CAMBRIDGE TRACTS IN MATHEMATICS

General Editors

237 Birational Geometry and Manin's Conjecture

Birational Geometry and Manin's Conjecture

SHO TANIMOTO
Nagoya University, Japan

CAMBRIDGE
UNIVERSITY PRESS

CAMBRIDGE
UNIVERSITY PRESS

Shaftesbury Road, Cambridge CB2 8EA, United Kingdom

One Liberty Plaza, 20th Floor, New York, NY 10006, USA

477 Williamstown Road, Port Melbourne, VIC 3207, Australia

314–321, 3rd Floor, Plot 3, Splendor Forum, Jasola District Centre, New Delhi – 110025, India

Cambridge University Press is part of Cambridge University Press & Assessment, a department of the University of Cambridge.

We share the University's mission to contribute to society through the pursuit of education, learning and research at the highest international levels of excellence.

www.cambridge.org
Information on this title: www.cambridge.org/9781009611633
DOI: 10.1017/9781009611640

When citing this work, please include a reference to the DOI 10.1017/9781009611640

First published 2026

A catalogue record for this publication is available from the British Library

A Cataloging-in-Publication data record for this book is available from the Library of Congress

ISBN 978-1-009-61163-3 Hardback

To my family, Mei, Yu, and Hal

Contents

Preface

In the late 1980s, Yuri Manin and his collaborators initiated a program that seeks geometric explanations for various asymptotic formulas for the counting functions of rational points on algebraic varieties defined over number fields. This program, which was later called *Manin's program*, leads to a conjecture which predicts the asymptotic formula for the counting function of rational points on a smooth Fano variety. This conjecture has been subsequently revised mainly by Victor Batyrev, Yuri Manin, Emmanuel Peyre, and Yuri Tschinkel, and it has been refined and formulated as *Manin's conjecture*.

At the beginning of the formulation of Manin's conjecture, it was understood that one cannot hope for the asymptotic formula, reflecting the global geometry of a polarized variety, for the counting function of the entire set of rational points. Indeed, it is possible that rational points are accumulating along some subvarieties such as lines on cubic surfaces. Thus it is important to consider *exceptional sets* and remove their contribution from the counting function of rational points. Originally, it was expected that an exceptional set should consist of rational points on a proper closed subset, and this prediction has been confirmed for various examples, including certain (almost) homogeneous spaces as well as some projective surfaces. However, Batyrev and Tschinkel found a counterexample to the closed set version of Manin's conjecture in the 1990s, and in their example, an exceptional set must be Zariski dense. Moreover, it seemed that such examples were quite common, so unfortunately Manin's conjecture was considered in arithmetic geometry as a guiding principle rather than a solid conjecture one could rely on. However, a new dawn for Manin's conjecture is emerging thanks to contributions of many people in recent years.

Around the beginning of the twenty-first century, Peyre suggested that any exceptional set in Manin's conjecture should be a *thin set*, a notion developed by Hilbert and Serre to study the inverse Galois problem. Recently, Le Rudulier and Browning–Heath-Brown proved the thin set version of Manin's conjecture for the Hilbert scheme of two points on a rational surface and a quadric bundle over the projective space respectively, and the reliability of Manin's conjecture has been significantly improved.

On the other hand, mainly together with Brian Lehmann, I have been studying the geometry of counterexamples for the closed-set version of Manin's conjecture using birational geometry such as the *minimal model program*, and this study leads to the formulation of exceptional sets using geometric invariants appearing in the asymptotic formula. Indeed, Brian Lehmann, Akash Kumar Sengupta, and I proposed a conjectural description of exceptional sets using birational geometry, and proved that this proposed set is thin. The proof uses geometric inputs, such as the minimal model program and the boundedness of singular Fano varieties (BAB conjecture), together with arithmetic inputs such as Hilbert's irreducibility theorem and systematic use of arithmetic étale fundamental groups. Moreover, our formulation is compatible with many existing results in the literature including Le Rudulier's example and Browning–Heath-Brown's example.

In this book, I explain this recent refinement of Manin's conjecture and the ideas behind the proofs of thinness of exceptional sets as well as its applications to the study of rational curves on Fano varieties leading to a conjecture called *geometric Manin's conjecture*, which predicts properties of the moduli space of rational curves on a Fano variety. This conjecture has been extensively studied by many mathematicians, and this book will include some foundational results on this conjecture as well as recent results confirming it in certain cases. It also includes several examples where exceptional sets are explicitly computed using geometric methods as well as geometric Manin's conjecture for higher genus curves. I hope that readers will find in this book a beautiful piece of mathematics, where arithmetic, geometry, and analysis are intertwined into one remarkable conjecture governing the asymptotic behavior of rational points and rational curves on a smooth Fano variety.

Acknowledgements

I would like to thank my colleagues for helping me throughout my career. In particular, I owe a debt of gratitude to Roya Beheshti, Margaret Bilu, Qile Chen, Dylon Chow, Ronno Das, Jordan Ellenberg, Kazuhiro Fujiwara, Runxuan Gao, Jakob Glas, Matthew Hase-Liu, Brendan Hassett, Tim Browning, Haruki Ito, Eric Jovinelly, Natsume Kitagawa, Aaron Landesman, Brian Lehmann, Dan Loughran, Fumiya Okamura, Emmanuel Peyre, Marta Pieropan, Eric Riedl, Will Sawin, Akash Kumar Sengupta, Arne Smeets, Jason Starr, Sam Streeter, Ramin Takloo-Bighash, Phil Tosteson, Yuri Tschinkel, Tony Várilly-Alvarado, Akshay Venkatesh, Melanie Matchett Wood, Huan Xiao, and Takehiko Yasuda for collaborations and discussions that helped shape my perspective on the birational geometry of Manin's conjecture as well as Manin's conjecture over function fields. In particular, I would like to thank Yuri for introducing me to this beautiful subject and Brian for many fruitful collaborations. I would also like to thank Lars Hesselholt and Tony Várilly-Alvarado for their enthusiastic support of this research.

I would like to thank Yoshinori Gongyo and Brian Lehmann for allowing me to include material in Section 4.4.2 which grew out of discussions between the three of us. I also would like to thank Chen Jiang for allowing me to include a proof of Lemma 6.5.9 which is originally due to him. I thank Qile Chen for pointing me to the reference [AV02] and answering my questions regarding stacks. I appreciate many detailed suggestions and comments on the manuscript from Brian Lehmann and Yuri Tschinkel, and these significantly improved the exposition. I also thank Jean-Louis Colliot-Thélène, Enhao Feng, Osamu Fujino, Kazuhiro Fujiwara, Runxuan Gao, Matthew Hase-Liu, Haruki Ito, Natsume Kitagawa, Aaron Landesman, Dan Loughran, Yutaro Naito, Yusuke Nakamura, Fumiya Okamura, Marta Pieropan, and Hiromu Tanaka for comments on this book. Finally, I thank anonymous reviewers for valuable comments which significantly improved the exposition of this book.

This book is based on a series of lectures I gave at Tohoku University, Beijing International Center for Mathematical Research, Saitama University, and Kyoto University. I thank Takehiko Yasuda, Zhiyu Tian, Takashi Kishimoto, Yuri Yatagawa, Osamu Fujino, Shu Kawaguchi, Yuji Odaka, and Hiromu Tanaka for opportunities to give lectures there and the audiences for various helpful questions and comments.

I deeply appreciate the various forms of financial support that made my research possible. During the preparation of this book, I was partially supported by JST FOREST program Grant number JPMJFR212Z, by JSPS KAKENHI Grant-in-Aid (B) 23H01067, by JSPS Bilateral Joint Research Projects Grant number JPJSBP120219935, by JSPS KAKENHI Early-Career Scientists Grant number 19K14512, by the Inamori Foundation, and by MEXT Japan, Leading Initiative for Excellent Young Researchers (LEADER).

Finally, I would like to express my gratitude to my family, especially Mei, Yu, and Hal, for giving me a joyful life outside of mathematics, as well as Shin, Mariko, and Ryo for their constant support throughout my life.

1
Introduction

One of the classical problems in mathematics is to study rational or integral solutions to a system of diophantine equations, that is, a system of polynomial equations with rational or integral coefficients. For example, if $f(x_1, \ldots, x_n)$ is a polynomial with rational coefficients, then one may ask for a description of the set of rational solutions:

$$\{(a_1, \ldots, a_n) \in \mathbb{Q}^n \mid f(a_1, \ldots, a_n) = 0\}.$$

The main questions regarding this set in diophantine geometry are:

(i) existence or non-existence of rational solutions;
(ii) finiteness of rational solutions, if they exist;
(iii) distribution of rational solutions, if there are infinitely many rational solutions;
(iv) explicit construction of some or all rational solutions.

Each question has been extensively studied, and all remain very active areas in mathematics.

A modern approach to studying the above questions is to consider rational solutions as points on an algebraic variety X defined by a system of polynomial equations, and this is why rational solutions are called *rational points*. It is customary to denote the set of rational points on X by $X(F)$ where F is the ground field under consideration. By considering rational solutions as points on a variety, we can apply techniques of algebraic geometry to the problems in diophantine geometry. For example, the above questions for higher dimensional varieties are almost completely open – there are many open conjectures on these questions: for example, Colliot-Thélène's conjecture, Bombieri–Lang's conjecture, Vojta's conjecture, Manin's conjecture, and so on. On the other hand, the period from 2005 to 2025 witnessed major developments in higher dimensional algebraic geometry, especially in the minimal model program, which aims to classify higher dimensional projective varieties up to birational equivalence. It is therefore natural to explore applications of techniques from the minimal model program to problems in diophantine geometry. Since birational equivalence identifies two varieties when they are isomorphic after removing proper closed sets, birational geometry will be best suited to study question (iii) on the distribution of rational points on an algebraic variety.

There are many questions one can ask about the distribution of rational points on algebraic varieties, but here we focus on two questions:

- the density of rational points on algebraic varieties in various topologies;
- the asymptotic formula for the counting function of rational points.

The first question concerns the Zariski density of rational points or weak approximation, that is, the density of rational points in the adelic space associated to a projective variety. Bombieri–Lang's conjecture predicts that the set of rational points on a smooth projective variety of general type is not Zariski dense. At the other extreme, it is expected that smooth Fano varieties satisfy potential density, that is, after some finite extension, the set of rational points becomes Zariski dense. See, for example, [BT99] and [HT00] for the first results in this direction, and [Cam04] and [Abr09] for the general framework on this problem. Weak approximation fails in general, but for smooth projective geometrically rationally connected varieties, all failures have so far been explained within the framework of Brauer–Manin obstructions. Colliot-Thélène's conjecture predicts that Brauer–Manin obstructions are only ones to weak approximation, and there is a vast amount of literature on this subject. See [CTS21] for more details.

The second question is the main theme of this book. We have the notion of a *height function*, which is a real-valued function on the set of rational points on a projective variety X:

$$\mathsf{H}_{\mathcal{L}} \colon X(F) \to \mathbb{R}_{>0},$$

where $\mathcal{L}$ is a line bundle L on X with an adelic metric. When L is ample, the set

$$\{x \in X(F) \mid \mathsf{H}_{\mathcal{L}}(x) \leq \mathsf{T}\}$$

is a finite set for any real number $\mathsf{T} > 0$. Thus, for any subset $Q \subset X(F)$, one can define the counting function as

$$N(Q, \mathcal{L}, \mathsf{T}) = \#\{x \in Q \mid \mathsf{H}_{\mathcal{L}}(x) \leq \mathsf{T}\}.$$

Manin's program, initiated in [FMT89] and [BM90], predicts the asymptotic formula of $N(Q, \mathcal{L}, \mathsf{T})$ as

$$N(Q, \mathcal{L}, \mathsf{T}) \sim c\mathsf{T}^{a(X,L)}(\log \mathsf{T})^{b(F,X,L)-1} \quad (\text{as } \mathsf{T} \to \infty), \tag{1.1}$$

for some appropriate set Q where $a(X, L)$, $b(F, X, L)$ are the geometric invariants of (X, L) introduced in [BM90] and c is some positive constant. Moreover, [Pey95], [BT98b], and [LS24] provide an adelic interpretation of c in terms of the Tamagawa number of X.

It was observed by Manin and his collaborators that, in the above asymptotic formula (1.1), one cannot take $Q = X(F)$ in general as it is possible that rational points are accumulating along some subvarieties. For example, for a smooth cubic surface S in $\mathbb{P}^3$, the asymptotic formula (1.1) will fail unless one removes contributions from lines on S. Hence it is important to consider an exceptional set $\mathsf{Z} \subset X(F)$, and set $Q = X(F) \setminus \mathsf{Z}$ in (1.1). Originally, it was expected that one could take our exceptional set Z as $V(F)$, where $V \subset X$ is a proper closed subset. Indeed, there are many examples satisfying this prediction (see Section 4.2 for some examples); however, Batyrev–Tschinkel found the first counterexample to this version of the conjecture in [BT96b]. After this counterexample, mathematicians in Manin's program found many counterexamples to the closed set version of the conjecture, e.g., [EJ06], [Els11], [BL17], [LR19], [BHB20], [Gao23], and [Gao24b].

In [Pey03], Emmanuel Peyre was the first to suggest that, instead of removing a proper closed subset, one should instead remove a *thin set*. Here a thin set $Z \subset X(F)$ is a subset of a finite union of sets of the form $f(Y(F))$ where $f: Y \to X$ is a generically finite morphism to the image without a rational section from X. This is the version now called *Manin's conjecture*. Manin's conjecture has received much attention, and this has been verified in a few cases such as [LR19] and [BHB20]. In [HTT15], with Brendan Hassett and Yuri Tschinkel, the author has started a systematic study of the exceptional set Z. This has been further developed mainly by Brian Lehmann and the author in [LT17] and [LTT18], also in [HJ17] and [Sen21] using the minimal model program. In [LST22], Brian Lehmann, Akash Kumar Sengupta, and the author proposed a purely geometric description of the exceptional set. The idea is that the exceptional set Z should consist of the contributions from generically finite morphisms $f: Y \to X$ such that

$$(a(X, L), b(F, X, L)) \leq (a(Y, f^*L), b(F, Y, f^*L))$$

in the lexicographic order. Moreover, using the minimal model program and the boundedness of singular Fano varieties, we showed that the exceptional set we proposed is indeed a thin set.

In this book, we explain the description of the exceptional set in [LST22] and some aspects of the proof of thinness of this set. In particular, we discuss how recent developments in the minimal model program lead to a finer refinement of Manin's conjecture that is compatible with many existing results in this field. One can also consider Manin's conjecture over global function fields, and this perspective enables us to translate our theorems on exceptional sets into results in the study of the space of rational curves on a smooth Fano variety X, that is,

$$\mathrm{Mor}(\mathbb{P}^1, X), \quad \overline{M}_{0,0}(X),$$

where the former space is the morphism scheme parametrizing rational curves $f\colon \mathbb{P}^1 \to X$ and the latter space is the moduli space of stable maps of genus 0 which provides a compactification for the space of rational curves. This leads to the formulation of geometric Manin's conjecture which predicts certain properties of these moduli spaces, and this is significantly inspired by Batyrev's heuristics of Manin's conjecture over global function fields in [Bat88]. We will discuss these exciting developments of Manin's conjecture in this book too.

Let me mention other survey papers and books in the literature. Manin's conjecture is at the interface of various subjects such as arithmetic geometry, analytic number theory, automorphic representation theory, ergodic theory, and birational geometry. However, this book is not a comprehensive introductory textbook on Manin's conjecture covering a wide spectrum of these subjects. Rather it is a monograph focusing on geometric aspects of Manin's conjecture, and it includes a recent refinement of Manin's conjecture using the notion of thin sets and its applications to the study of moduli spaces of rational curves on Fano varieties. Thus students who are considering doing research on Manin's conjecture should consult other sources in this field as well. One comprehensive survey paper on Manin's conjecture is Tschinkel's account [Tsc09], which discusses the formulation of Manin's conjecture as well as various methods of proofs for the conjecture. Browning's book [Bro09] treats topics from analytic number theory ranging from the Hardy–Littlewood circle method, dimension growth conjecture, to proofs of Manin's conjecture for certain (singular) del Pezzo surfaces. [Bro10] discusses a proof of Manin's conjecture for a certain smooth quartic del Pezzo surface which is established in [DlBB11]. [Bro21] also treats the circle method, focusing on its applications to cubic hypersurfaces. [Oh10] discusses results from [GMO08], which takes an ergodic approach to Manin's conjecture. [Jah14] includes various computational aspects of Manin's conjecture, in particular computations of Peyre's constants and experimental aspects. [ADHL15] is a research monograph specializing in Cox rings and their computations; it also uses the universal torsor method to prove Manin's conjecture for certain Mori dream spaces. [Pey21] discusses an appealing formulation of Manin's conjecture using the notion of slopes and freeness of rational points. Finally, there is an account by Margaret Bilu ([Bil23]) on the motivic version of Manin's conjecture.

Plan of this Book:

In Chapter 2, we introduce the notion of height functions and review their basic properties. Among them, the most important property is the Northcott property of height functions, which is a basis for the counting problem. Furthermore, in Example 2.1.8, we discuss a proof of Manin's conjecture for the projective space. This chapter is independent of others, so readers who wish to focus on the main theme of this book can safely skip this chapter.

In Chapter 3, we review techniques from higher dimensional algebraic geometry. After reviewing terminology from convex geometry in Section 3.1, we recall various notions of positivity of divisors in Section 3.2. Results in this section are mainly quoted from [Laz04]. In Section 3.3, we introduce various notions of singularities in the minimal model program. The main reference here is [KM98]. In Section 3.4, we discuss the minimal model program and results from [BCHM10]. This is one of the key inputs in our proof of thinness of the exceptional set. We also recall a few consequences of [BCHM10]: the canonical models, Zariski decomposition, and the relative minimal model program. In Section 3.5, the boundedness of singular Fano varieties, proved by Birkar in [Bir19] and [Bir21], is introduced. This is another key input in the proof of our main theorem. In Section 3.6, we review the notion of Mori dream spaces from [HK00] which are an important notion in modern algebraic geometry.

In Chapter 4, we discuss a modern formulation of Manin's conjecture; in particular, we develop conjectural constructions of exceptional sets proposed in [LST22]. In Section 4.1, we introduce the geometric invariants $a(X, L)$ and $b(F, X, L)$ appearing in the asymptotic formula of Manin's conjecture and we develop a birational-geometric study of these invariants. In Section 4.2, we discuss the formulation of Manin's conjecture and a key notion here is that of thin sets. Furthermore, we illustrate several examples from the literature that demonstrate subtleties of the formulation. In Section 4.3 we introduce conjectural constructions of exceptional sets from [LST22]. The main result of [LST22] is Theorem 4.3.5, which shows that the proposed exceptional sets are indeed thin. We discuss a few examples illustrating computations of the proposed exceptional sets at the end of Section 4.3 . In Section 4.4, some generalizations of Theorem 4.3.5 are discussed when a polarization is only big but not nef.

In Chapter 5, we discuss various ideas for proofs of Theorem 4.3.5. In Section 5.1, we describe the main result from [LT17, Theorem 1.5], Theorem 5.1.1, which asserts that the exceptional set formed by subvarieties violating geometric compatibility is a thin set. To prove this theorem, we first show that the exceptional set for the a-invariant is a proper closed subset (Theorem 5.1.6), and the main ingredient of this proof is weak BAB conjecture (Corollary 3.5.6). After studying some basic properties of families of adjoint rigid subvarieties, we prove Theorem 5.1.1 using Hilbert's irreducibility theorem (Theorem 5.1.10), which is one of the number-theoretic inputs in this proof. In Section 5.2, we prove the main result from [Sen21, Corollary 1.5], Theorem 5.2.3, which asserts that the exceptional set formed by adjoint rigid a-covers violating geometric compatibility is a thin set. The main technical input from the minimal model program is Theorem 5.2.7 which shows the finiteness of adjoint rigid a-covers up to birational equivalence over an algebraic closure. Combining it with an outcome of Hilbert's irreducibility theorem (Corollary 5.2.11), we prove Theorem 5.2.3. In Section 5.3, we explain the main ideas of the proof of Theorem 4.3.5. The core of the proof is a technical lemma (Lemma 5.3.5), whose proof relies on some arithmetic lifting properties of covering spaces with a rational point. Its proof is beyond the scope of this book: readers interested in it are encouraged to consult the original proof [LST22, Lemma 8.3]. Assuming Lemma 5.3.5, we prove Theorem 4.3.5.

In Chapter 6, we illustrate computations of exceptional sets constructed in Section 4.3 for several examples. Many results from adjunction theory will provide powerful tools to analyze specific examples. In Section 6.1, we study the simplest examples of homogeneous spaces, that is, projective spaces and smooth quadric hypersurfaces, using a geometric approach. In Section 6.2, we compute the exceptional set for the Hilbert scheme of length-2 subschemes on a smooth quadric surface, which is an example satisfying Manin's conjecture proved in [LR19], and confirm that our exceptional set coincides with the actual exceptional set used in [LR19]. To this end, first we recall several results from adjunction theory for lower dimensional varieties. Then we discuss a numerical description of subvarieties on the Hilbert scheme developed in [BC13, Example 18] and [LT17, Section 9.3]. Finally, we compute our exceptional set following the discussion of [LT17, Section 9.3.1] as well as [LST22, Example 5.13]. In Section 6.3, we compute our conjectural exceptional set for an example studied in [BHB20]. In this paper, Browning and Heath-Brown proved Manin's conjecture for a certain quadric surface bundle after removing a thin exceptional set, and we confirm that the exceptional set used in [BHB20] coincides with our conjectural exceptional set. This follows from the computation carried out in [LT19b, Section 12]. In Section 6.4, we discuss when our exceptional set is contained in a proper closed subset of smooth projective geometrically uniruled surfaces. In Section 6.4.1, we describe results in [LTT18, Proposition 5.9] and

[LT17, Theorem 1.8] showing that when our surface S is split, that is, when the Picard rank of S is equal to the geometric Picard rank of S, our exceptional set is indeed not Zariski dense. Then, in Section 6.4.2, we discuss non-split smooth del Pezzo surfaces. First, we provide the proof of [LT19b, Theorem 9.1], which shows that, for most smooth del Pezzo surfaces, our exceptional set is not Zariski dense. Then we turn to a surprising example in [Gao23], which is a del Pezzo surface of degree 1 and Picard rank 1 such that the exceptional set is Zariski dense, proving the first counterexample to the closed-set version of Manin's conjecture in dimension 2. In Section 6.5, we discuss our exceptional sets for smooth Fano threefolds of geometric Picard rank 1. In Section 6.5.1, we discuss del Pezzo threefolds. Then, in Section 6.5.2, we turn to smooth prime Fano threefolds of geometric index 1. In Section 6.6, we develop our conjectural exceptional sets for Fano hypersurfaces. A key to this problem is Conjecture 6.6.1, which we prove for cubic hypersurfaces and quartic hypersurfaces in Sections 6.6.1 and 6.6.2 respectively.

In Chapter 7, we turn to Manin's conjecture over global function fields and introduce a version of Manin's conjecture over $\mathbb{C}(t)$. First we introduce moduli spaces of rational curves and their deformation theory; then we discuss geometric Manin's conjecture, which was inspired by Victor Batyrev's heuristics of Manin's conjecture over finite fields ([Bat88]). This conjecture was proposed by Brian Lehmann and the author originally in [LT19a], and it has been further developed and refined in subsequent papers. In Section 7.1, we introduce two types of moduli spaces of rational curves on projective varieties: morphism schemes and moduli spaces of stable maps of genus 0. In Section 7.1.1, we discuss the morphism scheme parametrizing maps from a projective scheme to a quasi-projective scheme following the exposition of [Deb01], and in Section 7.1.2, we describe the coarse moduli space of stable maps of genus 0 on a projective variety. Then in Section 7.2, we develop deformation theory for both morphism schemes and coarse moduli spaces of stable maps of genus 0. In Section 7.3, Batyrev's heuristic argument for Manin's conjecture over finite fields is given following the exposition of [Tan21] and [LRT25c]. In Section 7.4, we introduce geometric Manin's conjecture, which is a refinement of predictions for moduli spaces by Victor Batyrev. This conjecture consists of three parts: (I) dimension, (II) number of irreducible components, and (III) homological stability. We discuss these three aspects in detail. In Section 7.5, we explore some arithmetic applications of geometric Manin's conjecture that were developed in [LRT25c].

In Chapter 8, we discuss proofs of geometric Manin's conjecture II for certain Fano varieties and Fano fibrations. In Section 8.1, we first discuss Mori's famous Bend and Break lemmas in various forms. The original lemmas of Mori assert that if you vary a curve while fixing a point on it, then it breaks into the union of a curve and rational curves. These lemmas are quoted from [KM98], and we also discuss a modified version of Bend and Break lemmas that are mainly from [LT24]. These techniques form the basis for our inductive approaches to the space of rational curves on Fano varieties. Then, in Section 8.2, we discuss the proofs of geometric Manin's conjecture II for del Pezzo threefolds (Theorem 8.2.1). This is by induction on the degrees of rational curves. The base cases are lines and conics, and the inductive step is by our improved version, the *movable Bend and Break lemma*. This is proved in Theorem 8.2.13. Our proofs are based on proofs in [LT19a] and [ST22]. In Section 8.3, we study the space of sections of del Pezzo fibrations over $\mathbb{P}^1$. After reviewing some preliminaries, we discuss (i) non-dominant families of sections, (ii) movable Bend and Break techniques for free sections, and (iii) examples. The main results are a certain characterization of non-dominant families of sections (Theorem 8.3.21) and movable Bend and Break technique (Theorem 8.3.30). Examples are general pencils of the fundamental linear system on del Pezzo threefolds. These results were mainly obtained in [LT24]. Finally, in Section 8.4, we discuss geometric Manin's conjecture for del Pezzo fibrations and Fano fibrations over higher genus curves. Results in this section are mainly from [LT22] and [LRT25c].

Prerequisites

I assume that readers have been exposed to some standard textbooks in algebraic geometry such as [Har77]. I also assume that readers possess a basic knowledge of algebraic number theory such as Galois theory, rings of integers, discrete valuations, adeles, and Galois cohomology (see, e.g., [CF10] for the necessary backgrounds). Some knowledge of higher dimensional algebraic geometry will be useful, but we summarize necessary definitions and results in Chapter 3.

Notations and Conventions

(i) Let $\xi(\mathsf{T}), \zeta(\mathsf{T})$ be two real-valued functions defined for positive real numbers $\mathsf{T} > 0$. We write $\xi(\mathsf{T}) = O(\zeta(T))$ if there exists $C > 0$ such that $|\xi(\mathsf{T})| < C|\zeta(\mathsf{T})|$ holds for sufficiently large $\mathsf{T} > 0$. We write $\xi(\mathsf{T}) \sim \zeta(\mathsf{T})$ when $\lim_{\mathsf{T}\to+\infty} \xi(\mathsf{T})/\zeta(\mathsf{T}) = 1$.

(ii) Through Chapters 1 to 6, F always denotes a field of characteristic 0 unless specified otherwise. Depending on the contexts, it may be algebraically closed or a number field. In Chapters 7 and 8, the ground field is denoted by k. This is usually an algebraically closed field or a finite field.

(iii) A variety defined over a field F is an integral separated scheme of finite type over F. Subvarieties are closed and integral. We denote the function field of a variety X by $F(X)$. For an extension F'/F we denote its base change by $X \otimes F'$. For an algebraic closure $\overline{F}$, the base change $X \otimes \overline{F}$ is denoted by $\overline{X}$.

(iv) Let F be a field and X be an F-scheme which is locally of finite type over F. A component of X means an irreducible component of X unless it is specified otherwise.

(v) Let X be a variety defined over a field F. For any sheaf $\mathcal{F}$ on X, its sheaf cohomology is denoted by $H^i(X, \mathcal{F})$ and its dimension is denoted by $h^i(X, \mathcal{F})$.

(vi) A *Cartier divisor* on X is a global section of the sheaf $\mathcal{K}_X^\times/\mathcal{O}_X^\times$ where $\mathcal{K}_X$ is the sheaf of rational functions on X. We denote the group of Cartier divisors by $\mathrm{CaDiv}(X)$. A $\mathbb{Q}$-*Cartier divisor* is a member of $\mathrm{CaDiv}(X) \otimes_{\mathbb{Z}} \mathbb{Q}$. A Cartier divisor L is effective if local equations for L are regular, and a $\mathbb{Q}$-Cartier divisor is effective if some multiple is an effective Cartier divisor.

(vii) Let L be a Cartier divisor on X. We denote the *complete linear series* of L by $|L|$. For a linear series M, we denote its *base locus* by $\mathrm{Bs}(M)$. We adopt Grothendieck's notation for projective bundles of locally free sheaves.

(viii) Let X be a variety defined over a field F with a geometric closed point $\overline{x} \to X$. We denote the *étale fundamental group* of X with the basepoint $\overline{x}$ by $\pi_1^{\text{ét}}(X, \overline{x})$.

(ix) For any generically finite and dominant morphism $f\colon Y \to X$ of F-varieties, $\mathrm{Bir}(Y/X)$ denotes the *group of birational X-automorphisms of Y* and $\mathrm{Aut}(Y/X)$ denotes the *group of X-automorphisms of Y*.

(x) A *general point* on a variety X means a closed point on some Zariski open subset of X; a *very general point* means a closed point on the complement of the union of countably many proper closed subvarieties in X. This makes sense only when the base field is uncountable. A *general member* of a family parametrizing schemes is a member of the family corresponding to a general point on the base. Similarly for a very general member as well.

(xi) For a quasi-projective variety X, a *smooth resolution* is a projective birational morphism $\beta\colon Y \to X$ such that Y is smooth. Its existence is guaranteed in characteristic 0 by the famous theorem of Hironaka ([Hir64]).

(xii) For a birational map $f: Y \dashrightarrow X$ and an irreducible subvariety $Z \subset X$, if f^{-1} is defined on some open subset $Z^\circ \subset Z$, then we call the closure of $f^{-1}(Z^\circ)$ the *strict transform* of Z and denote it by $f_*^{-1}(Z)$. For any birational map $f: Y \dashrightarrow X$ such that f^{-1} does not contract any divisor, we may define the strict transform of a Weil divisor $\sum_i a_i D_i$ on X as $\sum_i a_i f_*^{-1}(D_i)$.

(xiii) We use the definition of *strict normal crossings divisors* as in [Sta, Tag 0BI9].

(xiv) Let $f: X \to Y$ be a generically finite dominant morphism between varieties. The *exceptional locus of* $f: X \to Y$ is $\{x \in X \mid \dim_x f^{-1}(f(x)) > 0\}$. The *ramification locus* is the set of points $x \in X$ at which f is not étale. When X and Y are normal and f is birational, the ramification locus is equal to the exceptional locus.

(xv) A projective surjective morphism $f: X \to Y$ between varieties is an *algebraic fiber space* if $f_* \mathcal{O}_X = \mathcal{O}_Y$ holds. When f is an algebraic fiber space, every fiber of f is connected. Conversely, in characteristic 0, if Y is normal and every fiber of f is connected, then f is an algebraic fiber space. Also one should note that when X is normal and f is an algebraic fiber space, it follows from functoriality of normalizations that Y is normal.

(xvi) For a projective scheme X, $\mathrm{Hilb}(X)$ denotes the *Hilbert scheme* parametrizing flat families of closed subschemes of X. See [Kol96] for its constructions and their basic properties.

(xvii) For a quasi-projective variety X, $\mathrm{Sym}^{(n)}(X)$ denotes the *nth symmetric product of* X, which is the quotient of X^n by the action of $\mathfrak{S}_n$ permuting coordinates.

(xviii) For a Cartier divisor L on a projective variety X, an L-line, L-conic, or L-cubic means a rational curve C on X such that $L.C = 1, 2$, or 3 respectively.

2
Height Functions

2.1 Height Functions

One of the fundamental tools in diophantine geometry is the notion of a height function. This function measures the geometric and arithmetic complexity of rational points on a projective variety defined over a number field and is a key to a number of the fundamental finiteness results in arithmetic geometry, e.g., Mordell–Weil Theorem, Siegel's theorem, Mordell–Faltings' theorem and so on. Here is a basic example of height functions:

$$\begin{array}{cccc} \mathsf{H}\colon & \mathbb{P}^n(\mathbb{Q}) & \longrightarrow & \mathbb{R}_{>0} \\ & \rotatebox{90}{$\in$} & & \rotatebox{90}{$\in$} \\ & (x_0 : \ldots : x_n) & \longmapsto & \max\{|x_0|, \ldots, |x_n|\} \end{array} \tag{2.1}$$

Here we choose homogeneous coordinates x_i so that $x_i \in \mathbb{Z}$ for any i and $\gcd(x_0, \ldots, x_n) = 1$. Note that for any positive real number $\mathsf{T} > 0$, the set

$$\{P \in \mathbb{P}^n(\mathbb{Q}) \mid \mathsf{H}(P) \leq \mathsf{T}\}$$

is a finite set. This is the *Northcott property*, which allows one to consider the counting function of rational points of bounded height and try to identify the asymptotic formula for such a function as $\mathsf{T} \to \infty$.

In general, if one has the following data:

- a number field F;
- a projective variety X defined over F; and
- a line bundle L on X,

then one can associate a height function

$$\mathsf{H}_L\colon X(F) \to \mathbb{R}_{>0}$$

to the above data. Moreover, when L is ample, the height function enjoys the

Northcott property. There are two main constructions of the associated height functions to (F, X, L):

(i) the Weil height machine;
(ii) adelic metrizations.

Originally, height functions were developed by André Weil and Douglas Northcott in order to prove the Mordell–Weil theorem for abelian varieties defined over a number field. Today, this theory is summarized as the Weil height machine and a modern account of this theory is given in a beautiful book by Hindry and Silverman [HS00]. In the field of Manin's conjecture, the second construction is often used and we use this construction throughout this book. We closely follow the exposition of [CLT10]. Note that two constructions are essentially equivalent. See [HS00, Theorem B.3.2 (h) and Theorem B.10.7].

Let F be a number field. We denote the set of equivalence classes of absolute values of F by Ω_F. For each $v \in \Omega_F$, we consider the completion F_v of F with respect to v so that F_v is a local field. For each $v \in \Omega_F$, we have an absolute value

$$|\cdot|_v : F_v \longrightarrow \mathbb{R}_{\geq 0}.$$

We normalize this absolute value so that we have

$$|\cdot|_v = \begin{cases} \text{usual absolute value} & \text{if } F_v = \mathbb{R}; \\ (\text{usual absolute value})^2 & \text{if } F_v = \mathbb{C}; \\ |N_{F_v/\mathbb{Q}_p}(\cdot)|_p & \text{if } F_v \text{ is } p\text{-adic,} \end{cases}$$

where p is a prime number and $N_{F/\mathbb{Q}_p}(\cdot)$ is the norm of a finite extension $F/\mathbb{Q}_p$. Here we normalize $|\cdot|_p$ so that we have $|p|_p = 1/p$. As a result of these normalizations, we have the product formula, that is, for any $a \in F^\times$, we have

$$\prod_{v \in \Omega_F} |a|_v = 1. \tag{2.2}$$

For a non-archimedean place $v \in \Omega_F$, we denote the ring of integers for F_v by $\mathfrak{o}_v$. Let $S \subset \Omega_F$ be a finite set including all archimedean places. We denote the ring of S-integers by $\mathfrak{o}_{F,S}$.

Suppose now that we have a variety X defined over F_v. Then F_v-valued points $X(F_v)$ form a v-adic analytic space. When X is smooth, $X(F_v)$ is a v-adic manifold in the sense that, in the v-adic topology, any sufficiently small neighborhood in $X(F_v)$ is analytically isomorphic to an open set in $F_v^{\dim X}$.

When we further have a line bundle $L \to X$, we have a v-adic line bundle $L(F_v) \to X(F_v)$.

Next, we introduce the notion of v-adic metrics for analytic line bundles as above:

Definition 2.1.1 Let X be a variety defined over F_v and L a line bundle over X. A *metric* for $L(F_v) \to X(F_v)$ consists of norms $\| \cdot \|_v$ on $L_x(F_v)$ for each $x \in X(F_v)$, where L_x is a fiber at x, satisfying the following condition: for any Zariski open subset $U \subset X$ and a section $s \in H^0(U, L)$, the function

$$U(F_v) \ni x \mapsto \|s(x)\|_v \in \mathbb{R}_{\geq 0}$$

is continuous.

Here is a basic construction of metrics induced by integral models of varieties:

Example 2.1.2 (Metrics induced by integral models) Let v be a non-archimedean place. Suppose that we have a projective variety X defined over F_v and a line bundle L over X. We further assume that we are given a projective flat model $\mathcal{X}$ of X over $\operatorname{Spec} \mathfrak{o}_v$ with a line bundle $\mathcal{L}$ on $\mathcal{X}$ extending L. For each $x \in X(F_v) = \mathcal{X}(\mathfrak{o}_v)$, we define a norm $\| \cdot \|_v$ on $L_x(F_v)$ by

$$\|\ell\|_v \leq 1 \iff \ell \in \mathcal{L}(\mathfrak{o}_v) \subset L(F_v).$$

Then $\| \cdot \|_v$ is a metric for L, and we call it the metric induced by a model $(\mathcal{X}, \mathcal{L})$.

Here is the definition of adelic metrics, which is central to the construction of height functions:

Definition 2.1.3 (Adelic metrics) Let F be a number field and X be a projective variety defined over F. Suppose that we have a line bundle L on X. An *adelic metric* is a collection $\{\| \cdot \|_v\}_{v \in \Omega_F}$ where $\| \cdot \|_v$ is a metric on $L(F_v)$ with the following property: there exists a finite set $S \subset \Omega_F$ containing all archimedean places and a projective flat integral model $\mathcal{X} \to \operatorname{Spec} \mathfrak{o}_{F,S}$ of X with a line bundle $\mathcal{L}$ on $\mathcal{X}$ extending L such that for any place $v \notin S$, $\| \cdot \|_v$ is induced by $(\mathcal{X}, \mathcal{L})$. We also write $\mathcal{L} = (L, \{\| \cdot \|_v\}_{v \in \Omega_F})$ and call it an adelically metrized line bundle.

Using adelic metrics, one can introduce the notion of height functions:

Definition 2.1.4 (Height functions) Let F be a number field and X be a projective variety defined over F. Let $\mathcal{L} = (L, \{\|\cdot\|_v\}_v)$ be an adelically metrized line bundle on X. We define the *height function*

$$\mathsf{H}_{\mathcal{L}} \colon X(F) \to \mathbb{R}_{>0}$$

associated to $(X, \mathcal{L})$ in the following way: for each $x \in X(F)$, choose a non-zero vector $\ell \in L_x(F)$. We define

$$\mathsf{H}_{\mathcal{L}}(x) = \prod_{v \in \Omega_F} \|\ell\|_v^{-1}.$$

By the product formula (2.2), this definition is independent of the choice of ℓ.

One example of this construction is a naive height function on the projective space:

Example 2.1.5 Set $\mathcal{X} = \mathbb{P}^n_{\mathfrak{o}_F}$ and $\mathcal{L} = \mathcal{O}(1)$. Then $(\mathcal{X}, \mathcal{L})$ induces metrics $\|\cdot\|_v$ for all non-archimedean places. Let $\ell \in H^0(X, \mathcal{O}(1))$ be a linear form in $x_0, \ldots, x_n$. Then for any $(x_0 : \ldots : x_n) \in \mathbb{P}^n(F_v)$ and a non-archimedean place v, the induced metric $\|\cdot\|_v$ satisfies

$$\|\ell(x_0 : \ldots : x_n)\|_v = \frac{|\ell(x_0, \ldots, x_n)|_v}{\max\{|x_0|_v, \ldots, |x_n|_v\}}.$$

We use the same formula to define metrics at archimedean places. Then the height function associated to $\mathcal{L}$ is given by

$$\mathsf{H}_{\mathcal{L}}(x) = \prod_{v \in \Omega_F} \frac{\max\{|x_0|_v, \ldots, |x_n|_v\}}{|\ell(x_0, \ldots, x_n)|_v} = \prod_{v \in \Omega_F} \max\{|x_0|_v, \ldots, |x_n|_v\}.$$

Here the last equality follows from the product formula (2.2). Note that when $F = \mathbb{Q}$ and we choose homogeneous coordinates x_i so that

$$x_i \in \mathbb{Z} \text{ and } \gcd(x_i) = 1,$$

the above formula reduces to

$$\mathsf{H}_{\mathcal{L}}(x) = \max\{|x_0|, \ldots, |x_n|\},$$

which is a naive height function appearing in (2.1).

Remark 2.1.6 Height functions in this book correspond to *height functions relative to F* in [HS00]. The reference also discusses *absolute height functions* which are normalized by the degree of F over $\mathbb{Q}$. One reason for considering absolute height functions is that they are invariant under field extensions and therefore defined over the set of algebraic points on an algebraic variety. Such functions are employed in various contexts such as Vojta's conjecture, but in the field of Manin's conjecture, relative height functions are frequently used as the asymptotic formula takes a simpler form.

Next, we would like to state a few properties of height functions. Among them, the Northcott property is the most important and is central to counting problems of rational points on algebraic varieties.

Theorem 2.1.7 *Let F be a number field and X be a projective variety defined over F equipped with an adelically metrized line bundle $\mathcal{L} = (L, \{\|\cdot\|_v\}_v)$.*

(i) *For another adelically metrized line bundle $\mathcal{L}'$ whose underlying line bundle is L, there exist positive constants $C_1, C_2 > 0$ such that we have*

$$C_1 \mathsf{H}_{\mathcal{L}'}(x) \leq \mathsf{H}_{\mathcal{L}}(x) \leq C_2 \mathsf{H}_{\mathcal{L}'}(x)$$

for any $x \in X(F)$, that is, two height functions $\mathsf{H}_{\mathcal{L}}, \mathsf{H}_{\mathcal{L}'}$ are equivalent.

(ii) *Let $V = \mathrm{Bs}(|L|)$ be the base locus of the complete linear system of L. Then there exists a positive constant $C > 0$ such that*

$$\mathsf{H}_{\mathcal{L}}(x) \geq C$$

for any $x \in (X \setminus V)(F)$.

(iii) *When L is ample, for any positive number $\mathsf{T} > 0$ the set of rational points of bounded height*

$$\{x \in X(F) \mid \mathsf{H}_{\mathcal{L}}(x) \leq \mathsf{T}\}$$

is a finite set.

Proof The first statement follows from [HS00, Proposition B.10.6]. Note that if we have two integral models over $\operatorname{Spec} \mathfrak{o}_{F,S}$, then these two integral models are isomorphic to each other for almost all $v \notin S$. Next, since the height function associated to $(X, \mathcal{L})$ is equivalent to a height function defined by the Weil height machine by [HS00, Theorem B.3.2 (h) and Theorem B.10.7], the remaining statements follow from [HS00, Theorem B.3.2 (e) and (g)]. □

Property (iii) in Theorem 2.1.7 is called the *Northcott property*; it plays a central role in various proofs of finiteness results in diophantine geometry such as the Mordell–Weil theorem and Mordell–Faltings' theorem. Moreover, it is a key to the counting problem of rational points on a projective variety X when X admits infinitely many rational points.

Indeed, let F be a number field and X a projective variety defined over F. Let $\mathcal{L}$ be an ample line bundle on X with an adelic metrization. Let $Q \subset X(F)$ be a subset of the set of rational points. For any positive number $\mathsf{T} > 0$ we set the counting function as

$$N(Q, \mathcal{L}, \mathsf{T}) = \#\{x \in Q \mid \mathsf{H}_{\mathcal{L}}(x) \leq \mathsf{T}\}.$$

By the Northcott property, this function is well defined. Manin's conjecture predicts that if we choose Q in an appropriate way, then the asymptotic formula of $N(Q, \mathcal{L}, \mathsf{T})$ as $\mathsf{T} \to +\infty$ is given by an explicit formula in terms of the geometric and arithmetic invariants of $(X, \mathcal{L})$. A natural question is how one should choose Q and this is the main theme of this book. Before going to the next topic, let us explain how to establish the asymptotic formula for $X = \mathbb{P}^n$.

Example 2.1.8 ([Sch79]) The asymptotic formulas for the projective space over number fields have been obtained in [Sch79]. Over the field $\mathbb{Q}$ of rational numbers, one can give a simpler proof. The following proof is taken from [HS00, Theorem B.6.2].

Let $F = \mathbb{Q}$ and $X = \mathbb{P}^n$. We consider the height function $\mathsf{H}_{\mathcal{O}(1)}$ defined in Example 2.1.5. We would like to establish the asymptotic formula for

$$N(\mathbb{P}^n(\mathbb{Q}), \mathcal{O}(1), \mathsf{T}).$$

For $x = (x_0, \ldots, x_n) \in \mathbb{A}^{n+1}(\mathbb{Z})$ with $x \neq 0$, we set

$$\|x\| = \max\{|x_0|, \ldots, |x_n|\}, \quad \gcd(x) = \gcd(x_0, \ldots, x_n).$$

We define

$$\begin{aligned} M(\mathsf{T}) &:= \#\{x \in \mathbb{A}^{n+1}(\mathbb{Z}) \mid x \neq 0,\ \|x\| \leq \mathsf{T}\}, \\ M^*(\mathsf{T}, d) &:= \#\{x \in \mathbb{A}^{n+1}(\mathbb{Z}) \mid \gcd(x) = d,\ \|x\| \leq \mathsf{T}\}. \end{aligned}$$

Then we have

$$N(\mathbb{P}^n(\mathbb{Q}), \mathsf{H}_{\mathcal{O}(1)}, \mathsf{T}) = \frac{1}{2} M^*(\mathsf{T}, 1).$$

Thus we need to establish the asymptotic formula for $M^*(\mathsf{T}, 1)$. It is easy to see that $M^*(\mathsf{T}, d) = M^*(\mathsf{T}/d, 1)$. On the other hand, we have

$$M(\mathsf{T}) = \sum_{d \geq 1} M^*(\mathsf{T}, d) = \sum_{d \geq 1} M^*(\mathsf{T}/d, 1).$$

Thus, using the Möbius inversion formula, we obtain

$$M^*(\mathsf{T}, 1) = \sum_{d \geq 1} \mu(d) M(\mathsf{T}/d),$$

where μ is the Möbius function. Since we have $M(\mathsf{T}) = (2\lfloor \mathsf{T} \rfloor + 1)^{n+1} - 1$, we obtain

$$\begin{aligned}
M^*(\mathsf{T}, 1) &= \sum_{d \geq 1} \mu(d) M(\mathsf{T}/d) \\
&= \sum_{d \geq 1} \mu(d)((2\lfloor \mathsf{T}/d \rfloor + 1)^{n+1} - 1) \\
&= \sum_{1 \leq d \leq \mathsf{T}} \mu(d)((2\mathsf{T}/d + O(1))^{n+1} - 1) \\
&= \sum_{1 \leq d \leq \mathsf{T}} \mu(d) \left(2^{n+1} \frac{\mathsf{T}^{n+1}}{d^{n+1}} + O\left(\frac{\mathsf{T}^n}{d^n} \right) \right) \\
&= 2^{n+1} \mathsf{T}^{n+1} \left(\sum_{d \geq 1} \frac{\mu(d)}{d^{n+1}} \right) - (2\mathsf{T})^{n+1} \left(\sum_{d > \mathsf{T}} \frac{\mu(d)}{d^{n+1}} \right) + O\left(\mathsf{T}^n \sum_{1 \leq d \leq \mathsf{T}} d^{-n} \right).
\end{aligned}$$

In the last expression, the second term is $O(\mathsf{T})$ and the third term is $O(\mathsf{T}^n)$ if $n > 1$, and $O(\mathsf{T} \log \mathsf{T})$ if $n = 1$. Also we have $\sum_{d \geq 1} \frac{\mu(d)}{d^{n+1}} = \zeta(n+1)^{-1}$. Thus we conclude that

$$N(\mathbb{P}^n(\mathbb{Q}), \mathsf{H}_{O(1)}, \mathsf{T}) \sim \frac{2^n}{\zeta(n+1)} \mathsf{T}^{n+1}.$$

3

A Brief Introduction to Higher Dimensional Algebraic Geometry

From the beginning of Manin's conjecture, higher dimensional algebraic geometry plays important roles in its foundational aspects. This is already evident in early papers in this field, for example, [BT98b], [Tsc03], and it has been more systematically used in recent papers [HTT15], [LTT18], [HJ17], [LT17], [LT19a], [Sen21], [LST22], and [LT19b]. Indeed, the key inputs in the latter are the developments of higher dimensional algebraic geometry in the last decade, that is, the minimal model program [BCHM10] and the boundedness of singular Fano varieties (BAB conjecture) [Bir19] and [Bir21]. In this chapter, we review the necessary notions and results from higher dimensional algebraic geometry and the minimal model program. Readers who are familiar with the minimal model program can safely skip this chapter.

3.1 Convex Geometry

In higher dimensional algebraic geometry, certain convex cones play an important role in their study. Here we introduce some basic terminology and definitions from convex geometry.

Let V be a finite dimensional real vector space equipped with its Euclidean topology.

Definition 3.1.1 Let V be a finite dimensional real vector space and $\mathsf{C} \subset \mathsf{V}$ be a subset. We say C is a *cone* if for any $u \in \mathsf{C}$ and $\lambda \in \mathbb{R}_{\geq 0}$ we have $\lambda u \in \mathsf{C}$. We say a cone C is *convex* if for any $u, v \in \mathsf{C}$, we have $u + v \in \mathsf{C}$.

Let C be a convex cone. We say C is *strict* if C contains no 1-dimensional real subspace.

Next, we introduce the notion of faces:

Definition 3.1.2 Let V be a finite dimensional real vector space and $\mathsf{C} \subset \mathsf{V}$ be a closed convex cone. A *subcone* F of C is a closed convex cone contained in C. We say a subcone F is an *extremal face* if

$$u, v \in C \text{ and } u + v \in \mathsf{F} \implies u, v \in \mathsf{F}$$

holds. An extremal face F is called an *extremal ray* if the subspace $\langle \mathsf{F} \rangle$ spanned by F is 1-dimensional. Note that when the interior of C in V is non-empty, any element of an extremal face is contained in the boundary of C unless the face is equal to C itself.

A subset $\mathsf{F} \subset \mathsf{C}$ is a *supported face* if there exists a real linear map $\ell \colon \mathsf{V} \to \mathbb{R}$ that takes non-negative values on C such that we have

$$\mathsf{F} = \{v \in \mathsf{V} \mid \ell(v) = 0\} \cap \mathsf{C}.$$

Remark 3.1.3 Any supported face is an extremal face, but the converse is not true in general. To see this, consider a closed strict convex cone C in $\mathbb{R}^3$ such that the hyperplane defined by $z = 1$ cuts out the convex domain enclosed by the following three curves:

$$R_1 = \{(x, y, 1) \mid -1 \leq x \leq 0, y = 0\},$$
$$R_2 = \{(x, y, 1) \mid x^2 + (y-1)^2 = 1, 0 \leq x, y \leq 1\},$$
$$R_3 = \left\{(x, y, 1) \,\middle|\, y = \frac{1}{2}x + \frac{1}{2}, -1 \leq x \leq 1\right\}.$$

Then the ray $\mathsf{F} = \{(x, y, z) \mid x = y = 0, z \geq 0\}$ is an extremal ray for C, but it is not a supported face. Indeed, any supported face containing F contains the subcone generated by R_1.

When a convex cone C is finitely generated, that is, there are finitely many vectors $v_1, \ldots, v_r \in \mathsf{C}$ such that

$$\mathsf{C} = \sum_{i=1}^{r} \mathbb{R}_{\geq 0} v_i = \left\{ \sum_{i=0}^{r} a_i v_i \,\middle|\, a_i \in \mathbb{R}_{\geq 0} \right\},$$

then any extremal face of C is a supported face.

Finally, we introduce the dual cone:

Definition 3.1.4 Let V be a finite dimensional real vector space and $\mathsf{C} \subset \mathsf{V}$ be a closed convex cone. Let $\mathsf{V}^\vee$ be the dual vector space of V. Then the *dual cone* is given by

$$\mathsf{C}^\vee = \{\ell \in \mathsf{V}^\vee \mid \text{for any } u \in \mathsf{C}, \ell(u) \geq 0.\}.$$

This is a closed convex cone in $\mathsf{V}^\vee$. We have $\mathsf{C}^{\vee\vee} = \mathsf{C}$. Note that $\mathsf{C}^\vee$ has non-empty interior if and only if C is strict.

3.2 Divisors and Positivity

In this section, we recall several key definitions regarding divisors and various notions of their positivity. A basic reference is [Laz04], and we will cite relevant sections each time we mention results from this book. The results in [Laz04] are stated over an algebraically closed field of characteristic 0, but the adaptation to an arbitrary field of characteristic 0 is straightforward.

Let F be a field of characteristic 0. Let X be a normal variety defined over F. A $\mathbb{Q}$-*divisor* is a formal finite sum of prime divisors with $\mathbb{Q}$-coefficients:

$$D = \sum_{i=1}^{r} a_i D_i,$$

where $a_i \in \mathbb{Q}$ and D_i is a prime divisor, that is, a codimension-1 subvariety in X. When $a_i \geq 0$ for any i, we call D an effective $\mathbb{Q}$-divisor and write $D \geq 0$.

We say a $\mathbb{Q}$-divisor D is $\mathbb{Q}$-Cartier if there exists $m \in \mathbb{Z}_{>0}$ such that mD is integral and is associated to a Cartier divisor. Since X is normal, this Cartier divisor is uniquely determined. Thus we do not distinguish a $\mathbb{Q}$-Cartier divisor from the $\mathbb{Q}$-divisor associated to it in this case. See [Ful84, Section 2.1] for more details.

For each non-zero rational function $f \in F(X)$, we define the associated divisor by

$$\operatorname{div}(f) = \sum_{D:\text{ prime}} \operatorname{ord}_D(f) D,$$

where $\operatorname{ord}_D(f)$ is the order of f along D. We say two $\mathbb{Q}$-divisors D_1, D_2 are $\mathbb{Q}$-*linearly equivalent* if there exist rational functions f_i $(1 \leq i \leq r)$ with rational numbers $a_i \in \mathbb{Q}$ such that

$$D_1 - D_2 = \sum_{i=1}^{r} a_i \operatorname{div}(f_i).$$

When two $\mathbb{Q}$-divisors D_1, D_2 are $\mathbb{Q}$-linearly equivalent, we write

$$D_1 \sim_{\mathbb{Q}} D_2.$$

We define $\mathbb{R}$-*divisors* and $\mathbb{R}$-*linear equivalence* in a similar way. An $\mathbb{R}$-divisor D is $\mathbb{R}$-*Cartier* if D can be expressed as a linear combination of Cartier divisors over $\mathbb{R}$.

Suppose that X is a projective variety defined over F. Let D be a $\mathbb{Q}$-Cartier divisor and C be a curve on X. Then one can define the *intersection number*

$$D.C.$$

Indeed, one may take the normalization $\beta\colon C' \to C$ and define by

$$D.C = \deg_{C'} \beta^* D.$$

When $Z \subset X$ is an n-dimensional subvariety and $D_1, \ldots, D_n$ are $\mathbb{Q}$-Cartier divisors on X, we use the intersection number

$$(D_1. \ldots .D_n.Z).$$

There are several equivalent ways to define this. For a cohomological approach, one may consult [Deb01, Chapter 1] or the original source [Kle66]. For a general intersection theoretic approach, readers should consult [Ful84, Section 2.3] for example.

We say two $\mathbb{Q}$-Cartier divisors D_1, D_2 are *numerically equivalent* if

$$D_1.C = D_2.C,$$

for any curve $C \subset X$. When two $\mathbb{Q}$-Cartier divisors D_1, D_2 are numerically equivalent, we write

$$D_1 \equiv D_2.$$

Note that if two $\mathbb{Q}$-Cartier divisors are $\mathbb{Q}$-linearly equivalent, then they are numerically equivalent. We define the space of $\mathbb{Q}$-Cartier divisors modulo numerical equivalence by

$$N^1(X)_{\mathbb{Q}} = \{\mathbb{Q}\text{-Cartier divisors}\}/\equiv,$$

and we call it the *Néron–Severi space*. It is well known that this is a finite dimensional $\mathbb{Q}$-vector space. For smooth projective varieties, this is explained in [Ful84, Section 19.3]. We also define

$$N^1(X) = N^1(X)_{\mathbb{Q}} \otimes_{\mathbb{Q}} \mathbb{R}.$$

One can consider this space as the space of $\mathbb{R}$-Cartier divisors modulo numerical equivalence. We denote by $N^1(X)_{\mathbb{Z}}$ the lattice generated by Cartier divisors.

Next, *a* 1*-cycle* on X is a finite linear combination of integral curves on X. Here coefficients are either in $\mathbb{Z}, \mathbb{Q}$, or $\mathbb{R}$. Let α, α' be 1-cycles with real coefficients. They are numerically equivalent if for any Cartier divisor D we have $D.\alpha = D.\alpha'$, and in this situation we write

$$\alpha \equiv \alpha'.$$

We define the space of 1-cycles modulo numerical equivalence by

$$N_1(X) = \{\text{real 1-cycles}\}/\equiv .$$

By their constructions $N_1(X)$ is the dual vector space of $N^1(X)$. We also denote the lattice generated by integral 1-cycles by $N_1(X)_{\mathbb{Z}}$.

Let X be a normal projective variety defined over F, and L be a $\mathbb{Q}$-Cartier divisor on X. When mL is Cartier and the complete linear series $|mL|$ is non-empty where m is a positive integer, we denote the rational map associated to $|mL|$ by

$$\Phi_{|mL|} : X \dashrightarrow \mathbb{P}^n.$$

The following invariant is one of the most basic birational invariants in the classification theory of algebraic varieties:

Definition 3.2.1 (Iitaka dimension) Let X be a normal projective variety defined over F and L be a $\mathbb{Q}$-Cartier divisor on X. We define the *Iitaka dimension* $\kappa(L)$ by

$$\kappa(L) = \max\{\dim \Phi_{|mL|}(X) \mid m \in \mathbb{Z}_{>0},\ mL \text{ is Cartier and } |mL| \neq \emptyset\}.$$

When there is no positive integer m such that mL is Cartier and $|mL| \neq \emptyset$, we set $\kappa(L) = -\infty$. This is a birational invariant due to Zariski's main theorem and [Laz04, Lemma 2.1.13].

We say L is *rigid* when $\kappa(L) = 0$. In this case, L is $\mathbb{Q}$-linearly equivalent to a unique effective $\mathbb{Q}$-divisor. If we write this as $L \sim_{\mathbb{Q}} \sum_i a_i D_i$ with D_is prime and $a_i > 0$, then one can show that any non-negative linear combination of D_is that is $\mathbb{Q}$-Cartier is rigid.

When X is not normal, we take the normalization $\beta : X' \to X$ and define $\kappa(L) := \kappa(\beta^* L)$. This is well defined owing to birational invariance of the Iitaka dimension.

Next, we introduce the notion of nef divisors and curves, which satisfy certain positivity properties against effective classes. The word "nef" stands for numerically effective:

Definition 3.2.2 (Nef cones) Let X be a projective variety defined over F and let L be an $\mathbb{R}$-Cartier divisor on X. We say L is *nef* if for any curve $C \subset X$, $L.C \geq 0$.

Suppose that we have a real 1-cycle $\alpha \in N_1(X)$. Then we say α is nef if for any effective $\mathbb{Q}$-Cartier divisor D on X, $D.\alpha \geq 0$. We denote the cones of nef classes by

$$\mathrm{Nef}^1(X) \subset N^1(X), \quad \mathrm{Nef}_1(X) \subset N_1(X).$$

These are closed convex cones, and we call them the nef cone of divisors and curves respectively.

We also define the relative version of nefness: Let $f: X \to Y$ be a proper morphism. An $\mathbb{R}$-Cartier divisor is *relatively nef* or *f-nef* if for any curve $C \subset X$ such that $f(C)$ is a point we have $L.C \geq 0$.

Dually we have the notion of pseudo-effective classes, which are defined as limits of effective classes:

Definition 3.2.3 (Pseudo-effective cones) Let X be a projective variety defined over F. We define the *pseudo-effective cone of divisors* as the closure of the cone of effective $\mathbb{Q}$-Cartier divisors, that is,

$$\overline{\mathrm{Eff}}^1(X) = \overline{\text{the cone of effective } \mathbb{Q}\text{-Cartier divisors}} \subset N^1(X).$$

We also define the *pseudo-effective cone of curves* as the closure of the cone of effective real 1-cycles, that is,

$$\overline{\mathrm{Eff}}_1(X) = \overline{\text{the cone of effective real 1-cycles}} \subset N_1(X).$$

We call members of these cones *pseudo-effective classes.*

It is important to note that there are the following dualities between the nef cone of divisors and the pseudo-effective cone of curves, as well as the nef cone of curves and the pseudo-effective cone of divisors, respectively:

$$\mathrm{Nef}^1(X) \longleftrightarrow \overline{\mathrm{Eff}}_1(X),$$
$$\mathrm{Nef}_1(X) \longleftrightarrow \overline{\mathrm{Eff}}^1(X).$$

More explicitly, this means that we have the following identities of the cones:

$$\mathrm{Nef}^1(X) = \{D \in N^1(X) \mid D.\alpha \geq 0 \text{ for any } \alpha \in \overline{\mathrm{Eff}}_1(X)\},$$
$$\overline{\mathrm{Eff}}_1(X) = \{\alpha \in N_1(X) \mid D.\alpha \geq 0 \text{ for any } D \in \mathrm{Nef}^1(X)\}.$$

We have similar identities for the pair $\mathrm{Nef}_1(X)$ and $\overline{\mathrm{Eff}}^1(X)$. We record the following lemma for later use. A similar statement for a very general fiber is obtained in [Pet12, 6.2 Theorem]:

Lemma 3.2.4 *Let $\pi: X \to W$ be a morphism from a smooth projective variety to a normal projective variety with connected fibers. Let $W^\circ \subset W$ be a smooth Zariski open subset such that $\pi|_{\pi^{-1}(W^\circ)}: \pi^{-1}(W^\circ) \to W^\circ$ is smooth. Let $i: X_w \hookrightarrow X$ be the smooth fiber at a closed point $w \in W^\circ$. Then $i_*: N_1(X_w) \to N_1(X)$ maps nef classes on X_w to nef classes on X.*

Proof Any invertible sheaf $\mathcal{L}$ on X is flat over W°, hence the function

$$W^\circ \ni w \mapsto h^0(X_w, \mathcal{L}|_{X_w}) \in \mathbb{R}_{\geq 0}$$

satisfies upper semicontinuity. For any effective divisor D on X, its restriction to a general fiber over W° is effective. Thus $h^0(X_w, \mathcal{O}_X(D)|_{X_w})$ must be positive for general closed points $w \in W^\circ$, and therefore for all closed points $w \in W^\circ$ as well, by upper semicontinuity. Dually we conclude that nef classes on X_w are mapped to nef classes on X. □

The following theorem is quite foundational as well:

Theorem 3.2.5 ([Kle66]) *Let X be a projective variety defined over F. Let* $\mathrm{Amp}(X) \subset N^1(X)$ *be the ample cone of divisors, that is, the cone generated by ample $\mathbb{Q}$-Cartier divisors. Then the interior of* $\mathrm{Nef}^1(X)$ *coincides with* $\mathrm{Amp}(X)$*:*

$$\mathrm{int}(\mathrm{Nef}^1(X)) = \mathrm{Amp}(X).$$

Moreover, the closure of $\mathrm{Amp}(X)$ *is the nef cone of divisors:*

$$\overline{\mathrm{Amp}(X)} = \mathrm{Nef}^1(X).$$

Proof One may consult [Laz04, Theorem 1.4.23] or the original paper [Kle66]. □

Using the above theorem, one can deduce the following criterion of ampleness:

Corollary 3.2.6 ([Kle66]) *Let X be a projective variety defined over F and L be a $\mathbb{Q}$-Cartier divisor on X. Then L is ample if and only if, for any $\alpha \in$* $\overline{\mathrm{Eff}}_1(X) \setminus \{0\}$, $L.\alpha > 0$.

Proof One may consult [Laz04, Theorem 1.4.29] for a proof. □

Let X be a projective variety defined over F. Let C be a curve on X. We say C is *movable* if it is a member of a flat family $\pi: \mathcal{C} \to W$ of curves on X such that W is irreducible and the evaluation map $s: \mathcal{C} \to X$ is dominant. It is easy to verify that if a curve C is movable then it is nef. This results in the structural theorem for $\mathrm{Nef}_1(X)$:

Theorem 3.2.7 ([BDPP13]) *Suppose that $\overline{F} = F$. Let X be a smooth projective variety defined over F. Then we have*

$$\mathrm{Nef}_1(X) = \overline{\textit{the cone generated by movable curves}} \subset N_1(X).$$

The notion of big divisors is also important in modern algebraic geometry. In some classical references, this notion is called "almost ample":

Theorem 3.2.8 *Let X be a projective variety and L be a $\mathbb{Q}$-Cartier divisor on X. Then the following are equivalent:*

(i) *there exists $m \in \mathbb{Z}_{>0}$ such that mL is Cartier and $\Phi_{|mL|}\colon X \dashrightarrow \mathbb{P}^n$ is a generically finite map to the image;*
(ii) *one can write $L = A + N$ where A is an ample $\mathbb{Q}$-Cartier divisor and N be an effective $\mathbb{Q}$-Cartier divisor;*
(iii) $L \in \mathrm{int}(\overline{\mathrm{Eff}}^1(X))$.

Proof The equivalence (1) $\iff$ (2) follows from [Laz04, Corollary 2.2.7]. Finally the equivalence (1) $\iff$ (3) follows from [Laz04, Theorem 2.2.26]. □

Definition 3.2.9 (Big divisors) Let X be a projective variety defined over F, and L be an $\mathbb{R}$-Cartier divisor on X. We say L is *big* if $L \in \mathrm{int}(\overline{\mathrm{Eff}}^1(X))$.

We also need the relative notion of bigness:

Definition 3.2.10 Let $f\colon X \to Y$ be a projective morphism between quasi-projective varieties defined over F. Let L be a $\mathbb{Q}$-Cartier divisor on X. Then we say L is *relatively big* or *f-big* if the restriction of L to the generic fiber X_η is big.

Here is a characterization of relative bigness which is analogous to Theorem 3.2.8:

Proposition 3.2.11 *Let $f\colon X \to Y$ be a projective morphism between quasi-projective varieties defined over F. Let L be a $\mathbb{Q}$-Cartier divisor on X. Then the following statements are equivalent:*

(i) *L is f-big;*
(ii) *L is $\mathbb{Q}$-linearly equivalent to $A + E$, where A is an f-ample $\mathbb{Q}$-divisor and E is an effective $\mathbb{Q}$-divisor.*

Furthermore if Y is projective and M is a big $\mathbb{Q}$-divisor on Y, then the above statements are equivalent to

(iii) *$L + mf^*M$ is big for $m \gg 1$.*

Proof This can be found in [KM98, Lemma 3.23] . □

Let X be a normal projective variety defined over F and let L be a pseudo-effective $\mathbb{Q}$-Cartier divisor on X. Then the *stable base locus* of L is

$$\mathbf{B}(L) = \bigcap_{m\in\mathbb{Z}_{>0},\ mL \text{ is Cartier}} \mathrm{Bs}(|mL|),$$

where $\mathrm{Bs}(|mL|)$ is the base locus of the complete linear series $|mL|$.

Definition 3.2.12 (Augmented base locus) Let X be a normal projective variety defined over F and let L be a pseudo-effective $\mathbb{Q}$-Cartier divisor on X. The *augmented base locus of L* is

$$\mathbf{B}_+(L) = \bigcap_{A\,:\,\text{ample }\mathbb{Q}\text{-Cartier}} \mathbf{B}(L - A).$$

It follows from [ELM+06, Proposition 1.5] that one can find an ample $\mathbb{Q}$-Cartier divisor A such that $\mathbf{B}_+(L) = \mathbf{B}(L - A)$, so in particular, this is a closed subset. When L is big, for any subvariety $Y \not\subset \mathbf{B}_+(L)$, $L|_Y$ is big. Moreover, when L is nef, it is shown in [Nak00, Theorem 0.3] that we have

$$\mathbf{B}_+(L) = \bigcup_{Y \subset X,\; L^{\dim Y}.Y=0} Y,$$

where Y is a subvariety in X.

Using the notion of stable base loci, one can define the following notion of divisors, which is frequently used in modern higher dimensional algebraic geometry:

Definition 3.2.13 (Movable divisors) Let X be a normal projective variety defined over F. The *movable cone of divisors* $\overline{\mathrm{Mov}}^1(X) \subset N^1(X)$ is the closure of the cone of effective $\mathbb{Q}$-Cartier divisors whose stable base loci have codimension ≥ 2. A movable $\mathbb{R}$-Cartier divisor is any member of $\overline{\mathrm{Mov}}^1(X)$.

The volume functions of divisors are also essential in our applications:

Definition 3.2.14 (Volume functions) Let X be a normal projective variety of dimension n, and L be a Cartier divisor on X. The *volume of L* is

$$\mathrm{vol}(L) = \limsup_{m\to\infty} \frac{h^0(X, mL)}{m^n/n!}.$$

Here are some elementary properties of volume functions:

Lemma 3.2.15 *Let X be a normal projective variety of dimension n, and L be a Cartier divisor on X. Then the following statements hold:*

(i) *For any positive integer $a > 0$, we have* $\mathrm{vol}(aL) = a^n \mathrm{vol}(L)$. *Thus one can extend the definition of the volume to any $\mathbb{Q}$-Cartier divisor.*
(ii) *The volume function defines a continuous function $N^1(X)_{\mathbb{Q}} \to \mathbb{R}$, so in particular one can extend the definition of the volume to any $\mathbb{R}$-Cartier divisor.*
(iii) *We have* $\mathrm{vol}(L) > 0 \iff$ *L is big.*

(iv) *When L is nef, we have* $\mathrm{vol}(L) = L^n$.

Proof See [Laz04, Section 2.2.C] for proofs. □

Finally, we record the negativity lemma, which is quite useful in higher dimensional algebraic geometry:

Lemma 3.2.16 *Let $\beta\colon Y \to X$ be a birational projective morphism between normal quasi-projective varieties. Then the following statements are true:*

(i) *Let $E > 0$ be an exceptional effective $\mathbb{R}$-Cartier divisor on Y. Then there is a component F of E that is covered by curves Σ contracted by β such that $E.\Sigma < 0$.*
(ii) *If $\beta^* L \equiv M+G+E$, where L is a $\mathbb{R}$-Cartier divisor, M is a β-nef $\mathbb{R}$-Cartier divisor on Y, $G \geq 0$, E is β-exceptional, and E and G have no common components, then E is effective, that is, $E \geq 0$.*
(iii) *If X is $\mathbb{Q}$-factorial, then there is a β-exceptional divisor $E \geq 0$ such that $-E$ is ample over X. In particular, the exceptional locus of β is divisorial. (See Section 3.3 for the definition of $\mathbb{Q}$-factoriality.)*

Proof The proof of this lemma can be found in [BCHM10, Lemma 3.6.2]. □

3.3 Singularities in the Minimal Model Program

Even when we apply a minimal model program to a smooth projective variety, in general it will lead to a mildly singular projective variety as an outcome. Thus it is important to study how to deal with singularities in higher dimensional algebraic geometry. In this section, we recall various notions of singularities studied in the minimal model program. A basic reference is [KM98, Chapter 2] and we will cite this book each time we mention results from it. Again the results in [KM98] are stated over an algebraically closed field of characteristic 0, but their adaptation to an arbitrary field of characteristic 0 is straightforward.

Let F be a field of characteristic 0 and let X be a normal variety defined over F. First of all we define the notion of $\mathbb{Q}$-factoriality, which is preserved under any minimal model program: we say X is *$\mathbb{Q}$-factorial* if any prime divisor D on X is $\mathbb{Q}$-Cartier. We say X is *geometrically $\mathbb{Q}$-factorial* if $\overline{X}$ is $\mathbb{Q}$-factorial. Note that geometric $\mathbb{Q}$-factoriality implies $\mathbb{Q}$-factoriality but the converse fails in general.

Suppose that X is a smooth variety of dimension n. We denote the coherent sheaf of differential 1-forms by Ω_X^1: this is a locally free sheaf of rank n because of smoothness. Its determinant $\omega_X := \bigwedge^n \Omega_X^1$ is an invertible sheaf. This means that there is a Cartier divisor K_X such that $O_X(K_X) \cong \omega_X$. We call K_X the *canonical divisor of* X. Note that this divisor is not uniquely determined but we fix one divisor of this type each time throughout the discussion of the canonical divisor.

When X is only normal, we denote the smooth locus of X by X_{sm}, which is an open set of X. Moreover, since X is normal, the codimension of the closed set $X_{\mathrm{sing}} := X \setminus X_{\mathrm{sm}}$ is strictly greater than 1. This means that there is a unique Weil divisor K_X such that $K_X|_{X_{\mathrm{sm}}} = K_{X_{\mathrm{sm}}}$. We call this K_X the canonical divisor of X.

Let X be a normal variety and Δ be an effective $\mathbb{Q}$-divisor on X. We say a pair (X, Δ) is a *log pair* if $K_X + \Delta$ is $\mathbb{Q}$-Cartier.

Let (X, Δ) be a log pair. Suppose that we have a birational morphism,

$$f : Y \to X,$$

between normal varieties. Since $K_X + \Delta$ is $\mathbb{Q}$-Cartier, the pullback $f^*(K_X + \Delta)$ is well defined. Choose the canonical divisor K_Y so that $f_* K_Y = K_X$. Then one can uniquely write

$$K_Y = f^*(K_X + \Delta) + \sum_{E:\ \text{prime divisor}} a(E, X, \Delta)E,$$

where E runs over all prime divisors on Y and $a(E, X, \Delta) \in \mathbb{Q}$. We call $a(E, X, \Delta)$ the *discrepancy of* E. Note that, unless E is exceptional or the strict transform of a component of the support of Δ, we have $a(E, X, \Delta) = 0$. Thus the above summation is a finite sum.

The following definitions are central to the minimal model program and crucial to Manin's conjecture too:

Definition 3.3.1 Let (X, Δ) be a log pair.

We say (X, Δ) has only *terminal* (resp. *canonical*) singularities if for any birational morphism $f : Y \to X$ between normal varieties and any exceptional prime divisor E on Y, we have $a(E, X, \Delta) > 0$ (resp. ≥ 0).

We say (X, Δ) has only *kawamata log terminal* (resp. *log canonical*) singularities if for any birational morphism $f : Y \to X$ between normal varieties and any prime divisor E on Y, we have $a(E, X, \Delta) > -1$ (resp. ≥ -1). We use the notation klt for kawamata log terminal and lc for log canonical. Note that when (X, Δ) is klt, the above definition implies that $\lfloor \Delta \rfloor = 0$ where $\lfloor \Delta \rfloor$ is the *round down* given by $\lfloor \Delta \rfloor = \sum \lfloor a_i \rfloor D_i$ when $\Delta = \sum a_i D_i$.

Fix $\epsilon > 0$. We say (X, Δ) has only *ϵ-lc* singularities if for any birational morphism $f: Y \to X$ between normal varieties and any prime divisor E on Y, we have $a(E, X, \Delta) \geq -1 + \epsilon$.

We say X has only terminal/canonical/klt/lc/ϵ-lc singularities if $(X, 0)$ does.

Here are some examples of these singularities:

Example 3.3.2 It is an easy exercise to prove that smooth varieties have only terminal singularities. See, for example, [KM98, Corollary 2.31] for a proof.

Example 3.3.3 When S is a surface, S has only terminal singularities if and only if S is smooth. See [KM98, Theorem 4.5] for a proof.

The surface S has only canonical singularities if and only if S has du Val singularities. Readers may consult [KM98, Theorem 4.5] for a proof. These singularities are classified analytically and well understood when the ground field is algebraically closed. See [KM98, Theorem 4.22] for more details.

The following lemma follows from the definition of singularities:

Lemma 3.3.4 *Let (X, Δ) be a log pair and $\Delta' \geq 0$ be an effective $\mathbb{Q}$-Cartier divisor on X. Then the following statements are true:*

(i) *If $(X, \Delta + \Delta')$ has only terminal/canonical/klt/lc singularities, then so has (X, Δ);*
(ii) *if (X, Δ) has only terminal/klt singularities, then for any sufficiently small positive rational number ϵ, $(X, \Delta + \epsilon\Delta')$ has only terminal/klt singularities.*

Proof This is easy to prove, but one might find it in [KM98, Corollary 2.35]. □

The following notion of singularities is also important in the minimal model program and a large part of the MMP is established for this category:

Definition 3.3.5 Let X be a normal variety defined over F, and let $\Delta \geq 0$ be an effective $\mathbb{Q}$-divisor $\sum_i a_i D_i$ such that D_is are distinct prime divisors and $0 \leq a_i \leq 1$. Assume that (X, Δ) is a log pair. We say a pair (X, Δ) has only *divisorial log terminal* (dlt for short) singularities if there exists a closed set $V \subset X$ such that

(i) $X \setminus V$ is smooth and the support of $\Delta|_{X \setminus V}$ is a strict normal crossings divisor; and
(ii) for any birational morphism $\beta: Y \to X$ from a normal variety and any prime divisor $E \subset Y$ with $f(E) \subset V$, we have $a(E, X, \Delta) > -1$.

3.4 The Minimal Model Program

The Italian school in algebraic geometry completed the minimal model program for smooth projective surfaces in the early twentieth century. The main result is Castelnuovo's theorem on contraction of (-1)-curves. Originally, it was considered to be impossible to generalize this story to higher dimension, but Mori took the first step towards this generalization and established the cone and contraction theorems for smooth projective varieties in [Mor82], which were generalizations of Castelnuovo's theorem in higher dimension. The cone and contraction theorems were established for mildly singular projective varieties by the contributions of many mathematicians including Kawamata, Kollár, Shokurov, Reid, and others, (see, e.g., [Kaw84], [Sho85], and [KMM87]), and Mori completed the minimal model program in dimension 3 in the celebrated paper [Mor88] by proving the existence of 3-dimensional flips. After this paper, there were many developments regarding the log minimal model program in dimension 3, 4, and so on, but the attempts to generalize the minimal model program in higher dimension were not so successful until [BCHM10]. In [BCHM10] and [HM10], Birkar, Cascini, Hacon, and McKernan completed a large part of the (log) minimal model program using complicated inductive arguments in dimension, and these results are essential for our study of Manin's conjecture. Here we recall some results from [BCHM10].

Let F be a field of characteristic 0. Suppose that $f\colon X \dashrightarrow Y$ is a birational map between normal projective varieties defined over F. We say that f is a *birational contraction* if the inverse map $f^{-1}\colon Y \dashrightarrow X$ does not contract any prime divisor on Y. We will use the following form of results from [BCHM10]:

Theorem 3.4.1 ([BCHM10]) *Let X be a smooth projective variety defined over F, and L be a big and nef $\mathbb{Q}$-divisor on X. Assume that $L + K_X$ is pseudo-effective. After running a $(K_X + L)$-minimal model program, we achieve a birational contraction $\phi\colon X \dashrightarrow X'$ such that*

(i) *X' is a $\mathbb{Q}$-factorial terminal projective variety;*
(ii) *$K_{X'} + \phi_* L$ is semi-ample.*

Here a $\mathbb{Q}$-Cartier divisor D is *semi-ample* if some multiple of D has the basepoint-free complete linear series.

A birational contraction $\phi\colon X \dashrightarrow X'$ is an outcome of a process called a *$(K_X + L)$-MMP* which stands for a $(K_X + L)$-minimal model program, and it can be obtained as the composition of a sequence of birational maps

$$\phi_i\colon X_i \dashrightarrow X_{i+1},$$

which is either a $(K_X + L)$-*divisorial contraction* or $(K_X + L)$-*flip*. A $(K_X + L)$-divisorial contraction is a certain projective birational morphism whose exceptional locus consists of a unique prime divisor. Examples of such operations include the contraction of a (-1)-curve in 2-dimensional geometry, as well as blowups along smooth centers on smooth projective varieties. A $(K_X + L)$-flip is a birational map which is isomorphic in codimension 1, that is, outside of codimension ≥ 2 loci, $\phi_i : X_i \dashrightarrow X_{i+1}$ induces an isomorphism of Zariski open subsets. Such an operation only appears in dimension ≥ 3. The existence of a $(K_X + L)$-MMP is guaranteed by the contributions of several people which is finalized in [BCHM10]. Its termination is still open in general, but [BCHM10] confirms termination of special MMPs called the *minimal model program with scaling*. An outcome of a $(K_X + L)$-MMP is called a $(K_X + L)$-*minimal model.* Even when $L + K_X$ is not pseudo-effective, one can still perform a $(K_X + L)$-MMP, but an outcome will be a *Mori fiber space* $\pi : X' \to B$ with a birational contraction $\phi : X \dashrightarrow X'$ in the sense that

(i) X' is a $\mathbb{Q}$-factorial terminal projective variety;
(ii) $\pi : X' \to B$ is an algebraic fiber space;
(iii) $-(K_{X'} + \phi_* L)$ is relatively π-ample; and
(iv) for any two curves C and C' contracted by π, there exists some rational numbers $c > 0$ such that for any Cartier divisor D on X', we have

$$D.C = cD.C'.$$

Actual steps of the MMP do not matter in most proofs in this book, so we omit details and refer the reader to [KM98] for the definitions and basic properties.

To deduce Theorem 3.4.1 from [BCHM10], we will need the following lemma:

Lemma 3.4.2 *Let X be a projective variety defined over F. and L be a big and nef $\mathbb{Q}$-Cartier divisor on X. Then there exists an effective divisor $E \geq 0$ such that for any sufficiently small rational $t > 0$, $L - tE$ is ample.*

Proof The following proof is taken from [KM98, Proposition 2.61]. It follows from a characterization of bigness (Theorem 3.2.8) that there exist some positive integer m, an ample divisor A, and an effective divisor E such that $mL = A + E$. Then one can write

$$L = t(A + (1/t - m)L) + tE.$$

When $1/t - m \geq 0$, $t(A + (1/t - m)L)$ is ample by Corollary 3.2.6 because L is nef. Thus our assertion follows. □

Now let us deduce Theorem 3.4.1 from [BCHM10]:

Proof of Theorem 3.4.1 Let E be an effective divisor as in Lemma 3.4.2. By Lemma 3.3.4 and Lemma 3.4.2, for a sufficiently small rational $t > 0$, (X, tE) is terminal and $L - tE$ is ample. By Bertini's theorem, (see [Kol97, Lemma 4.7 and Theorem 4.8]), there exists $0 \leq A_t \sim_{\mathbb{Q}} L - tE$ such that $(X, A_t + tE)$ has only terminal singularities. By [BCHM10, Corollary 1.4.2], we run a $(K_X + A_t + tE)$-minimal model program and obtain a birational contraction $\phi\colon X \dashrightarrow X'$ such that $(X', K_{X'} + \phi_* A_t + t\phi_* E)$ has only klt singularities and $K_{X'} + \phi_* L$ is semi-ample. Thus we need to show that X' has only terminal singularities. Since the minimal model program is numerical, for sufficiently small $t' > 0$, $\phi\colon X \dashrightarrow X'$ is a $(K_X + L - t'E)$-MMP. Then it follows from [KM98, Corollaries 3.42 and 3.43] that there exists $0 \leq A_{t'} \sim_{\mathbb{Q}} L - t'E$ such that $(X', \phi_* A_{t'})$ is terminal. By Lemma 3.3.4, X' has only terminal singularities. □

Remark 3.4.3 In the above proof, one may choose t'' sufficiently small so that $(X', t''\phi_* E)$ is terminal. Then by [KM98, Corollaries 3.42 and 3.43], there exists $0 \leq A_{t''} \sim_{\mathbb{Q}} L - t''E$ such that $(X', \phi_* A_{t''} + t''\phi_* E)$ is terminal. The upshot is that one can find $0 \leq L' \sim_{\mathbb{Q}} L$ such that $(X', \phi_* L')$ is a terminal pair.

3.4.1 Canonical Maps Associated with Adjoint Divisors

In the setting of Theorem 3.4.1, $K_{X'} + \phi_* L$ is semi-ample, so it defines the semi-ample fibration $\pi'\colon X' \to W$ associated to $K_{X'} + \phi_* L$ (see cite[Theorem 2.1.27]Laz04). Its composition $\pi = \pi' \circ \phi\colon X \dashrightarrow W$ has the following concrete description: let m be a positive integer such that mL is Cartier. Then we consider a graded algebra

$$R = \bigoplus_{n \geq 0} H^0(X, \mathcal{O}(n(mK_X + mL))).$$

It follows from [BCHM10, Corollary 1.1.2] that this algebra is finitely generated. Thus we can construct

$$\psi\colon X \dashrightarrow \operatorname{Proj}(R),$$

and we can show that this is isomorphic to $\pi\colon X \dashrightarrow W$. We call this map the *canonical map* or the *canonical model of* (X, L). This is also birational to the Iitaka fibration associated to $K_X + L$, as in [Laz04, Theorem 2.1.33], so in particular, for a general closed point $w \in W$, the restriction to the fiber X_w, that is, $(K_X + L)|_{X_w}$ is rigid.

3.4.2 Zariski Decomposition

The following notion is also useful in our study:

Definition 3.4.4 ([KMM87, Definition 7-3-5]) Let F be a field of characteristic 0. Let X be a normal projective variety defined over F. Let $D \sim_{\mathbb{Q}} P + N$ be an expression of $\mathbb{Q}$-Cartier divisors on X. This expression is called the *Zariski decomposition of D in the sense of Cutkosky–Kawamata–Moriwaki* (we write C–K–M for short) if the following conditions hold:

(i) P is nef;
(ii) N is effective; and
(iii) for a sufficiently divisible m such that mP and mD are Cartier, the natural map

$$H^0(X, \mathcal{O}_X(mP)) \to H^0(X, \mathcal{O}_X(mD))$$

is an isomorphism.

The divisor P is called the *positive part* and N is called the *negative part*. If they exist, then N is unique and P is unique up to $\mathbb{Q}$-linear equivalence.

In general it is not true that every pseudo-effective $\mathbb{Q}$-Cartier divisor D admits a Zariski decomposition in the sense of Definition 3.4.4. However, this is true after a birational modification when the MMP with respect to D runs and terminates:

Theorem 3.4.5 ([Kaw87, Proposition 5]) *Let F be a field of characteristic* 0. *Let X be a smooth projective variety defined over F, and L be a big and nef $\mathbb{Q}$-divisor on X. Assume that $K_X + L$ is pseudo-effective. Let $\phi\colon X \dashrightarrow X'$ be the outcome of a $(K_X + L)$-MMP whose existence is guaranteed by Theorem 3.4.1. Let $\widetilde{X}$ be a common resolution of ϕ with projections $q\colon \widetilde{X} \to X$ and $p\colon \widetilde{X} \to X'$. Let $D = K_X + L$ and $P = K_{X'} + \phi_* L$. The divisor P is nef by Theorem 3.4.1. Then there exists an effective p-exceptional $\mathbb{Q}$-divisor E on $\widetilde{X}$ such that we have*

$$q^* D \sim_{\mathbb{Q}} p^* P + E,$$

which is the Zariski decomposition in the sense of C–K–M.

Proof One may consult [FS04, Theorem 5.4]. In [FS04], the statement is given for toric varieties; however, the proof works as soon as a $(K_X + L)$-MMP runs and terminates, which is guaranteed by Theorem 3.4.1. □

3.4.3 The Relative Minimal Model Program

In this section, we assume that our ground field F is an algebraically closed field of characteristic 0. The minimal model program can be formulated in the relative setting $f\colon X \to S$ where f is a projective morphism between normal quasi-projective varieties. Readers interested in its formulation should consult [KM98, Section 3.31] for details. When X is smooth and L is f-big and f-nef, the relative $(K_X + L)$-MMP over S always exists. Moreover, by [BCHM10, Corollary 1.4.2], any relative $(K_X + L)$-MMP over S with scaling always terminates.

In particular, one can use this to perform the MMP in a family: let $f\colon X \to Y$ be a smooth projective morphism between smooth quasi-projective varieties. Let L be an f-big and f-nef $\mathbb{Q}$-divisor on X. In [KM92, Proposition 12.2.5], a local system $\mathcal{G}N^1(X/Y)$ over Y in the étale topology with a finite monodromy is constructed. In [KM92], it is constructed in the analytic topology over $\mathbb{C}$, but one can apply the same construction in étale topology using the comparison theorem. See [LT17, Remark 3.3] for more details. When $H^1(X_t, \mathcal{O}_{X_t}) = 0$ for any $t \in Y$, we have an isomorphism $\mathcal{G}N^1(X/Y)|_t \cong N^1(X_t)$ by Hodge theory. Here is one corollary of the relative MMP that we will use later:

Theorem 3.4.6 ([dFH11, Theorem 4.1]) *Let $f\colon X \to Y$ be a smooth projective morphism between smooth quasi-projective varieties defined over F, and let L be an f-big and f-nef $\mathbb{Q}$-divisor on X. Assume that $H^1(X_t, \mathcal{O}_{X_t}) = 0$ for any $t \in Y$ and the fiber X_t of f at t, and that the monodromy action on $\mathcal{G}N^1(X/Y)$ is trivial. We run a relative $(K_X + L)$-MMP over Y and the outcome is a Y-birational contraction $\phi\colon X \dashrightarrow X'$ with the structure morphism $f'\colon X' \to Y$, which is projective. Then for a general $t \in Y$, $\phi|_{X_t}\colon X_t \dashrightarrow X'_t$ is the outcome of an absolute $(K_{X_t} + L|_{X_t})$-MMP, where X_t and X'_t denote fibers of f and f' at t respectively.*

Proof This follows from the proof of [dFH11, Theorem 4.1]. □

We also recall the following theorem on deformation invariance of log plurigenera:

Theorem 3.4.7 ([HMX13, Theorem 1.8]) *Let $f\colon X \to Y$ be a projective morphism of smooth varieties defined over F. Let (X, L) be a klt pair that has simple normal crossings over Y, and for which L is big. Let m be a positive integer such that mL is integral. Then $h^0(X_y, \mathcal{O}_{X_y}(m(K_{X_y} + L|_{X_y})))$ is independent of $y \in Y$.*

Proof This is proved in (1) of [HMX13, Theorem 1.8]. □

3.5 The Boundedness of Singular Fano Varieties

The main protagonists of this book are (log) Fano varieties. Here is their definition:

Definition 3.5.1 Let (X, Δ) be a log pair defined over a field F of characteristic 0.

We say that (X, Δ) is *log Fano* if the following conditions hold:

(i) X is projective;
(ii) (X, Δ) has only klt singularities; and
(iii) the anti-log canonical divisor $-(K_X + \Delta)$ is ample.

We say X is *Fano* if $(X, 0)$ is log Fano, and *weak Fano* if X is projective, X has only klt singularities, and $-K_X$ is big and nef.

Here are some examples of Fano varieties:

Example 3.5.2 (Fano hypersurfaces) Let $X \subset \mathbb{P}^n$ be a smooth hypersurface of degree d. Let H be the hyperplane class on X. Then it follows from the adjunction formula that $-K_X \sim (n+1-d)H$. Thus we conclude that X is Fano if and only if $d \leq n$.

Example 3.5.3 (Del Pezzo surfaces) A smooth Fano surface is called a *del Pezzo surface*. Let S be a del Pezzo surface defined over a field of characteristic 0. Then the degree $(-K_S)^2$ is an integer between 1 and 9, and there are 10 deformation types.

Indeed, assume that $\overline{F} = F$. Then we have the following classification:

(i) when $(-K_S)^2 = 9$, we have $S \cong \mathbb{P}^2$;
(ii) when $(-K_S)^2 = 8$, we have $S \cong \mathbb{P}^1 \times \mathbb{P}^1$ or $\mathrm{Bl}_x(\mathbb{P}^2)$;
(iii) when $(-K_S)^2 = 9 - r$ with $(r \geq 2)$, we have $S \cong \mathrm{Bl}_{x_1,\ldots,x_r}(\mathbb{P}^2)$ where $x_1, \ldots, x_r$ are r general points on $\mathbb{P}^2$.

Here $\mathrm{Bl}_{x_1,\ldots,x_r}(\mathbb{P}^2)$ denotes the blowup of $\mathbb{P}^2$ at $x_1, \ldots, x_r$.

A smooth weak Fano surface is called a *weak del Pezzo surface*. Such a class of surfaces has a similar classification to the one above.

Example 3.5.4 (Smooth Fano threefolds) Suppose that F is an algebraically closed field of characteristic 0. Then there exists a complete classification of smooth Fano threefolds defined over F. Smooth Fano threefolds of Picard rank 1 are classified by Fano and Iskovskikh ([Isk77], [Isk78], and [Isk79]), and there are 17 deformation types. Smooth Fano threefolds of Picard rank ≥ 2 are classified by Mori–Mukai ([MM82], [MM83], and [MM03]), and there are 88 deformation types. [IP99] includes many classification results on Fano varieties. The website [Bel25] is also useful and accessible.

It is shown in [KMM92a] that smooth Fano varieties of fixed dimension are *bounded*, that is, they are parametrized by finitely many schemes of finite type, using the geometry of rational curves on Fano varieties. This was extended to singular toric Fano varieties by Borisov–Borisov in [BB92] and singular log del Pezzo surfaces by Alexeev in [Ale94], and the boundedness of singular Fano varieties was formulated as the Borisov–Alexeev–Borisov conjecture (BAB conjecture). This conjecture was proved by Birkar in [Bir19] and [Bir21]:

Theorem 3.5.5 ([Bir21, Theorem 1.1]) *Let F be an algebraically closed field of characteristic* 0. *Let n be a positive integer and $\epsilon > 0$ be a positive real number. Then projective varieties X of dimension n defined over F such that*

(i) *there exists an effective $\mathbb{Q}$-divisor $\Delta \geq 0$ such that (X, Δ) has only ϵ-lc singularities, and*
(ii) $-(K_X + \Delta)$ *is big and nef,*

form a bounded family.

As a corollary, we have the following result.

Corollary 3.5.6 ([Bir19, Theorem 1.6]) *Let F be a field of characteristic* 0. *Let n be a positive integer and $\epsilon > 0$ be a positive real number. There exists a positive constant $\mathsf{C} = \mathsf{C}(n, \epsilon) > 0$ such that any geometrically integral ϵ-lc weak Fano variety X of dimension n defined over F satisfies*

$$\mathrm{vol}(-K_X) = (-K_X)^n \leq \mathsf{C}(n, \epsilon).$$

This corollary is one of the main inputs in the applications of birational geometry to Manin's conjecture in [LTT18], [HJ17], [LT17], [LT19a], [Sen21], and [LST22], and we will use it later.

Finally, we record a cone theorem that follows from Theorem 3.5.5:

Lemma 3.5.7 ([Ara10]) *Let X be a smooth projective variety defined over a field F of characteristic* 0, *and Δ be an effective $\mathbb{Q}$-divisor on X such that (X, Δ) is klt. Fix an ample $\mathbb{Q}$-divisor A on X. Then the cone*

$$\overline{\mathrm{Eff}}_1(X)_{K_X+\Delta\geq 0} + \mathrm{Nef}_1(X)$$

has only finitely many extremal rays with negative intersection with $K_X + \Delta + A$. Moreover, these rays are generated by movable curves. Here the cone $\overline{\mathrm{Eff}}_1(X)_{K_X+\Delta\geq 0}$ is defined as

$$\overline{\mathrm{Eff}}_1(X)_{K_X+\Delta\geq 0} = \{\alpha \in \overline{\mathrm{Eff}}_1(X) \mid (K_X + \Delta).\alpha \geq 0\}.$$

For a proof, readers might consult [LST22, Lemma 3.7].

3.6 Mori Dream Spaces

In this section, we review the notion of Mori dream spaces, which is introduced in [HK00]. We assume that our ground field F is a field of characteristic 0. First let us define the following notion:

Definition 3.6.1 ([HK00, 1.8 Definition]) Let X be a normal projective variety defined over F. A birational contraction $f\colon X \dashrightarrow X'$ to a normal projective variety is a *small $\mathbb{Q}$-factorial modification* (SQM) if X' is $\mathbb{Q}$-factorial and f is an isomorphism in codimension 1.

The following definition is quite influential in modern algebraic geometry and it also plays important roles in arithmetic geometry:

Definition 3.6.2 ([HK00, 1.10 Definition]) Let F be an algebraically closed field of characteristic 0. Let X be a normal projective variety defined over F. We say X is a *Mori dream space* if the following conditions hold:

(i) X is $\mathbb{Q}$-factorial and $\mathrm{Pic}(X)_{\mathbb{Q}} = N^1(X)_{\mathbb{Q}}$;
(ii) $\mathrm{Nef}^1(X)$ is the affine hull of finitely many semi-ample line bundles; and
(iii) there are finitely many SQMs $f_i\colon X \dashrightarrow X_i$ such that each X_i satisfies (ii) and $\overline{\mathrm{Mov}}^1(X)$ is the union of the $f_i^*(\mathrm{Nef}^1(X_i))$.

When F is a field of characteristic 0, we say X is a Mori dream space if $\overline{X}$ is.

A reason why this is called a Mori dream space is given by the following proposition:

Proposition 3.6.3 ([HK00, 1.11 Proposition (1)]) *Let X be a Mori dream space defined over F. Then, for any divisor D, any D-MMP runs and terminates, that is, all necessary contractions and flips exist, any sequence terminates, and if the strict transform of D becomes nef in any D-MMP, then it is semi-ample.*

Being a Mori dream space is closely related to the finite generation of a ring called the Cox ring, which is the subject of intensive studies. One can interpret the Cox ring as the multi-graded ring encoding all projective embeddings of the underlying projective variety at once. See, for example, [ADHL15] for more information.

Definition 3.6.4 ([HK00, 2.6 Definition]) Let F be an algebraically closed field of characteristic 0. Let X be a $\mathbb{Q}$-factorial normal projective variety defined over F such that $\mathrm{Pic}(X)_{\mathbb{Q}} = N^1(X)_{\mathbb{Q}}$. We choose a basis $\mathcal{O}_X(D_1), \dots, \mathcal{O}_X(D_r)$ for $\mathrm{Pic}(X)/\mathrm{Pic}(X)_{\mathrm{tor}}$ where $\mathrm{Pic}(X)_{\mathrm{tor}}$ is the torsion part of $\mathrm{Pic}(X)$. The *Cox ring* *of* X is the following multi-graded ring:

$$\mathrm{Cox}(X) = \bigoplus_{(n_1,\ldots,n_r)\in\mathbb{Z}^r} H^0(X, \mathcal{O}_X(n_1D_1 + \cdots + n_rD_r)).$$

When $\mathrm{Pic}(X)$ is torsion free, this definition does not depend on the choice of our basis. In general this ring does depend on the choice of a basis, but certain properties such as finite generation do not. There is a more sophisticated way to define the Cox ring in general. See [ADHL15, Section 1.4.2], for example.

One of the foundational results in [HK00] is the following theorem, which is proved using Geometric Invariant Theory (GIT):

Theorem 3.6.5 ([HK00, 2.9 Proposition]) *Let X be a normal $\mathbb{Q}$-factorial projective variety defined over F such that* $\mathrm{Pic}(\overline{X})_{\mathbb{Q}} = N^1(\overline{X})_{\mathbb{Q}}$. *Then X is a Mori dream space if and only if* $\mathrm{Cox}(\overline{X})$ *is finitely generated.*

One example of Mori dream spaces is any toric variety. Indeed, for a $\mathbb{Q}$-factorial normal projective toric variety X, $\mathrm{Cox}(\overline{X})$ is a polynomial ring. This result is due to Cox [Cox95], hence its name. Other classical examples are certain almost homogeneous spaces such as spherical varieties. See [BK94] for more details.

The following theorem is quite important in higher dimensional algebraic geometry:

Theorem 3.6.6 ([BCHM10, Corollary 1.3.2]) *Let X be a log Fano variety defined over F. Then X is a Mori dream space.*

This was proved in [HK00] for dimension ≤ 3 using the low dimensional minimal model program. In higher dimension, this is an outcome of [BCHM10]. We will use this result later in our study of Manin's conjecture.

4
Manin's Conjecture

Manin's program, initiated in [FMT89] and [BM90], provides a way to understand the asymptotic formula for the counting function of rational points on a Fano variety, and this program leads to Manin's conjecture, which was further revised in a series of papers [Pey95], [BT98b], [Pey03], [Pey17], [LST22], and [LS24]. In this chapter, we discuss the formulation of Manin's conjecture and examples that have been verified in the literature. In Manin's conjecture it is important to remove the contribution of the exceptional set so that the asymptotic formula of the counting function reflects the global geometry of a projective variety. In Section 4.3 we discuss a conjectural construction of exceptional sets in Manin's conjecture which was proposed in [LST22] by Lehmann, Sengupta, and the author.

4.1 Geometric Invariants

In this section, we discuss the birational invariants appearing in Manin's conjecture. They first appeared in this context in [BM90], and [HTT15] provided the first systematic study of these invariants. Their geometry has been further developed in [LTT18], [HJ17], [LT17], [Sen21], and [LST22] to study the exceptional sets in Manin's conjecture. The basic references for this section are [HTT15] and [LST22].

4.1.1 Fujita Invariants

Let F be a field of characteristic 0. The following invariant is one of the main protagonists in this book:

Definition 4.1.1 (Fujita invariants) Let X be a normal projective variety defined over F with canonical singularities and L be a big $\mathbb{Q}$-Cartier divisor. The *Fujita invariant* or the *a-invariant* is the following geometric invariant:

$$a(X, L) := \min\{t \in \mathbb{R} \mid tL + K_X \in \overline{\mathrm{Eff}}^1(X)\}.$$

When L is not big, we formally set $a(X, L) = +\infty$.

Remark 4.1.2 When L is big, by [BDPP13], the following statements are equivalent:

(i) $a(X, L) > 0$;
(ii) K_X is not pseudo-effective;
(iii) X is *geometrically uniruled*, that is, there exists a dominant and generically finite rational map $Y \times \mathbb{P}^1 \dashrightarrow \overline{X}$ where Y is a variety of dimension $\dim X - 1$.

In this book, we mainly concern ourselves with projective varieties with positive a-invariants, and these are exactly the geometrically uniruled projective varieties.

The Fujita invariant is related to other invariants in birational geometry. The invariant

$$\kappa\epsilon(X, L) = -a(X, L)$$

was introduced and studied by Fujita under the name *Kodaira energy* ([Fuj87], [Fuj92], [Fuj96], and [Fuj97]). A related invariant

$$\sigma(X, L) = \dim X + 1 - a(X, L)$$

was studied by Sommese in [Som86] under the name *spectral value*. The Fujita invariant is also known as the *pseudo-effective threshold* (the threshold of the pseudo-effective cone of divisors). While the nef threshold (the threshold of the nef cone of divisors) plays a central role in the minimal model program, the pseudo-effective threshold plays an important role in the classification of polarized varieties, that is, the study of a pair (X, L) where X is a smooth projective variety and L is an ample divisor on X. This is the main motivation of Fujita and Sommese in their study of Fujita invariants. The Fujita invariant also plays a central role in the study of cylinders on Fano varieties. See a survey paper [CPPZ21] for details of this application.

The asymptotic formula in Manin's conjecture should be invariant under the pullback of a birational morphism because of the functoriality of height functions. Indeed, one of the basic properties of the Fujita invariant is its birational invariance:

Proposition 4.1.3 ([HTT15, Proposition 2.7]) *Let X be a normal projective variety defined over F with canonical singularities, and L be a big $\mathbb{Q}$-Cartier divisor on X. Let $\beta\colon X' \to X$ be a birational morphism with X' smooth and projective. Then we have*

$$a(X', \beta^* L) = a(X, L).$$

Proof Since X has only canonical singularities, we can write

$$K_{X'} \sim_{\mathbb{Q}} \beta^* K_X + E$$

with $E \geq 0$ an exceptional effective $\mathbb{Q}$-divisor. Let m be a positive integer such that mK_X is Cartier and n be an integer such that nL is Cartier. Then we have

$$\begin{aligned} h^0(X, \mathcal{O}_X(nL + mK_X)) &= h^0(X', \mathcal{O}_{X'}(n\beta^* L + m\beta^* K_X)) \\ &= h^0(X', \mathcal{O}_{X'}(n\beta^* L + mK_{X'})). \end{aligned} \tag{4.1}$$

The last equality follows from Lemma 3.2.16 (ii). Indeed, since we have

$$(n\beta^* L + mK_{X'}) - (n\beta^* L + m\beta^* K_X) \sim mE \geq 0,$$

there is a natural injection

$$H^0(X', \mathcal{O}_{X'}(n\beta^* L + m\beta^* K_X)) \hookrightarrow H^0(X', \mathcal{O}_{X'}(n\beta^* L + mK_{X'})).$$

Hence, to prove that this is bijective, we need to show that any effective divisor $0 \leq D \sim n\beta^* L + mK_{X'}$, we have $D - mE \geq 0$. Note that

$$D - \beta^* \beta_* D = E',$$

is a divisor supported on the exceptional locus. Moreover we have $E' \sim mE$. By Lemma 3.2.16 (ii), we conclude $E' - mE \geq 0$ as well as $mE - E' \geq 0$ showing that $mE = E'$. Therefore, our claim follows.

Back to the original question, (4.1) shows that

$$h^0(X, \mathcal{O}_X(nL + mK_X)) > 0 \iff h^0(X', \mathcal{O}_{X'}(n\beta^* L + mK_{X'})) > 0$$

This implies that for any rational number $t \in \mathbb{Q}$ there exists an effective $\mathbb{Q}$-Cartier divisor D on X, which is $\mathbb{Q}$-linearly equivalent to $K_X + tL$, if and only if there exists an effective $\mathbb{Q}$-Cartier divisor D' on X', which is $\mathbb{Q}$-linearly equivalent to $K_{X'} + t\beta^* L$. Hence we have

$$a(X, L) = a(X', \beta^* L).$$

Thus our assertion follows. □

This property naturally leads us to the following definition:

Definition 4.1.4 Let X be a projective variety defined over F, and L be a $\mathbb{Q}$-Cartier divisor on X. Let $\beta\colon X' \to X$ be a birational morphism such that X' is smooth and projective. We define the Fujita invariant $a(X, L)$ by

$$a(X, L) := a(X', \beta^* L).$$

This is well defined by Proposition 4.1.3.

Sheaf cohomology is invariant under base change of the ground field, and this leads to the following lemma:

Lemma 4.1.5 ([LST22, Corollary 4.5]) *Let X be a projective variety defined over F, and L be a $\mathbb{Q}$-Cartier divisor on X. Then for any extension F'/F and for any irreducible component $X' \subset X \otimes F'$, we have*

$$a(X', L \otimes F'|_{X'}) = a(X, L).$$

Proof After applying a resolution, we may assume that X is smooth. When L is not big, we have $a(X', L \otimes F'|_{X'}) = a(X, L) = \infty$. Thus we may assume that L is big. It suffices to show the statement when F'/F is an algebraic extension or a purely transcendental extension as any extension can be realized as the composition of two such extensions.

When F'/F is algebraic, we may take a Galois closure $\widetilde{F}/F'$ over F. Then it suffices to show our assertion for $\widetilde{F}/F'$ and $\widetilde{F}/F$. Thus we may assume that F'/F is Galois. Then the Galois group $\mathrm{Gal}(F'/F)$ acts on the components of $X \otimes F'$ transitively so that we have

$$X \otimes F' = \bigsqcup_{i=1}^{r} \sigma_i(X'),$$

where $\sigma_i \in \mathrm{Gal}(F'/F)$ with $\sigma_1 = \mathrm{id}$. Let m be an integer such that mL is Cartier, and n be any positive integer. Since sheaf cohomology is invariant under flat base change, we have

$$h^0(X, mL + nK_X) > 0 \iff h^0(X \otimes F', mL \otimes F' + nK_{X\otimes F'}) > 0.$$

Then since we have

$$\begin{aligned} h^0(X \otimes F', mL \otimes F' + nK_{X\otimes F'}) &= \sum_{i=1}^{r} h^0(\sigma_i(X'), (mL \otimes F' + nK_{X\otimes F'})|_{\sigma_i(X')}) \\ &= r h^0(X', (mL \otimes F' + nK_{X'})|_{X'}), \end{aligned}$$

our assertion follows.

When F'/F is purely transcendental, $X \otimes F'$ is still irreducible. Thus our assertion again follows from the invariance of sheaf cohomology under flat base change. □

Rationality of the nef thresholds plays a crucial role in the formulation of the minimal model program. (See [KM98, Theorem 3.5] for more details of Rationality theorem.) Analogously, the following theorem is also crucial in our study of the adjoint divisor $a(X, L)L + K_X$:

Theorem 4.1.6 *Let X be a smooth projective geometrically uniruled variety defined over F, and L be a big and nef $\mathbb{Q}$-divisor on X. Then $a(X, L)$ is a rational number.*

Proof When L is ample, this theorem is established in [Bat92] and [Ara10] for threefolds and in [BCHM10, Cor 1.1.7] for higher dimensional varieties. When L is only big and nef, the corresponding theorem is stated in [HTT15, Theorem 2.16]. This follows from the work [Leh12]. Its refinement is provided in [LST22, Lemma 3.7].

Let us deduce this statement from [BCHM10, Corollary 1.1.5]. In view of Lemma 4.1.5, we may assume that F is algebraically closed. After rescaling we may assume that $a(X, L) < 1$ and we pick a positive rational number $t_0 < a(X, L)$. By Lemma 3.4.2, there exists an effective $\mathbb{Q}$-divisor $E \geq 0$ such that for a sufficiently small rational $t > 0$, $A = L - tE$ is ample and there exist two general effective $\mathbb{Q}$-divisors $A_1' \sim A_2' \sim A$ such that $(X, A_1' + A_2' + tE)$ is a terminal pair and A_1', A_2', E have no common component. Let V be the $\mathbb{R}$-vector space in the space $\mathrm{WDiv}_{\mathbb{R}}(X)$ of Weil divisors generated by A_2' and E. We define the set $\mathcal{E}$ by

$$\mathcal{E} = \{\Delta = t_0 A_1' + B \in \mathrm{WDiv}_{\mathbb{R}}(X) \mid B \in \mathsf{V}, B \geq 0,\ (X, \Delta) \text{ satisfies } (*)\},$$

where $(*)$ represents the conditions (X, Δ) is lc and $K_X + \Delta$ is pseudo-effective. It follows from [BCHM10, Corollary 1.1.5] that $\mathcal{E}$ is a rational polytope. Then note that the divisor

$$\Delta_t = t_0 A_1' + (a(X, L) - t_0) A_2' + a(X, L) t E$$

is in the boundary of this polytope, showing that $a(X, L)$ is rational. □

Remark 4.1.7 If L is only big but not nef, then the above theorem fails in general. See Example 4.4.2 for examples.

The following notion is a crucial ingredient in the applications of the BAB conjecture (Corollary 3.5.6):

Definition 4.1.8 (Adjoint rigid pairs) Let X be a geometrically uniruled projective variety defined over F, and L be a big and nef $\mathbb{Q}$-Cartier divisor on X. We say (X, L) is *adjoint rigid* if there exists a resolution $\beta \colon X' \to X$ with

X' smooth and projective such that $a(X, L)\beta^* L + K_{X'}$ is rigid, that is, the Iitaka dimension of this divisor is 0. Note that this definition does not depend on the choice of a resolution. We should also emphasize that, by Theorem 3.4.1, the Iitaka dimension of $a(X, L)\beta^* L + K_{X'}$ is always non-negative.

The following proposition suggests that adjoint rigid varieties are closely related to weak Fano varieties, and this observation is a key in our applications of the BAB conjecture (Corollary 3.5.6) to Manin's conjecture.

Proposition 4.1.9 ([LT17, Theorem 3.5]) *Let X be a geometrically uniruled projective variety with terminal singularities defined over F, and L be a big and nef $\mathbb{Q}$-Cartier divisor on X. Assume that (X, L) is adjoint rigid. Then there exists a birational contraction $\phi\colon X \dashrightarrow X'$ such that*

(i) *X' is a $\mathbb{Q}$-factorial terminal weak Fano variety;*
(ii) $a(X, L)\phi_* L + K_{X'} \sim_{\mathbb{Q}} 0$.

In particular, if we assume X is geometrically irreducible, it must be geometrically rationally connected, that is, there is a family of rational curves on $\overline{X}$ such that any two general closed points can be connected by a member of that family on $\overline{X}$.

Proof After taking a resolution, we may assume that X is smooth. After rescaling, we may also assume that $a(X, L) = 1$. We apply a $(K_X + L)$-MMP (Theorem 3.4.1) and obtain a birational contraction

$$\psi\colon X \dashrightarrow \widetilde{X}.$$

By Remark 3.4.3, we can find $0 \leq L' \sim_{\mathbb{Q}} L$ such that

$$(\widetilde{X}, \psi_* L')$$

is a terminal pair. Since the Iitaka dimension of $K_X + L$ is 0, there exists a unique effective $\mathbb{Q}$-divisor $E \sim_{\mathbb{Q}} K_X + L'$ such that the exceptional divisors of ψ coincide with the support of E. This means that, after taking pushforward, we have

$$K_{\widetilde{X}} + \psi_* L' \sim_{\mathbb{Q}} 0.$$

Since $\psi_* L'$ is big, we can write $\psi_* L' = A + E'$ where A is an ample $\mathbb{Q}$-divisor and E' is an effective $\mathbb{Q}$-divisor on $\widetilde{X}$. For a sufficiently small rational $0 < \epsilon \ll 1$,

$$(\widetilde{X}, (1 - \epsilon)\psi_* L' + \epsilon E')$$

is terminal. Moreover, we have

$$-(K_{\widetilde{X}} + (1 - \epsilon)\psi_* L' + \epsilon E') \sim_{\mathbb{Q}} \epsilon A,$$

which is ample. Thus the pair $(\widetilde{X}, (1-\epsilon)\psi_* L' + \epsilon E')$ is a log Fano variety. It follows from Theorem 3.6.6 that $\widetilde{X}$ is a Mori dream space.

On the other hand, since L' is nef, its pushforward $\psi_* L'$ is movable. According to the definition of Mori dream spaces, one can find an SQM $f\colon \widetilde{X} \dashrightarrow X'$ such that $-K_{X'} \sim_{\mathbb{Q}} f_*\psi_* L'$ is big and nef. Thus we may define

$$\phi = f \circ \psi\colon X \dashrightarrow X'.$$

To prove that X' has only terminal singularities, let us fix an ample divisor $H \geq 0$ on X' and denote its strict transform on $\widetilde{X}$ by $\widetilde{H}$. We fix a sufficiently small $\epsilon' > 0$ such that $(\widetilde{X}, \psi_* L' + \epsilon' \widetilde{H})$ is a terminal pair. We apply a $(K_{\widetilde{X}} + \psi_* L' + \epsilon' \widetilde{H})$-MMP, and the outcome will be $f\colon \widetilde{X} \dashrightarrow X'$ because H is ample and X' is $\mathbb{Q}$-factorial. Since f is an isomorphism in codimension 1, f decomposes as a composition of $(K_{\widetilde{X}} + \psi_* L' + \epsilon' \widetilde{H})$-flips. It follows from [KM98, Corollary 3.42] that $(X', f_*\psi_* L' + \epsilon' H)$ is a terminal pair. Thus our claim follows. The last assertion on rational connectedness follows from [HM07]. □

One can show that adjoint rigid subvarieties form building blocks of geometrically uniruled varieties using the minimal model program. To that end, let X be a smooth projective geometrically uniruled variety defined over F and L be a big and nef $\mathbb{Q}$-divisor on X. Let $\pi\colon X \dashrightarrow W$ be the canonical map of $(X, a(X, L)L)$ constructed in Section 3.4.1. Let $\beta\colon X' \to X$ be a resolution with X' smooth and projective such that $\pi \circ \beta\colon X' \dashrightarrow W$ is a morphism. For any general closed point $y \in W$, the fiber $X'_y = (\pi \circ \beta)^{-1}(y)$ satisfies $a(X'_y, \beta^* L|_{X'_y}) = a(X, L)$ and $(X'_y, \beta^* L|_{X'_y})$ is adjoint rigid because $\pi \circ \beta$ is the Iitaka fibration for $a(X, L)\beta^* L + K_{X'}$. Thus general fibers are adjoint rigid subvarieties. This observation often enables us to reduce a problem of geometrically uniruled varieties to the case of adjoint rigid varieties.

Finally, we discuss how the Fujita invariant behaves in a smooth family:

Theorem 4.1.10 ([LT17, Theorem 4.3]) *Let $\pi\colon \mathcal{X} \to W$ be a smooth family of geometrically uniruled projective varieties defined over F. Let L be a relatively big and relatively nef $\mathbb{Q}$-Cartier divisor on $\mathcal{X}$. Then there exists a non-empty Zariski open subset $W^\circ \subset W$ such that the Fujita invariant $a(X_y, L|_{X_y})$ of the fiber X_y is constant for any closed point $y \in W^\circ$. Moreover, after shrinking W° further, the Iitaka dimension of $a(X_y, L|_{X_y})L|_{X_y} + K_{X_y}$ is also constant.*

Proof In view of Lemma 4.1.5, we may assume that F is algebraically closed. It follows from a relative version of Lemma 3.4.2 that there exists an effective divisor E on $\mathcal{X}$ such that, for any sufficiently small $t > 0$, $L - tE$ is π-ample.

After replacing L by $L + \pi^*N$ where N is a suitable ample divisor on W, we may assume that $A = L - tE$ is ample. After applying a log resolution to (X, E), we may assume that E has simple normal crossings. By Bertini's theorem for a basepoint-free linear system, we may assume that $L = A + tE$ has a simple normal crossings support and $(\mathcal{X}, L)$ is a klt pair. Then after shrinking W°, we may assume that $(\mathcal{X}, L)$ has simple normal crossings over W°. Then our assertion follows from Theorem 3.4.7. □

4.1.2 *b*-invariants

Let F be a field of characteristic 0. Another main protagonist in this book is the following invariant:

Definition 4.1.11 Let X be a $\mathbb{Q}$-factorial geometrically uniruled projective variety defined over F with terminal singularities, and L be a big $\mathbb{Q}$-divisor on X. We define the face of the nef cone of curves as

$$\mathcal{F}(X, L) = \mathrm{Nef}_1(X) \cap \{\alpha \in N_1(X) \mid (a(X, L)L + K_X).\alpha = 0\}.$$

Note that since $a(X, L)L + K_X$ is pseudo-effective, this divisor is non-negative on $\mathrm{Nef}_1(X)$, so that $\mathcal{F}(X, L)$ is a supported face in the sense of Definition 3.1.2.

We define the *b-invariant* of (X, L) by

$$b(F, X, L) := \dim\langle \mathcal{F}(X, L)\rangle,$$

where $\langle \mathcal{F}(X, L)\rangle$ is the subspace in $N_1(X)$ spanned by $\mathcal{F}(X, L)$.

Remark 4.1.12 Another definition of the b-invariant that is in the literature is the codimension of the minimal supported face of $\overline{\mathrm{Eff}}^1(X)$ containing $a(X, L)L + K_X$. Since $\mathrm{Nef}_1(X)$ and $\overline{\mathrm{Eff}}^1(X)$ are dual to each other, the two definitions are equivalent.

Here are basic examples of computations of the a, b-invariants:

Example 4.1.13 Let X be a smooth Fano variety and $L = -K_X$. Then we have

$$a(X, L) = 1, \quad b(F, X, L) = \dim N_1(X) = \dim N^1(X).$$

Example 4.1.14 The following computations of a, b-invariants are frequently used in the literature: let X be a smooth projective variety defined over F such that $-K_X$ is big and the pseudo-effective cone $\overline{\mathrm{Eff}}^1(X)$ is a simplicial cone generated by effective divisors $\{D_\alpha\}_{\alpha \in \mathcal{A}}$.

Let $L = \sum_{\alpha \in \mathcal{A}} \lambda_\alpha D_\alpha$ be a big $\mathbb{Q}$-divisor on X with $\lambda_\alpha > 0$ for any $\alpha \in \mathcal{A}$. We also write

$$-K_X \sim_{\mathbb{Q}} \sum_{\alpha \in \mathcal{A}} \kappa_\alpha D_\alpha,$$

where κ_α is a positive rational number. Then we have the following formulas for a, b-invariants:

$$a(X, L) = \max_{\alpha \in \mathcal{A}} \left\{ \frac{\kappa_\alpha}{\lambda_\alpha} \right\}, \quad b(F, X, L) = \# \left\{ \alpha \in \mathcal{A} \,\middle|\, a(X, L) = \frac{\kappa_\alpha}{\lambda_\alpha} \right\}.$$

Remark 4.1.15 In contrast to Lemma 4.1.5, the invariant $b(F, X, L)$ does depend on F. For example, let S be a non-split smooth quadric surface defined over F in $\mathbb{P}^3$, that is, two rulings of lines are not defined over F. Let $L = -K_S$ be the anticanonical class. While we have $b(F, S, L) = 1$, for a quadratic extension F'/F such that two rulings are defined over F', we have $b(F', S \otimes F', L \otimes F') = 2$ because $N^1(S \otimes F')$ has rank 2. In general, for any finite extension F'/F, it follows from the definition that $b(F, X, L) \leq b(F', X \otimes F', L \otimes F')$.

The b-invariant is also a birational invariant:

Proposition 4.1.16 ([HTT15, Proposition 2.10]) *Let X be a $\mathbb{Q}$-factorial geometrically uniruled projective variety defined over F with terminal singularities, and L be a big $\mathbb{Q}$-divisor on X. Suppose that we have a birational morphism $\beta \colon X' \to X$ with X' smooth and projective. Then the pushforward $\beta_* \colon N_1(X') \to N_1(X)$ induces an isomorphism*

$$\beta_* \colon \mathcal{F}(X', \beta^* L) \cong \mathcal{F}(X, L).$$

In particular, we have $b(F, X', \beta^* L) = b(F, X, L)$.

Proof Let $E_1, \ldots, E_r$ be irreducible components of the exceptional divisor of β. The negativity lemma (Lemma 3.2.16 (ii)) and $\mathbb{Q}$-factoriality of X imply that

$$N^1(X') = \beta^* N^1(X) \oplus \mathbb{R}E_1 \oplus \cdots \oplus \mathbb{R}E_r.$$

Since X has only terminal singularities, we can write

$$K_{X'} \sim_{\mathbb{Q}} \beta^* K_X + \sum_{i=1}^{r} a_i E_i$$

with $a_i \in \mathbb{Q}_{>0}$. This implies that elements in $\mathcal{F}(X', \beta^* L)$ have vanishing intersection with $\beta^*(a(X, L)L + K_X)$ and all E_is. Hence β_* maps $\mathcal{F}(X', \beta^* L)$ to $\mathcal{F}(X, L)$. On the other hand, there is a natural pullback $\beta^* \colon N_1(X) \to N_1(X')$ that maps $\mathcal{F}(X, L)$ to $\mathcal{F}(X', \beta^* L)$. Indeed, for $\alpha \in N_1(X)$ one may define $\beta^* \alpha$

by $\beta^* D.\beta^*\alpha = D.\alpha$ for any $D \in N^1(X)$ and $E_i.\beta^*\alpha = 0$ for any $i = 1, \ldots, r$. Then we need to show that

$$\beta_* \circ \beta^*|_{\mathcal{F}(X,L)} = \mathrm{id}_{\mathcal{F}(X,L)}, \quad \beta^* \circ \beta_*|_{\mathcal{F}(X',\beta^* L)} = \mathrm{id}_{\mathcal{F}(X',\beta^* L)}.$$

The first one holds in general. The second one also holds because of vanishing properties of elements in $\mathcal{F}(X', \beta^* L)$. Hence the claim. □

This proposition leads us to the following definition:

Definition 4.1.17 Let X be a geometrically uniruled projective variety defined over F, and L be a big $\mathbb{Q}$-Cartier divisor on X. Let $\beta\colon X' \to X$ be a birational morphism with X' smooth and projective. We define the b-invariant of (X, L) by

$$b(F, X, L) := b(F, X', \beta^* L).$$

This is again well defined by Proposition 4.1.16.

The following lemma will be useful to compute the b-invariants:

Lemma 4.1.18 ([LTT18, Lemma 3.5]) *Let X be a geometrically uniruled smooth projective variety and L a big and nef $\mathbb{Q}$-divisor on X. Let $\phi\colon X \dashrightarrow X'$ be a minimal model for a pair $(X, a(X, L)L)$ constructed by Theorem 3.4.1. Then we have*

$$a(X, L) = a(X', \phi_* L), \quad b(F, X, L) = b(F, X', \phi_* L).$$

Proof Since any minimal model program preserves the canonical fibration, it follows that $a(X, L)\phi_* L + K_{X'}$ is pseudo-effective but not big. Thus we have $a(X, L) = a(X', \phi_* L)$. After rescaling we may assume that

$$a(X, L) = a(X', \phi_* L) = 1.$$

Let $p\colon W \to X$ and $q\colon W \to X'$ be a common smooth resolution of ϕ. Then we know by Proposition 4.1.16 that

$$b(F, X, L) = b(F, W, p^* L), \quad b(F, X', \phi_* L) = b(F, W, q^* \phi_* L).$$

Thus it suffices to show that

$$\mathcal{F}(W, p^* L) = \mathcal{F}(W, q^* \phi_* L).$$

By the negativity lemma (Lemma 3.2.16 (ii)), there exists an effective q-exceptional $\mathbb{Q}$-divisor $E \geq 0$ such that

$$p^* L + E = q^* \phi_* L.$$

It follows from this that $\mathcal{F}(W, q^*\phi_* L) \subset \mathcal{F}(W, p^* L)$.

On the other hand, by Remark 3.4.3, there exists $0 \le L' \sim_{\mathbb{Q}} L$ such that $(X', \phi_* L')$ is a terminal pair. Then there exists a q-exceptional $\mathbb{Q}$-divisor $N \ge 0$ such that

$$K_W + p^* L' \sim_{\mathbb{Q}} q^*(\phi_* L' + K_{X'}) + N.$$

Thus we conclude that $\mathcal{F}(W, p^* L) \subset \mathcal{F}(W, q^*\phi_* L)$. Our assertion follows. □

Corollary 4.1.19 ([LTT18, Lemma 3.10]) *Let X be a geometrically uniruled smooth projective variety defined over F, and L be a big and nef $\mathbb{Q}$-divisor on X. Assume that (X, L) is adjoint rigid so that there exists a unique effective $\mathbb{Q}$-divisor $D = \sum_{i=1}^n a_i D_i$ such that $a_i > 0$, the D_is are distinct prime divisors, and $a(X, L)L + K_X \sim_{\mathbb{Q}} D$. Let $\mathsf{V} = \langle D_i \rangle \subset N^1(X)$ be the subspace generated by D_is. Then we have $b(F, X, L) = \dim N^1(X)/\mathsf{V} = \dim N^1(X) - n$.*

Proof This immediately follows from Lemma 4.1.18. Note that the rigidity of D implies that the D_is are linearly independent in $N^1(X)$. Thus the last assertion follows as well. □

We also record the following lemma for a later application:

Lemma 4.1.20 ([LST22, Lemma 4.15]) *Let X be a geometrically uniruled geometrically integral smooth projective variety defined over F, and let L be a big and nef $\mathbb{Q}$-divisor on X. Let $\pi : X \dashrightarrow W$ be the canonical model for $K_X + a(X, L)L$.*

Suppose that there is a Zariski open subset $W^\circ \subset W$ such that: (i) W° is smooth, (ii) π is well defined and smooth over W°, and (iii) every fiber X_w over a closed point $w \in W^\circ$ satisfies $a(X_w, L_{X_w}) = a(X, L)$. Then we have

(i) *for any closed point $w \in W^\circ$, we denote the inclusion $X_w \subset X$ by i. Then we have $i_*(\mathcal{F}(X_w, L|_{X_w})) \subset \mathcal{F}(X, L)$; and*
(ii) *if $w \in W^\circ$ is a general point, then the map $i_* : \mathcal{F}(X_w, L|_{X_w}) \to \mathcal{F}(X, L)$ is a surjection.*

Proof We may resolve the indeterminacy of π without affecting $\pi^{-1}(W^\circ)$. By Proposition 4.1.16 we can deduce our assertion from a statement for this blowup model. So let us assume that π is a morphism.

(i) This follows from Lemma 3.2.4 and the fact that

$$(a(X, L)L + K_X)|_{X_w} \sim_{\mathbb{Q}} a(X_w, L|_{X_w})L|_{X_w} + K_{X_w}.$$

(ii) By Lemma 3.4.2, $a(X, L)L$ is $\mathbb{Q}$-linearly equivalent to $\Delta + A$ such that Δ is an effective $\mathbb{Q}$-divisor on X, A is an ample $\mathbb{Q}$-divisor on X, and $(X, \Delta + A)$ is terminal. It follows from Lemma 3.5.7 that $\mathcal{F}(X, L)$ contains only finitely many extremal rays, and those rays are generated by movable curves C_js. Since our canonical model $\pi\colon X \to W$ is a morphism, there exists an ample $\mathbb{Q}$-Cartier divisor H on W such that $a(X, L)L + K_X - \pi^* H$ is $\mathbb{Q}$-linearly equivalent to an effective $\mathbb{Q}$-divisor. Since C_j has vanishing intersection with $a(X, L)L + K_X$, this implies that C_js are π-vertical. Thus for a general $w \in W^\circ$, the fiber X_w contains an irreducible deformation of C_j which itself is movable in X_w. Thus our assertion follows. □

Finally, we discuss the behavior of the b-invariant in a smooth family. While the behavior is very subtle over a non-algebraically closed field, the b-invariant behaves well over an algebraic closure.

Theorem 4.1.21 ([Sen19, Theorem 1.2]) *Suppose that $\overline{F} = F$. Let $\pi\colon \mathcal{X} \to W$ be a smooth family of projective geometrically uniruled varieties defined over F, and let L be a relatively big and relatively nef $\mathbb{Q}$-Cartier divisor on $\mathcal{X}$. Then there exists a non-empty Zariski open subset $W^\circ \subset W$ such that $b(F, X_y, L)$ is constant for any closed point $y \in W^\circ$.*

4.2 The Statement and Examples of Manin's Conjecture

In this section, we introduce a modern formulation of Manin's conjecture that is developed in a series of papers [BM90], [Pey95], [BT98b], [Pey03], [LST22], and [LS24]. One of the key ingredients in its formulation is the notion of thin sets, which we will discuss below.

4.2.1 Thin Sets

As mentioned in the introduction, it is important to remove the contribution of the exceptional sets from the counting function of rational points so that one can aim for a precise asymptotic formula in terms of the global geometry of a polarized variety. In its original formulation in [BM90], it was expected that the exceptional set should be contained in a proper closed subset of the underlying variety. However, the first counterexample appeared in [BT96b]. Peyre was the first to suggest that the exceptional set in Manin's conjecture should be a thin set. (See [Pey03], for example.) In this section, we introduce the notion of thin sets, which was first introduced by Hilbert and Serre to study the inverse Galois problem:

Definition 4.2.1 (Thin sets) Let F be a field of characteristic 0, and X be a variety defined over F. For any proper closed subset $V \subset X$, the set

$$V(F) \subset X(F)$$

is called a *type I thin set*. For any generically finite dominant morphism $f: Y \to X$ from an irreducible variety Y such that the degree of f is > 1, we call

$$f(Y(F)) \subset X(F)$$

a *type II thin set*.

A thin set of X is any subset Z of $X(F)$ such that Z is contained in a finite union of type I and II thin sets.

We say a morphism between irreducible varieties $f: Y \to X$ is a thin map if it is generically finite to the image and admits no rational section from X.

Using this notion, one can rephrase the definition of thin sets in the following way: the set $Z \subset X(F)$ is thin if Z is contained in a finite union of sets of the form $f(Y(F))$ where $f: Y \to X$ is a thin map.

Here is a basic example of thin sets:

Example 4.2.2 Let F be a field of characteristic 0. Let $n > 1$ and consider $f: \mathbb{A}^1 \to \mathbb{A}^1$ defined by

$$x \mapsto x^n.$$

Then

$$f(\mathbb{A}^1(F)) = \{x^n \mid x \in F\}$$

is a type II thin set.

Note that when F is algebraically closed, the notion of thin sets is meaningless as any subset of $X(F)$ is a thin set. However, this notion makes sense when F is a number field:

Proposition 4.2.3 (Hilbert) *Let F be a number field. Then the subset*

$$\mathbb{P}^1(F) \subset \mathbb{P}^1$$

is not thin.

The following proof can be found in [Ser92, Proposition 3.4.2]:

Proof Let $\mathsf{H}: \mathbb{P}^1(F) \to \mathbb{R}_{>0}$ be the height function associated to the hyperplane class. Then [Sch79] and [Pey95] show that

$$N(\mathbb{P}^1(F), \mathsf{H}, \mathsf{T}) = \#\{x \in \mathbb{P}^1(F) \mid \mathsf{H}(x) \leq \mathsf{T}\} \sim c\mathsf{T}^2 \quad (\mathsf{T} \to +\infty)$$

for some positive constant $c > 0$. Let $f : C \longrightarrow \mathbb{P}^1$ be any finite morphism from a smooth projective curve such that $\deg f > 1$. Let $g(C)$ denote the genus of C. There are three possible cases we need to consider.

Case 1: $g(C) = 0$. We may assume that $C(F)$ is not empty. Then C is isomorphic to $\mathbb{P}^1$. [Sch79] and [Pey95] show that

$$N(\mathbb{P}^1(F), \mathsf{H} \circ f, \mathsf{T}) \sim c\mathsf{T}^{\frac{2}{\deg f}} \quad (\mathsf{T} \longrightarrow +\infty)$$

for some constant $c > 0$.

Case 2: $g(C) = 1$. Again we may assume that $C(F)$ is not empty. Then C becomes an elliptic curve. By the Mordell–Weil theorem, $C(F)$ is a finitely generated group. Let r be the rank of $C(F)$. It is shown by Néron (see, e.g., [HS00, Theorem B.6.3]) that

$$N(C(F), \mathsf{H} \circ f, \mathsf{T}) \asymp (\log \mathsf{T})^{\frac{r}{2}} \quad (\mathsf{T} \longrightarrow +\infty),$$

that is, there exist positive constants c_1, c_2 such that

$$c_1(\log \mathsf{T})^{\frac{r}{2}} \leq N(C(F), \mathsf{H} \circ f, \mathsf{T}) \leq c_2(\log \mathsf{T})^{\frac{r}{2}} \quad (\mathsf{T} \longrightarrow +\infty).$$

Case 3: $g(C) \geq 2$. In this situation, it follows from Mordell–Faltings' theorem that $C(F)$ is finite.

The upshot is that any finite sum of three types of asymptotic formulas cannot form cT^2, thus $\mathbb{P}^1(F)$ cannot be covered by finitely many sets of the form $f(C(F))$. Hence the claim. □

Next, let us introduce the following definition:

Definition 4.2.4 Let F be a number field and X be a variety defined over F. We say X satisfies the *weak weak approximation* if there exists a finite set $S_0 \subset \Omega_F$ such that, for any finite set $S \subset \Omega_F$ with $S \cap S_0 = \emptyset$, $X(F)$ is dense in $\prod_{v \in S} X(F_v)$.

Assume that X is projective. We say that X satisfies the *weak approximation* if X satisfies the weak weak approximation with $S_0 = \emptyset$.

Using this notion, we have the following theorem which is due to Ekedahl and Colliot-Thélène:

Theorem 4.2.5 ([Eke90] and [CT]) *Let F be a number field. Suppose that a variety X defined over F satisfies the weak weak approximation property. Then $X(F) \subset X$ is not thin.*

For a proof, see [Ser92, Theorem 3.5.7]. This theorem leads to the following conclusion.

Example 4.2.6 The projective space $\mathbb{P}^n$ satisfies the weak approximation property. Thus Theorem 4.2.5 implies that $\mathbb{P}^n(F)$ is not thin. Since the weak approximation property is a birational property ([Ser92, Lemma 3.5.5]), any F-rational variety X satisfies that $X(F)$ is not thin.

Remark 4.2.7 Let X be a smooth projective geometrically rationally connected variety defined over a number field F. Colliot-Thélène's conjecture predicts that the set of rational points is dense inside the *Brauer–Manin set*, that is,

$$\overline{X(F)} = X(\mathbb{A}_F)^{\mathrm{Br}(X)}.$$

See [CTS21] for a formulation of the Brauer–Manin set. If this conjecture is true, then X satisfies the weak weak approximation property whenever $X(F)$ is not empty. Thus Colliot-Thélène's conjecture predicts that the set $X(F)$ is not thin if it has a rational point.

Finally, we have the following lemma:

Lemma 4.2.8 *Let F be a field of characteristic* 0 *and X be a geometrically integral variety defined over F. Suppose that F'/F is a finite extension. If the set $\mathsf{Z} \subset X(F')$ is a thin set of $X(F')$, then $\mathsf{Z} \cap X(F)$ is a thin set of $X(F)$.*

Proof This can be proved using the Weil restriction of scalars functor. See [Ser92, Proposition 3.2.1] for a proof. □

4.2.2 Manin's Conjecture: the Statement and Examples

Finally, we state Manin's conjecture as formulated in a series of papers [FMT89], [BM90], [Pey95], [BT98b], [Pey03], and [LS24]:

Conjecture 4.2.9 (Manin's conjecture) Let F be a number field and X be a smooth projective geometrically rationally connected variety defined over F. Let $\mathcal{L}$ be a big and nef $\mathbb{Q}$-divisor with an adelic metrization. Suppose that the set $X(F)$ is not thin. Then there exists a thin set $\mathsf{Z} \subset X(F)$ such that

$$N(X(F) \setminus \mathsf{Z}, \mathcal{L}, \mathsf{T}) \sim c(F, X(F) \setminus \mathsf{Z}, \mathcal{L})\mathsf{T}^{a(X,L)}(\log \mathsf{T})^{b(F,X,L)-1} \quad (\mathsf{T} \to +\infty),$$

where $c(F, X(F) \setminus \mathsf{Z}, \mathcal{L})$ is Peyre's constant introduced for the anticanonical divisor in [Pey95], extended to any big divisor in [BT98b], and further revised in [LS24].

Remark 4.2.10 When $a(X, L)L + K_X$ is rigid, Peyre's constant,

$$c(F, X(F) \setminus \mathsf{Z}, \mathcal{L}),$$

does not depend on Z. In this situation we write $c(F, X, \mathcal{L})$ instead of the above.

Remark 4.2.11 By Remark 4.2.7, it is expected that $X(F)$ is not thin whenever there is a rational point. Thus the set of rational points we count, that is, $X(F)\backslash \mathsf{Z}$, is not thin under the assumption of $X(F) \neq \emptyset$, so in particular, it is expected to be Zariski dense.

In Conjecture 4.2.9, the set Z is called an *exceptional set* and it is important to remove the contribution from this set to have the asymptotic formula predicted in this conjecture. Here is a series of examples illustrating this feature:

Example 4.2.12 ([Fou98]) Let S be a singular cubic surface in $\mathbb{P}^3$ defined by

$$xyz = w^3.$$

Manin's conjecture for this particular cubic surface has been established in [Fou98] for the anticanonical height. Note that S is a toric surface whose open torus part is

$$U = \{xyz \neq 0\} \subset S,$$

which is isomorphic to $\mathbb{G}_m^2$. The boundary divisor $D = \{xyz = w = 0\}$ is the union of three lines. Let $\beta\colon \widetilde{S} \to S$ be the minimal resolution. Since S has only canonical singularities, we conclude that $-\beta^* K_S = -K_{\widetilde{S}}$. Moreover, we have $\dim N^1(\widetilde{S}) = 7$, thus it follows that

$$a(S, -K_S) = 1, b(\mathbb{Q}, S, -K_S) = 7.$$

Indeed, Fouvry established that

$$N(U(\mathbb{Q}), -K_S, \mathsf{T}) \sim c\mathsf{T}(\log \mathsf{T})^6 \quad (\mathsf{T} \to \infty),$$

for some positive constant $c > 0$. On the other hand, let ℓ be one of the lines in the boundary. Since $-K_S.\ell = 1$, we have $a(\ell, -K_S|_\ell) = 2/(-K_S.\ell) = 2$ and $b(\mathbb{Q}, \ell, -K_S|_\ell) = 1$. Thus we have

$$N(\ell(\mathbb{Q}), -K_S, \mathsf{T}) \sim c'\mathsf{T}^2$$

for some positive constant $c' > 0$. Hence it is important to remove the contributions from these lines; otherwise the asymptotic formula will be dominated by T^2.

In Example 4.2.12, the exceptional set is $D(\mathbb{Q})$, which is contained in a proper closed subset D. Originally, it was predicted in [BM90] that an exceptional set in Manin's conjecture would not be Zariski dense. Indeed, the closed set version of Manin's conjecture has been confirmed for many examples:

Example 4.2.13 (Generalized flag varieties) Let G be a connected semi-simple linear algebraic group defined over a number field F. Let P be a parabolic subgroup of G. Manin's conjecture for $X = P\backslash G$ with arbitrary polarization has been confirmed in [FMT89] using results from the Langlands program.

One can define a Dirichlet series called the *height zeta function*, and by the Tauberian theorem, establishing the asymptotic formula is reduced to a problem of a meromorphic continuation of this height zeta function. It turns out that the height zeta function for $X = P\backslash G$ coincides with a specialization of an Eisenstein series associated to $P\backslash G$, thus one can appeal to the Langlands program for such a meromorphic continuation. In this setting, one can choose an exceptional set Z to be the empty set for any polarization.

This example has been generalized to examples of equivariant compactifications of homogeneous spaces:

Example 4.2.14 (Equivariant compactifications) Let G be a connected linear algebraic group defined over a number field F. Let X be a smooth equivariant compactification of G defined over F, that is, X is a smooth projective variety defined over F, and G acts on X with a unique open dense orbit isomorphic to G. Let $\mathcal{L}$ be a big divisor with an adelic metrization. To address Manin's conjecture for $(X, \mathcal{L})$, one may use the following strategy called the *height zeta function method*: first we extend the domain of the height function $\mathsf{H}_{\mathcal{L}}$ to the adelic space:

$$\mathsf{H}_{\mathcal{L}} \colon G(\mathbb{A}_F) \to \mathbb{R}.$$

This is a continuous function in the adelic topology. Then for any complex number s such that the real part $\Re(s)$ is sufficiently large, we consider the following Dirichlet series:

$$\mathcal{Z}(s, g) = \sum_{\gamma \in G(F)} \mathsf{H}_{\mathcal{L}}(\gamma g)^{-s},$$

where g is an element of the adelic group $G(\mathbb{A}_F)$ over F. This converges to a continuous L^2-function in g and a holomorphic function in s when $\Re(s)$ is sufficiently large. This is an example of height zeta functions. If one obtains a meromorphic continuation of this zeta function in s, then Manin's conjecture for $(X, \mathcal{L})$ will be verified by using Tauberian theorem. Thus our goal is to obtain a meromorphic continuation of this zeta function. To this end, the height zeta function is an element of the Hilbert space

$$\mathsf{L}^2(G(F)\backslash G(\mathbb{A}_F))$$

so one hopes to use spectral theory/harmonic analysis of this space to obtain a meromorphic continuation. This program has been carried out for several examples including:

- toric varieties ([BT98a] and [BT96a]);
- equivariant compactifications of vector groups ([CLT02]);
- wonderful compactifications of semi-simple groups of adjoint type ([STBT07]);
- bi-equivariant compactifications of unipotent groups ([ST16]).

In all examples of this type, the exceptional set Z is $D(F)$ where D is the boundary divisor. Readers who are interested in more detail should consult an excellent survey paper by Tschinkel [Tsc09, Section 6].

The circle method is a very powerful tool to analyze smooth projective varieties of low degree:

Example 4.2.15 (The circle method) Hardy–Littlewood circle method is a technique in analytic number theory that was invented by Hardy and Littlewood to study Waring's problem. Using this method, [Bir62], [BHB17], and [FM17] settled Manin's conjecture for smooth complete intersections of low degree. Their results are as follows: let X be a smooth projective variety in $\mathbb{P}^n$ defined over a number field F such that we have

$$\dim X > (\deg(X) - 1)2^{\deg(X)}.$$

Then Manin's conjecture is true with $\mathsf{Z} = \emptyset$. It follows from [BEL91] that such a variety is a complete intersection in $\mathbb{P}^n$. Readers who are interested in more detail should consult a beautiful book by Browning [Bro09].

Another prominent method to establish Manin's conjecture for projective varieties is the method of *universal torsors*:

Example 4.2.16 (Universal torsors and lattice points counting) The counting arguments in Example 2.1.8 have been generalized to the strategy called the universal torsor method, which reduces a problem of counting rational points on a geometrically rationally connected projective variety X to a problem of counting integral points on a universal torsor that is an open variety above X. A universal torsor can be realized as a certain open subset of $\mathrm{Spec}(\mathrm{Cox}(X))$ when $\mathrm{Cox}(X)$ is finitely generated. In the case of Example 2.1.8, a universal torsor for $\mathbb{P}^n$ is $\mathbb{A}^{n+1} \setminus \{0\}$.

This method is particularly appropriate to apply for toric varieties as in this case, the Cox ring is a polynomial ring. See [Sal98] for a treatment of Manin's conjecture for toric varieties using the universal torsors. Moreover, the universal torsor method has been quite successful in treating Manin's conjecture for geometrically rational projective surfaces including some (singular) del Pezzo surfaces. In particular, this method has been executed when the relation of the

Cox ring among generators is given by one equation. Readers who are interested in this method should consult a book on Cox rings [ADHL15].

However, as mentioned before, there are many counterexamples to the closed set version of Manin's conjecture. The first counterexample was found by Batyrev–Tschinkel in [BT96b]:

Example 4.2.17 ([BT96b]) Let F be a number field and $X \subset \mathbb{P}^3_x \times \mathbb{P}^3_y$ be a smooth hypersurface defined by

$$\sum_{i=0}^{3} x_i y_i^3 = 0.$$

The fivefold X is a smooth Fano variety and we let $L = -K_X$ with an adelic metrization. It follows from the Lefschetz hyperplane theorem that $\dim N^1(X) = 2$. Thus we have

$$a(X, L) = 1, \quad b(F, X, L) = 2.$$

Hence if the closed set version of Manin's conjecture holds for X, then there exists a non-empty Zariski open subset $X^\circ \subset X$ such that

$$N(X^\circ(F), \mathcal{L}, \mathsf{T}) \sim c\mathsf{T} \log \mathsf{T} \quad (\mathsf{T} \to \infty)$$

for some positive constant $c > 0$.

However, let $\pi\colon X \to \mathbb{P}^3_x$ be the first projection. Then this is a cubic surface fibration. For a rational point $P = (x_0 : x_1 : x_2 : x_3) \in \mathbb{P}^3_x(F)$ with $x_i \in (F^\times)^3$ for any i, we have

$$b(F, X_P, L) = \begin{cases} 7 & \text{if } \sqrt{-3} \in F, \\ 4 & \text{if } \sqrt{-3} \notin F, \end{cases}$$

where $X_P = \pi^{-1}(P)$ is the fiber at P. Batyrev and Tschinkel showed that for any non-empty Zariski open subset $U \subset X_P$ we have

$$N(U(F), \mathcal{L}, \mathsf{T}) \gg \mathsf{T}(\log \mathsf{T})^3 \quad (\mathsf{T} \to \infty)$$

under the assumption of $\sqrt{-3} \in F$. (Later this condition is removed in [FLS18].) This yields a contradiction. Thus X cannot satisfy the closed set version of Manin's conjecture. Note that Manin's conjecture for X has not yet been proved.

Another surprising counterexample has been found in [LR19]:

Example 4.2.18 ([LR19]) Let $S = \mathbb{P}^1 \times \mathbb{P}^1$ defined over $\mathbb{Q}$. We define X as $\mathrm{Hilb}^{[2]}(S)$, that is, the Hilbert scheme of length-2 zero-dimensional subschemes on S. Then X is a smooth weak Fano variety. Let $L = -K_X$, which is a big and nef divisor on X. Then one can conclude that

$$a(X, L) = 1, \quad b(\mathbb{Q}, X, L) = \dim N^1(X) = 3.$$

We consider the following diagram

$$\begin{array}{ccc} W = \mathrm{Bl}_\Delta(S \times S) & \xrightarrow[\psi]{} & S \times S \\ \Big\downarrow f & & \Big\downarrow g \\ X & \xrightarrow[\phi]{} & \mathrm{Sym}^{(2)}(S) \end{array}$$

where $\mathrm{Bl}_\Delta(S \times S)$ is the blowup of $S \times S$ along the diagonal and $\mathrm{Sym}^{(2)}(S)$ is the symmetric product of S.

Now $\phi\colon X \longrightarrow \mathrm{Sym}^{(2)}(S)$ is a crepant resolution, that is, $K_X \sim \phi^* K_{\mathrm{Sym}^{(2)}(S)}$ holds. Thus we conclude that

$$f^* L = -\psi^* g^* K_{\mathrm{Sym}^{(2)}(S)}.$$

Since g is étale in codimension 1, we have

$$a(W, f^* L) = a(S \times S, -K_{S\times S}) = 1, \quad b(\mathbb{Q}, W, f^* L) = b(\mathbb{Q}, S \times S, -K_{S\times S}) = 4.$$

Now Manin's conjecture for $(W, f^* L)$ is reduced to Manin's conjecture for $(S \times S, -K_{S\times S})$, which is established in [FMT89]. Thus X does not satisfy the closed set version of Manin's conjecture. However, [LR19] shows that Manin's conjecture (Conjecture 4.2.9) holds after removing the contribution from a thin set. See [LR19] for more details.

4.3 Geometric Constructions of Exceptional Sets

In Examples 4.2.17 and 4.2.18, the closed set version of Manin's conjecture fails for the following reasons:

- There is a Zariski dense set of subvarieties Y such that

  $$(a(X, L), b(F, X, L)) < (a(Y, L|_Y), b(F, Y, L|_Y))$$

  in the lexicographic order.
- There is a dominant generically finite morphism $f\colon W \longrightarrow X$ such that

  $$(a(X, L), b(F, X, L)) < (a(W, f^* L), b(F, W, f^* L))$$

  in the lexicographic order.

Thus it is natural to consider all thin maps $f\colon Y \to X$ such that

$$(a(X,L), b(F,X,L)) < (a(Y,f^*L), b(F,Y,f^*L))$$

and look at

$$\bigcup_f f(Y(F)).$$

This is the motivation for the conjectural construction of exceptional sets in [LST22]. However, to form an exceptional set, one also needs to consider thin maps $f\colon Y \to X$ such that

$$(a(X,L), b(F,X,L)) = (a(Y,f^*L), b(F,Y,f^*L))$$

and decide whether we should remove the contribution from such a thin map. This is done in [LST22] using the notion of *face-contracting morphisms*.

4.3.1 Face-contracting Morphisms

Let F be a field of characteristic 0. Let X be a projective variety defined over F, and L be a big and nef $\mathbb{Q}$-Cartier divisor on X. Assume that X is geometrically uniruled. We take a smooth resolution $\beta\colon \widetilde{X} \to X$ and define the *fiber dimension of* (X,L) as

$$d(X,L) = \dim X - \kappa(a(X,L)\beta^*L + K_{\widetilde{X}}).$$

In other words, $d(X,L)$ is the dimension of a general fiber of the canonical fibration for $(\widetilde{X}, a(X,L)\beta^*L)$. This definition does not depend on the choice of the resolution $\beta\colon \widetilde{X} \to X$.

Now to define the notion of face-contracting morphisms, we consider the following two situations:

Assumptions 4.3.1 Let X be a smooth projective geometrically uniruled variety defined over F, and L be a big and nef $\mathbb{Q}$-divisor on X. Let $f\colon Y \to X$ be a thin map from a smooth projective variety defined over F such that $a(Y,f^*L) = a(X,L)$. We consider two assumptions on f:

(i) f is dominant; or
(ii) we have $d(Y,f^*L) = d(X,L)$ and there exists a commutative diagram

$$\begin{array}{ccc} Y & \xrightarrow{f} & X \\ {\scriptstyle \pi_Y}\downarrow & & \downarrow{\scriptstyle \pi_X} \\ T & \longrightarrow & W \end{array}$$

where π_X, π_Y are the canonical maps for $(X, a(X,L)L)$ and $(Y, a(Y,f^*L)f^*L)$

respectively. Moreover, we assume that a general fiber of π_Y maps to a fiber of π_X with the same a-invariant as X and is contained in the locus where π_X is smooth.

Under these assumptions we have the following property:

Lemma 4.3.2 ([LST22, Lemma 4.25]) *Under Assumptions 4.3.1, the pushforward $f_*: N_1(Y) \to N_1(X)$ maps $\mathcal{F}(Y, f^*L)$ to $\mathcal{F}(X, L)$.*

Proof In the situation of assumption (i), f_* maps nef classes to nef classes. Moreover, by the ramification formula, there exists $R \geq 0$ such that

$$a(X, L)f^*L + K_Y \sim_{\mathbb{Q}} f^*(a(X, L)L + K_X) + R.$$

Hence elements in $\mathcal{F}(Y, f^*L)$ have vanishing intersection with

$$f^*(a(X, L)L + K_X) \text{ and } R.$$

Hence the claim.

For assumption (ii), let Y_t be a general fiber of π_Y, and X_w be its image on X. Then it follows from Lemma 4.1.20 that $\mathcal{F}(Y_t, f^*L|_{Y_t})$ maps surjectively to $\mathcal{F}(Y, f^*L)$, and similarly $\mathcal{F}(X_w, L|_{X_w})$ maps surjectivly to $\mathcal{F}(X, L)$. (Note that because of Assumption 4.3.1, X_w has the same a-invariant as X and π_X is smooth at w.) Since we have $a(Y_t, f^*L|_{Y_t}) = a(X_w, L|_{X_w})$, by (i), $\mathcal{F}(Y_t, f^*L|_{Y_t})$ maps to $\mathcal{F}(X_w, L|_{X_w})$. Together these imply that $\mathcal{F}(Y, f^*L)$ maps to $\mathcal{F}(X, L)$. Hence the claim. □

Finally, we define the notion of face-contracting morphisms:

Definition 4.3.3 Let X, Y be smooth projective geometrically uniruled varieties defined over F, and L be a big and nef $\mathbb{Q}$-divisor on X. Let $f: Y \to X$ be a thin map.

Suppose that we have a commutative diagram

$$\begin{array}{ccc} Y' & \xrightarrow{f'} & X' \\ {\scriptstyle \phi_Y}\downarrow & & \downarrow{\scriptstyle \phi_X} \\ Y & \xrightarrow[f]{} & X \end{array}$$

where ϕ_X, ϕ_Y are birational projective morphisms such that f' satisfies Assumptions 4.3.1. By Lemma 4.3.2, $f_*: N_1(Y) \to N_1(X)$ maps $\mathcal{F}(Y, f^*L)$ to $\mathcal{F}(X, L)$.

We say that f is face-contracting if $f_*: \mathcal{F}(Y, f^*L) \to \mathcal{F}(X, L)$ is not injective.

Remark 4.3.4 Under the assumption of Definition 4.3.3, if

$$b(F, X, L) < b(F, Y, f^*L),$$

then f is automatically face-contracting.

4.3.2 Constructions of Exceptional Sets

In this section, we recall a conjectural construction of exceptional sets from [LST22]. Let F be a field of characteristic 0, and X be a geometrically uniruled smooth projective variety defined over F. Let L be a big and nef $\mathbb{Q}$-divisor on X. We consider the augmented base locus $\mathbf{B}_+(L)$, which is a proper closed subset of X. Any height function associated to L satisfies the Northcott property outside of $\mathbf{B}_+(L)$.

Let $\pi\colon X \dashrightarrow W$ be the canonical map for $(X, a(X, L)L)$. After applying a resolution we may assume that π is a morphism. Let $V \subset W$ be a proper closed subset such that π is a smooth morphism above $W \setminus V$ and all fibers above $W \setminus V$ have the same a-invariant as X.

We define Z_0 as

$$\mathsf{Z}_0 = (\mathbf{B}_+(L) \cup \pi^{-1}(V))(F) \subset X(F).$$

Note that this is a type I thin set of X.

Next, we let $f\colon Y \to X$ run over all thin maps defined over F such that

- Y is geometrically integral, smooth, and projective;
- $d(Y, f^*L) < d(X, L)$; and
- the inequality

$$(a(X, L), b(F, X, L)) \leq (a(Y, f^*L), b(F, Y, f^*L))$$

  holds in the lexicographic order.

We define Z_1 as

$$\mathsf{Z}_1 = \bigcup_f f(Y(F)) \subset X(F).$$

Next, we let $f\colon Y \to X$ run over all thin maps defined over F such that

- Y is geometrically integral, smooth, and projective;
- the image $f(Y)$ meets the open set $\pi^{-1}(W \setminus V)$;
- $d(Y, f^*L) = d(X, L)$;

- the inequality

$$(a(X,L), b(F,X,L)) \leq (a(Y,f^*L), b(F,Y,f^*L))$$

  holds in the lexicographic order; and
- f is face-contracting when the above inequality is an equality.

We define Z_2 as

$$\mathsf{Z}_2 = \bigcup_f f(Y(F)) \subset X(F).$$

Finally, we let $f\colon Y \to X$ run over all thin maps defined over F such that

- Y is geometrically integral, smooth, and projective;
- $d(Y, f^*L) > d(X, L)$; and
- the inequality

$$(a(X,L), b(F,X,L)) \leq (a(Y,f^*L), b(F,Y,f^*L))$$

  holds in the lexicographic order.

We define Z_3 as

$$\mathsf{Z}_3 = \bigcup_f f(Y(F)) \subset X(F).$$

It turns out that the set Z_3 is contained in a proper closed subset. See Lemma 5.3.1.

The main theorem of [LST22] is the following:

Theorem 4.3.5 ([LST22, Theorem 5.6]) *The set* $\mathsf{Z}_0 \cup \mathsf{Z}_1 \cup \mathsf{Z}_2 \cup \mathsf{Z}_3$ *is a thin set.*

This naturally leads to the following conjecture:

Conjecture 4.3.6 ([LST22, Conjecture 5.1]) In Conjecture 4.2.9 (Manin's conjecture), one can take the exceptional set Z as

$$\mathsf{Z} = \mathsf{Z}_0 \cup \mathsf{Z}_1 \cup \mathsf{Z}_2 \cup \mathsf{Z}_3.$$

Remark 4.3.7 When (X, L) is adjoint rigid, we have $d(X, L) = \dim X$. Thus in this situation we have $\mathsf{Z}_3 = \emptyset$. Also in the definition of Z_2, one only needs to consider dominant thin maps $f\colon Y \to X$ such that (Y, f^*L) is adjoint rigid.

4.3.3 Why Face-contracting?

In the construction of Z_2 in Section 4.3.2, it is important to insist that $f\colon Y \to X$ is face-contracting when the a, b-invariants are equal. Let us show an example justifying this. To understand this example, we need to introduce the idea of twists. We will discuss this notion briefly here. Readers interested in more detail should consult a beautiful book by Poonen [Poo17, Section 4.5].

Let F be a field of characteristic 0. Let $f\colon Y \to X$ be a generically finite dominant proper morphism between irreducible varieties defined over F. We denote the group of automorphisms over X as

$$\mathrm{Aut}(Y/X).$$

Suppose that we have another generically finite cover $f'\colon Y' \to X$ defined over F. When there exists an $\overline{X}$-isomorphism between $\overline{Y}$ and $\overline{Y}'$, that is, there exists an isomorphism $\phi\colon \overline{Y} \to \overline{Y'}$ such that $\overline{f}' \circ \phi = \overline{f}$, we say that $f'\colon Y' \to X$ is a twist of $f\colon Y \to X$. Then there is a one-to-one correspondence:

$$\{\text{twists of } f\}/X\text{-isomorphisms} \longleftrightarrow H^1(\mathrm{Gal}(\overline{F}/F), \mathrm{Aut}(\overline{Y}/\overline{X})).$$

Here is a description of this correspondence:

$(\longrightarrow)$ Suppose that we have a twist $f'\colon Y' \to X$ of f. Then there exists some finite Galois extension F' and an $X \otimes F'$-isomorphism $\phi\colon Y \otimes F' \to Y' \otimes F'$. Then we may construct a 1-cocycle σ by

$$\mathrm{Gal}(F'/F) \ni s \mapsto \phi^{-1} \circ \phi^s \in \mathrm{Aut}(Y \otimes F'/X \otimes F'),$$

where ϕ^s is the conjugate of ϕ by s.

$(\longleftarrow)$ Suppose that we have a 1-cocycle

$$\sigma\colon \mathrm{Gal}(F'/F) \ni s \mapsto \sigma_s \in \mathrm{Aut}(Y \otimes F'/X \otimes F'),$$

where F' is some finite Galois extension. We define the action of $s \in \mathrm{Gal}(F'/F)$ on $Y \otimes F'$ by $\sigma_s \circ (\mathrm{id}_Y \otimes s)$ and take the quotient by this action

$$Y^\sigma := (Y \otimes F')/\mathrm{Gal}(F'/F).$$

Then $f^\sigma\colon Y^\sigma \to X$ is a twist of $f\colon Y \to X$.

Here is an example demonstrating the necessity of the face-contracting condition:

Example 4.3.8 ([LT17, Example 8.7]) The following example is originally due to [San95]. Let F be a number field. Let $Y = (\mathbb{P}^1)^3$. We consider the following involution on Y:

$$\iota\colon [(x_0 : x_1), (y_0 : y_1), (z_0 : z_1)] \mapsto [(-x_0 : x_1), (-y_0 : y_1), (-z_0 : z_1)].$$

We define the variety X by taking the quotient

$$X := Y/\langle \iota \rangle.$$

Then X is a non-Gorenstein $\mathbb{Q}$-factorial terminal Fano threefold. The quotient map $f\colon Y \to X$ has eight points as the ramification locus, so in particular, f is étale in codimension 1. Let $L = -K_X$. Then we have

$$a(X, L) = 1, \quad b(F, X, L) = 3,$$
$$a(Y, f^*L) = 1, \quad b(F, Y, f^*L) = 3.$$

Since the action of ι on $N^1(\overline{Y})$ is trivial, for any twist $f^\sigma : Y^\sigma \to X$, we have

$$a(Y^\sigma, (f^\sigma)^*L) = 1, \quad b(F, Y^\sigma, (f^\sigma)^*L) = 3.$$

Moreover, it follows from [Poo17, Theorem 8.4.1] that

$$X(F) = \bigcup_\sigma f^\sigma(Y^\sigma(F)),$$

which is not thin because X is F-rational. On the other hand, for any twist σ, $f^\sigma : Y^\sigma \to X$ is not face-contracting, so these thin maps do not contribute to Z_2.

4.3.4 Some Examples

Let us illustrate some computations of the conjectural exceptional set

$$\mathsf{Z} = \mathsf{Z}_0 \cup \mathsf{Z}_1 \cup \mathsf{Z}_2 \cup \mathsf{Z}_3.$$

Example 4.3.9 ([LST22, Example 5.11]) Let $X = P\backslash G$ be a generalized flag variety defined over a number field F as in Example 4.2.13. It is shown in [Bor96] that the Brauer–Manin obstructions are the only ones for weak approximation, so in particular, X satisfies the weak weak approximation property whenever there is a rational point. Thus, when $X(F)$ is non-empty, the set $X(F)$ is not a thin set of X. On the other hand, it is easy to see that the conjectural exceptional set $\mathsf{Z}_0 \cup \mathsf{Z}_1 \cup \mathsf{Z}_2 \cup \mathsf{Z}_3$ is invariant under the action of $G(F)$, and the action of $G(F)$ on $X(F)$ is transitive. Altogether we conclude that

$$\mathsf{Z}_0 \cup \mathsf{Z}_1 \cup \mathsf{Z}_2 \cup \mathsf{Z}_3 = \emptyset,$$

because this set is a thin set by Theorem 4.3.5. This prediction of Manin's conjecture has been confirmed in [FMT89].

Other similar examples are the following:

Example 4.3.10 Let X be an equivariant compactification of a connected linear algebraic group G defined over a number field F as in Example 4.2.14. [Bor96] still applies to G so, in particular, G satisfies weak weak approximation. Thus $G(F)$ is not thin in G. Therefore, arguing as in Example 4.3.9, we conclude that the set

$$\mathsf{Z}_0 \cup \mathsf{Z}_1 \cup \mathsf{Z}_2 \cup \mathsf{Z}_3$$

is contained in the boundary.

We will discuss further examples of computations of the exceptional set in Chapter 6.

4.4 Possible Generalizations of Theorem 4.3.5

The Northcott property holds over some open subset of the underlying variety for any height function associated to a big $\mathbb{Q}$-divisor. Thus it is natural to wonder whether Manin's conjecture (Conjecture 4.2.9) holds for a polarization which is only big, not nef. It turns out that such a statement has been proved for various equivariant compactifications of connected linear algebraic groups ([BT96a], [BT98a], [CLT02], and [STBT07]).

However, [LST22, Section 5.1] shows that generalizing Theorem 4.3.5 to an arbitrary big $\mathbb{Q}$-divisor is subtle. In Section 4.4.1, we discuss various examples illustrating this subtlety. On the other hand, when the underlying variety is a smooth Fano variety, this subtlety disappears; we discuss this generalization in Section 4.4.2.

4.4.1 Counterexamples to Generalizations for Arbitrary Big Divisors

In this section, we illustrate a few examples demonstrating some pathological behavior of a, b-invariants when our polarization L is only big, not nef. Here we focus on examples whose anticanonical class is big. We use the following construction in all examples below, which we found in [Cut86]: let W be any smooth projective variety defined over a field F of characteristic 0, and H be an ample divisor on W such that $H - K_W$ is ample. Let $X = \mathbb{P}_W(\mathcal{O} \oplus \mathcal{O}(-H))$ with the projection $\pi\colon X \to W$. We denote the class $\mathcal{O}(1)$ by ξ. Then we have

$$-K_X \sim 2\xi + \pi^*(H - K_W).$$

From this one can prove that, for any $\alpha \in \mathrm{Nef}_1(X) \setminus \{0\}$, $-K_X.\alpha > 0$, showing that $-K_X$ is big. In this case, we always have

$$a(X, -K_X) = 1, \quad b(F, X, -K_X) = \dim N^1(X) = 1 + \dim N^1(W).$$

Example 4.4.1 ([LST22, Example 5.14]) Let W be the Craighero–Gattazzo surface in [CG94] defined over a number field $\mathbb{Q}(\zeta)$, where ζ satisfies

$$\zeta^3 + \zeta^2 - 1 = 0.$$

Then W is a surface of general type with $H^1(W, \mathcal{O}_W) = H^2(W, \mathcal{O}_W) = 0$. By [RTU17, Theorem 6.2], W is simply connected, thus $\mathrm{Pic}(W)_{\mathrm{tor}} = 0$ because $H_1(W(\mathbb{C}), \mathbb{Z})$ vanishes.

Using the above construction, we find a projective bundle X over W which is almost Fano in the sense that $-K_X$ is big, $H^i(X, \mathcal{O}_X) = 0$ for $i > 0$, and $\mathrm{Pic}(X)$ is torsion free. The vanishing of cohomology follows from the Leray spectral sequence, e.g., [Voi07, Theorem 4.11]. The torsion-freeness of the Picard group follows from [Har77, Chapter II Exercise 7.9].

The pair $(X, -K_X)$ satisfies $a(X, -K_X) = 1$, however, Bombieri–Lang's conjecture predicts that the set of rational points on W is not Zariski dense even after taking some finite extension. Thus it is difficult to imagine that Manin's conjecture (Conjecture 4.2.9) holds for this example.

Here is an example showing that in contrast to Theorem 4.1.6, $a(X, L)$ may not be rational if we assume L is only big, not nef:

Example 4.4.2 ([Cut86, Example 1.6]) Let W be the self-product of an elliptic curve defined over a number field F without complex multiplication; we construct a projective bundle X by applying the above construction to W. The computations in [Cut86, Example 1.6] show that if L is a big $\mathbb{Q}$-divisor, but not nef, then it is possible that $a(X, L)$ is irrational. Thus Theorem 4.1.6 can fail if we assume that L is only big, not nef. To see this, let D be a rigid section of $\pi\colon X \to W$, that is, a unique member of $|\xi|$. Then the cone of pseudo-effective divisors on X is generated by D and $\pi^*\overline{\mathrm{Eff}}^1(W)$. Let $L = tD + \pi^*M$ where $t \in \mathbb{Q}_{>0}$ and M is a big $\mathbb{Q}$-divisor on W. When t is sufficiently large, $a := a(X, L)$ is characterized by the fact that $aM - H \in \partial\overline{\mathrm{Eff}}^1(W)$. However, $\partial\overline{\mathrm{Eff}}^1(W)$ is a quadric cone, so in general we have $a(X, L) \notin \mathbb{Q}$.

In this situation the set of rational points on X is thin by the weak Mordell–Weil theorem ([HS00, Theorem C.0.2]).

Of course, when we consider Manin's conjecture (Conjecture 4.2.9), it is natural to restrict ourselves to geometrically rationally connected varieties.

However, even in this setting, the a, b-invariants can exhibit some pathological behavior:

Example 4.4.3 ([LST22, Example 5.16]) Choose a pencil of cubics on $\mathbb{P}^2$ defined over $\mathbb{Q}$ whose generic fiber is a smooth elliptic curve with positive Mordell–Weil rank. (For the construction of such examples see [Shi92] and [Kur14].) Let W be the blowup of the base locus of this pencil. Then W admits infinitely many (-1)-curves defined over $\mathbb{Q}$ which are sections of the elliptic fibration coming from the pencil.

Let H be an arbitrary ample divisor on W. Then since $-K_W$ is nef, $H - K_W$ is ample. Thus we apply the above construction to obtain a projective bundle X over W such that $-K_X$ is big. Let T_i be any (-1)-curve on W. Set $S_i = \pi^{-1}(T_i)$. Then S_i is the Hirzebruch surface and Manin's conjecture is known for any big divisor on S_i by [BT98a] and [BT96a]. However, since we have

$$-K_X|_{S_i} + K_{S_i} \sim S_i|_{S_i} \sim \pi^*\mathcal{O}_{T_i}(-1),$$

which is antinef, we conclude that $a(S_i, -K_X|_{S_i}) > a(X, -K_X) = 1$. In particular, Theorem 5.1.6, which is an important ingredient for Theorem 4.3.5, fails if we assume only that L is big, not nef. Hence the weak Manin's conjecture, which predicts that there exists a non-empty Zariski open subset $U \subset X$ such that $N(U(F), \mathcal{L}, T) \ll_\epsilon T^{a(X,L)+\epsilon}$ for any $\epsilon > 0$, fails if we assume only that the polarization L is big.

Remark 4.4.4 On the other hand, if we fix $\epsilon > 0$, then the S_is such that $a(S_i, -K_X|_{S_i}) > 1 + \epsilon$ are finite. Indeed, if such an inequality holds, then there is a movable curve C on S_i such that

$$(-(1 + \epsilon)K_X + K_{S_i}).C < 0.$$

Then C must map to T_i dominantly and we denote its degree by d. The above inequality implies that

$$0 > -\epsilon K_X.C + \pi^*\mathcal{O}_{T_i}(-1).C \geq d\epsilon(H - K_W).T_i - d,$$

which implies that

$$(H - K_W).T_i < \frac{1}{\epsilon}$$

holds. Since $H - K_W$ is ample, there are only finitely many such T_is. Thus our claim follows.

In particular, this example is not a counterexample to the following conjecture: for any $\epsilon > 0$, there exists a non-empty open subset $U_\epsilon \subset X$ depending on ϵ such that $N(U_\epsilon(F), \mathcal{L}, \mathsf{T}) \ll_\epsilon \mathsf{T}^{a(X,L)+\epsilon}$.

In contrast to big divisors, this accumulation phenomenon of the a-invariant never happens when L is big and nef, and this is known as Fujita's spectrum conjecture, which was proved by Di Cerbo using Birkar's solution to the BAB conjecture ([Bir19]). See his original paper [DC17, Theorem 1.2] as well as its generalizations in [HJ17, Corollary 2.15] and [HL20, Theorem 1.3].

In Example 4.4.3, we do not know whether the set of rational points on the S_is is thin or not. While the set of rational points on sections of an elliptic fibration is thin by a version of the weak Mordell–Weil theorem, these sections may not be the only source of (-1)-curves. Thus it is natural to ask the following question:

Question 4.4.5 Let W be a rational surface defined over a number field such that W admits infinitely many (-1)-curves defined over the ground field. Is the set of rational points on (-1)-curves thin?

4.4.2 Big Divisors on Fano Varieties with Terminal Singularities

In this section, we prove the following partial generalization of Theorem 4.3.5. The following theorem grew out of discussion between Yoshinori Gongyo, Brian Lehmann, and the author.

Theorem 4.4.6 *Let X be a $\mathbb{Q}$-factorial Fano variety with terminal singularities defined over a field F of characteristic 0. Let L be an arbitrary big $\mathbb{Q}$-divisor on X. Let $f: Y \to X$ run over all F-thin maps from a smooth projective geometrically integral variety defined over F such that*

$$(a(X, L), b(F, X, L)) < (a(Y, f^*L), b(F, Y, f^*L))$$

in the lexicographic order. Then the union

$$\bigcup_f f(Y(F)) \subset X(F)$$

is a thin set of X.

Proof According to Theorem 3.6.6, a $\mathbb{Q}$-factorial Fano variety with terminal singularities is a Mori dream space. In particular, an L-MMP exists and it terminates by Proposition 3.6.3. Let $\phi: X \dashrightarrow X'$ be an outcome of L-MMP. Then ϕ_*L is big and nef. Let $\widetilde{X}$ be a common resolution of ϕ with projections $p: \widetilde{X} \to X$ and $q: \widetilde{X} \to X'$. To prove our assertion, it suffices to prove the claim for $(\widetilde{X}, p^*L)$. By applying the negativity lemma (Lemma 3.2.16) to each step of L-MMP, one can prove that there exists an effective ϕ-exceptional $\mathbb{Q}$-divisor $E_1 \geq 0$ such that

$$L = p_*q^*\phi_*L + E_1.$$

Then again by the negativity lemma (Lemma 3.2.16) and the fact that $q^*\phi_*L$ is nef, there exists an effective p-exceptional $\mathbb{Q}$-divisor $E_2 \geq 0$ such that

$$q^*\phi_*L + E_2 = p^*p_*q^*\phi_*L.$$

Let $E = p^*E_1 + E_2$ be an q-exceptional $\mathbb{Q}$-divisor. The above two equalities imply that

$$p^*L = p^*p_*q^*\phi_*L + p^*E_1 = q^*\phi_*L + E. \tag{4.2}$$

Let $f: Y \to \widetilde{X}$ be any thin map defined over F such that the image $f(Y)$ is not contained in the support of E. Then it follows from (4.2) that

$$(a(Y, f^*p^*L), b(F, Y, f^*p^*L)) \leq (a(Y, f^*q^*\phi_*L), b(F, Y, f^*q^*\phi_*L)) \tag{4.3}$$

in the lexicographic order.

Next, we show that

$$(a(\widetilde{X}, p^*L), b(F, \widetilde{X}, p^*L)) = (a(\widetilde{X}, q^*\phi_*L), b(F, \widetilde{X}, q^*\phi_*L))$$

holds. Since $-K_X$ is nef, it follows from the negativity lemma (Lemma 3.2.16) that there exists an effective q-exceptional $\mathbb{Q}$-divisor N such that

$$p^*K_X \sim_{\mathbb{Q}} q^*K_{X'} + N.$$

Also since X has only terminal singularities, there exists an effective p-exceptional $\mathbb{Q}$-divisor E' such that

$$K_{\widetilde{X}} \sim_{\mathbb{Q}} p^*K_X + E'.$$

The upshot is that for any positive rational number a we have

$$ap^*L + K_{\widetilde{X}} \sim_{\mathbb{Q}} q^*(a\phi_*L + K_{X'}) + aE + E' + N.$$

It follows from the negativity lemma (Lemma 3.2.16) that

$$a(\widetilde{X}, p^*L) = a(\widetilde{X}, q^*\phi_*L).$$

Now since $-K_X$ is ample, any ϕ-exceptional divisor appears in the support of N. Since X has only terminal singularities, any p-exceptional divisor appears in the support of E' as well. Thus the support of $aE + E' + N$ coincides with the support of $E' + N$. We also have

$$a(X, L)q^*\phi_*L + K_{\widetilde{X}} \sim_{\mathbb{Q}} q^*(a(X, L)\phi_*L + K_{X'}) + E' + N,$$

so $\mathcal{F}(\widetilde{X}, p^*L)$ and $\mathcal{F}(\widetilde{X}, q^*\phi_*L)$ must coincide. We conclude that

$$b(F, \widetilde{X}, p^*L) = b(F, \widetilde{X}, q^*\phi_*L).$$

Our claim follows.

Finally, applying Theorem 4.3.5 to $(\widetilde{X}, q^*\phi_*L)$, our assertion follows due to (4.3).

□

5
Proofs of Main Results

In this chapter, we explain the main ideas of the proof of Theorem 4.3.5. This proof has been developed in a series of papers: [LTT18], [HJ17], [LT17], [Sen21], and [LST22]. We now explain the history of this theorem. In [LTT18], Lehmann, Tschinkel, and the author proved that the exceptional set for the a-invariant is a proper closed subset assuming the weak BAB conjecture (Corollary 3.5.6). At the time of writing [LTT18], the BAB conjecture had not yet been proved, so our result was unconditional only in dimension ≤ 4. In [HJ17], Christopher Hacon and Chen Jiang made this result unconditional in arbitrary dimension using the first paper [Bir19] on BAB conjecture proved by Birkar. In [LT17], Lehmann and the author showed a version of Theorem 4.3.5 when the Ys forming the exceptional set Z_1 are only subvarieties using the boundedness of singular Fano varieties together with Hilbert's irreducibility theorem (due to Hilbert and Serre). Moreover, we conjectured the finiteness of adjoint rigid a-covers up to birational equivalence over an algebraic closure aiming to prove a version of Theorem 4.3.5 when $f\colon Y \to X$ is an adjoint rigid a-cover. This was achieved by Sengupta in [Sen21]. Finally, Lehmann, Sengupta, and the author jointly proved Theorem 4.3.5 in [LST22] building on early papers mentioned above as well as using intricate arguments involving twists and the arithmetic fundamental groups of algebraic varieties defined over a non-closed field. Its core technical lemma is Lemma 5.3.5, whose proof is beyond the scope of this book. Readers interested in its proof are encouraged to consult the original paper [LST22, Lemma 8.3].

5.1 Subvarieties

One of the main results in [LT17] is the following theorem, which shows that subvarieties violating geometric compatibility form only a thin set:

Theorem 5.1.1 ([LT17, Theorem 1.5]) *Let F be a field of characteristic 0, and X be a geometrically uniruled smooth projective variety defined over F. Let L be a big and nef $\mathbb{Q}$-divisor on X. Assume that*

(i) $N^1(X) = N^1(\overline{X})$*; and*
(ii) $a(X, L)L + K_X$ *is rigid.*

Let Y run over all proper geometrically integral subvarieties $Y \subset X$ such that

$$(a(X, L), b(F, X, L)) \leq (a(Y, L|_Y), b(F, Y, L|_Y))$$

holds in the lexicographic order. Then the set

$$\bigcup_Y Y(F) \subset X(F)$$

is a thin set of X.

Our goal of this section is to prove the above theorem. The main ingredients are BAB conjecture (Corollary 3.5.6) and Hilbert's irreducibility theorem (Theorem 5.1.10), which we will explain later.

First, we prove that the exceptional set for the a-invariant is a proper closed subset. The following lemma shows that this is indeed true if one fixes one family.

Lemma 5.1.2 ([LTT18, Proposition 4.1]) *Let F be a field of characteristic 0. Let X be a smooth projective geometrically uniruled variety defined over F, and L be a big and nef $\mathbb{Q}$-divisor on X. Let $p\colon \mathcal{U} \to W$ be a family of subvarieties of X such that the evaluation map $s\colon \mathcal{U} \to X$ is dominant. Then, for a general member $Y \subset \mathcal{U}$, we have $a(Y, L|_Y) \leq a(X, L)$.*

Proof In view of Lemma 4.1.5, we may assume that $\overline{F} = F$ without loss of generality. Let $\mathcal{U}'$ be a smooth resolution of $\mathcal{U}$ with the family map

$$p'\colon \mathcal{U}' \to W$$

and the evaluation map

$$s'\colon \mathcal{U}' \to X.$$

It follows from Lemma 3.2.4 that, for a general fiber Y' of p', the pushforward of any nef 1-cycle on Y' is nef in $\mathcal{U}'$. Hence, for any $\alpha \in \mathrm{Nef}_1(Y')$, $s'_*\alpha$ is also nef on X. We may further assume that $s\colon \mathcal{U}' \to X$ is smooth at a general point on Y' and that $w \in W$, which is a point corresponding to Y, is a smooth point of W.

Let $d = \dim Y$, $n = \dim X$. Let $T \subset W$ be a general complete intersection of dimension $n - d$ containing w. We denote the base change $\mathcal{U}'$ by $T \subset W$ by $\mathcal{U}'_T$. Then, by generality, $s'|_{\mathcal{U}'_T} : \mathcal{U}'_T \to X$ is dominant and generically finite. By the ramification formula we have

$$K_{\mathcal{U}'_T} \sim s'^* K_X + E,$$

where $E \geq 0$ is an effective divisor on $\mathcal{U}'_T$. By our construction, the support of E does not contain Y'. Thus we have

$$\begin{aligned}(K_{Y'} + a(X,L)s'^*L).\alpha &= (K_{\mathcal{U}'_T} + a(X,L)s'^*L).i_*\alpha \\ &= (s'^*K_X + E + a(X,L)s'^*L).i_*\alpha \\ &\geq (K_X + a(X,L)L).s_*\alpha \geq 0,\end{aligned}$$

where $i\colon Y' \hookrightarrow \mathcal{U}'_T$ is the inclusion. Thus $K_{Y'} + a(X,L)s'^*L|_{Y'}$ is pseudo-effective. This implies that $a(Y, L|_Y) \leq a(X, L)$. □

The following lemma enables us to reduce our problem of geometrically uniruled varieties to a problem for adjoint rigid subvarieties:

Lemma 5.1.3 ([LTT18, Theorem 4.5] and [LST22, Lemma 4.15]) *Let F be a field of characteristic 0. Let X be a geometrically uniruled geometrically integral smooth projective variety defined over F, and L be a big and nef $\mathbb{Q}$-divisor on X.*

Let $\pi\colon X \dashrightarrow W$ be the canonical map for $(X, a(X,L)L)$ as constructed in Section 3.4.1. Then for a general fiber X_w of π at a general closed point $w \in W$, we have $a(X_w, L|_{X_w}) = a(X, L)$ and $(X_w, L|_{X_w})$ is adjoint rigid. After taking a resolution, we may assume that π is a morphism. Then the pushforward map $i_*\colon N_1(X_w) \to N_1(X)$ induces

$$i_*\colon \mathcal{F}(X_w, L|_{X_w}) \to \mathcal{F}(X, L),$$

and it is surjective where $i_*\colon X_w \hookrightarrow X$ is the inclusion. In particular, we have

$$b(F, X, L) \leq b(F, X_w, L|_{X_w}).$$

Proof Here we sketch the proof. Readers interested in more detail should consult [LST22, Lemma 4.15]. The first statement follows from basic properties of Iitaka fibrations.

After taking a resolution, we may assume that $\pi\colon X \to W$ is a morphism. It follows from Lemma 3.2.4 that i_* maps nef classes on X_w to nef classes on X. Thus we have the well-defined map

$$i_* : \mathcal{F}(X_w, L|_{X_w}) \to \mathcal{F}(X, L).$$

The cone theorem for nef curves (Lemma 3.5.7), which was developed in [Ara10] and [Leh12], shows that $\mathcal{F}(X, L)$ contains finitely many extremal rays and they are generated by movable curves C_i. Since we have

$$(a(X, L)L + K_X).C_i = 0,$$

C_i is π-vertical. Thus we conclude that $C_i \subset X_w$ is a movable curve in X_w for a general $w \in W$. This completes the proof. □

The main input of the next theorem is the BAB conjecture (Corollary 3.5.6):

Theorem 5.1.4 ([LT17, Corollary 4.8]) *Assume $\overline{F} = F$. Let X be a smooth uniruled projective variety defined over F, and L be a big and nef $\mathbb{Q}$-divisor on X. We fix a positive real number $\tau > 0$. Then in the Hilbert scheme* Hilb(X), *the following set is bounded:*

$$\{Y \subset X \textit{ subvarieties } \mid Y \not\subset \mathbf{B}_+(L),\ a(Y, L|_Y) \geq \tau, \textit{ and } (Y, L|_Y) \textit{ is adjoint rigid}\}$$

To prove this theorem we need to record the following lemma:

Lemma 5.1.5 ([LTT18, Lemma 4.7]) *Assume $\overline{F} = F$. Let X be a smooth projective variety defined over F and L be a big and nef $\mathbb{Q}$-divisor on X. Fix a positive constant $\mathsf{T} > 0$. Then the subset of* Hilb(X) *parametrizing subvarieties Y that are not contained in $\mathbf{B}_+(L)$ and are of L-degree at most T is bounded.*

Proof Due to properties of augmented base loci, there exist an ample $\mathbb{Q}$-divisor A and an effective $\mathbb{Q}$-divisor E such that $L = A + E$ and $\mathbf{B}_+(L)$ is the stable base locus of E. Let $Y \not\subset \mathbf{B}_+(L)$. Then there exists an effective $\mathbb{Q}$-divisor E' such that $E' \sim_{\mathbb{Q}} E$ and $Y \not\subset \mathrm{Supp}(E')$. An easy computation shows that for $d = \dim Y$,

$$L^d - (L - E)^d = \sum_{i=0}^{d-1} E'.L^i.(L - E)^{d-1-i}.$$

Since $E' \cap Y$ is an effective class of dimension $d - 1$, and L and $L - E$ are nef, the intersection of the right-hand side with Y is non-negative. This implies that

$$\mathsf{T} \geq L^d.Y \geq (L - E)^d.Y = A^d.Y.$$

Thus our assertion follows from the corresponding statement for an ample $\mathbb{Q}$-divisor. Indeed, such a statement follows from the corresponding statement for the Chow variety [Kol96, Chapter I, 3.21 Theorem] and the comparison theorem between the Hilbert scheme and the Chow variety [Kol96, Chapter I, 6.3 Theorem]. □

Now let us prove Theorem 5.1.4 using the BAB conjecture (Corollary 3.5.6):

Proof of Theorem 5.1.4 Let Y be a subvariety such that $Y \not\subset \mathbf{B}_+(L)$, $a(Y, L|_Y) \geq \tau$, and $(Y, L|_Y)$ is adjoint rigid. Let $\beta\colon \widetilde{Y} \to Y$ be a smooth resolution. It follows from Proposition 4.1.9 that there exists a birational contraction $\phi\colon \widetilde{Y} \dashrightarrow Y'$ such that Y' is a weak Fano variety with terminal singularities and we have $a(Y, L|_Y)\phi_*\beta^* L + K_{Y'} \equiv 0$. Then we have

$$L^{\dim Y}.Y = \beta^* L^{\dim Y}.\widetilde{Y} = \mathrm{Vol}(\beta^* L) \leq \mathrm{Vol}(\phi_*\beta^* L)$$
$$= \frac{\mathrm{Vol}(-K_{Y'})}{a(Y, L|_Y)^{\dim Y}} \leq \frac{\mathrm{Vol}(-K_{Y'})}{\tau^{\dim Y}}.$$

The first inequality follows from the negativity lemma (Lemma 3.2.16). Thus our assertion follows from Corollary 3.5.6 and Lemma 5.1.5. □

The following theorem has been proved in [LTT18] assuming BAB conjecture. The first unconditional proof was provided in [HJ17]:

Theorem 5.1.6 ([HJ17] and [LT19a, Theorem 3.3]) *Let F be a field of characteristic 0. Let X be a smooth geometrically uniruled projective variety defined over F, and L be a big and nef $\mathbb{Q}$-divisor on X. Let $Y \subset X$ run over all subvarieties such that $a(Y, L|_Y) > a(X, L)$. Then the union*

$$\bigcup_Y Y \subset X$$

is a proper closed subset.

Proof Let V be the Zariski closure of the above union. Our goal is to prove that this is a proper subset. By Lemma 4.1.5, we may assume that $\overline{F} = F$. It follows from Lemma 5.1.3 that we only need to show that V is proper when Y is adjoint rigid. It follows from Theorem 5.1.4 that such Ys are bounded. Thus our claim follows from Lemma 5.1.2.

Finally, let V_i be an irreducible component of V. Any irreducible component N of $\mathbf{B}_+(L)$ has $a(N, L|_N) = \infty$, thus we may assume that $a(V_i, L|_{V_i}) < \infty$. Then V_i is geometrically uniruled and Lemma 5.1.2 implies that $a(V_i, L|_{V_i}) > a(X, L)$. Hence the claim. □

Let us discuss the above theorem when X is weak Fano and the Ys are 1-dimensional:

Example 5.1.7 ($-K_X$-lines) Let X be a smooth weak Fano variety defined over $F = \overline{F}$ with $L = -K_X$. Let $C \subset X$ be an irreducible curve not contained in $\mathbf{B}_+(L)$. Then $a(C, L|_C) > 0$ if and only if C is rational. For such a rational curve C we have

$$a(C, L|_C) = \frac{2}{L.C}.$$

Thus $a(C, L|_C) > a(X, L) = 1$ if and only if $L.C = 1$. A rational curve C with $L.C = 1$ is called a $-K_X$-line, and Theorem 5.1.6 shows that such lines are contained in a proper closed subset.

When X is a smooth del Pezzo surface, $-K_X$-lines are (-1)-curves except when X has degree 1. For a degree-1 del Pezzo surface, $-K_X$-lines are either (-1)-curves or rational members of $|-K_X|$. It is well known that such lines are contained in a proper closed subset.

The following proposition is crucial to our understanding of families of adjoint rigid subvarieties:

Proposition 5.1.8 ([LT17, Proposition 4.14]) *Let F be a field of characteristic 0. Let X be a smooth geometrically uniruled projective variety of dimension n defined over F, and L be a big and nef $\mathbb{Q}$-divisor on X. Suppose that we have a dominant family $p\colon \mathcal{U} \to W$ of integral subvarieties such that a general fiber Y satisfies $a(Y, L|_Y) = a(X, L)$ and $(Y, L|_Y)$ is adjoint rigid. Assume that the induced rational map $W \dashrightarrow \mathrm{Hilb}(X)$ is quasi-finite. Then the evaluation map $s\colon \mathcal{U} \to X$ is generically finite.*

Proof We may assume that $\overline{F} = F$. First, let us assume that (X, L) is adjoint rigid. Let d be the dimension of a general fiber Y of $p\colon \mathcal{U} \to W$. We assume that $s\colon \mathcal{U} \to X$ is not generically finite. After taking a resolution and shrinking W, we may assume that $p\colon \mathcal{U} \to W$ is smooth and W is smooth. After taking an étale cover, we may assume that the monodromy on $N^1(\mathcal{U}_w)$ is trivial.

We apply a relative $(a(X, L)s^*L + K_{\mathcal{U}})$-MMP over W and obtain a relative minimal model $\phi\colon \mathcal{U} \dashrightarrow \widetilde{\mathcal{U}}$ as explained in Section 3.4.3. Let $E_1, \ldots, E_r \subset \mathcal{U}$ be exceptional prime divisors of $\phi\colon \mathcal{U} \dashrightarrow \widetilde{\mathcal{U}}$ and set $E = \sum_{i=1}^r E_i$. Then since $(Y, L|_Y)$ is adjoint rigid and Y is general, there exists a unique effective $\mathbb{Q}$-divisor D_Y on Y such that $a(Y, L)s^*L|_Y + K_Y \sim_{\mathbb{Q}} D_Y$ and $\mathrm{Supp}(D_Y) = \mathrm{Supp}(E \cap Y)$. Indeed, this follows from Theorem 3.4.6.

Let $V \subset X$ be a general complete intersection of dimension $n - d - 1$. Then our assumption implies that $\dim s^{-1}(V) \geq n - d$. Since V is general, we have $s^{-1}(V) \not\subset E$ and

$$\overline{s(p^{-1}(p(s^{-1}(V))))} = X.$$

Indeed, for a general point $x \in X$, there exists a one-parameter family of Y

passing through x. Then such Ys sweep out a $(d + 1)$ dimensional locus in X and this locus meets with V because V has codimension $d + 1$ and V is a complete intersection. Thus our claim follows. In particular, this implies that

$$\dim p(s^{-1}(V)) \geq n - d.$$

Pick a general subvariety $W' \subset p(s^{-1}(V))$ of dimension $n - d$ such that

- $s \colon \mathcal{U}' = p^{-1}(W') \to X$ is dominant and generically finite; and
- for a general fiber $Y' \subset \mathcal{U}'$ we have $s^{-1}(V) \cap Y' \not\subset E$.

Note that our construction implies that, after shrinking W', W' and $\mathcal{U}'$ are smooth.

Now since (X, L) is adjoint rigid, there exists a unique effective $\mathbb{Q}$-divisor D such that $a(X, L)L + K_X \sim_{\mathbb{Q}} D$. For a general member Y' of $p \colon \mathcal{U}' \to W'$, we have

- $Y' \not\subset \mathrm{Supp}(s^*D)$;
- Y' is not contained in the ramification locus of $s \colon \mathcal{U}' \to X$; and
- there exists $y' \in s^{-1}(V) \cap Y'$ such that $y \notin E \cap Y'$.

On the other hand, the ramification formula tells us that

$$a(X, L)s^*L + K_{\mathcal{U}'} \sim_{\mathbb{Q}} s^*D + R,$$

where R is the ramification divisor. Since the locus $p^{-1}(W') \cap s^{-1}(V)$ has dimension at least $n-d$ and this locus maps to V which has dimension $n-d-1$, we conclude that y' is in the exceptional locus of $s \colon \mathcal{U}' \to X$. This implies that y' is contained in the support of R. However, this implies that $y' \in \mathrm{Supp}(E \cap Y)$, a contradiction. Thus our assertion holds when (X, L) is an adjoint rigid pair.

In general let $\pi \colon X \dashrightarrow B$ be the canonical map for $(X, a(X, L)L)$. After applying a resolution, we may assume that π is a morphism. We claim that a general member Y of p is contracted by π. Let $\mathcal{U}'$ be a resolution of $\mathcal{U}$. After shrinking W, we may assume that W is smooth and $p \colon \mathcal{U}' \to W$ is smooth. Taking a general complete intersection $T \subset W$, we may assume that its base change $\mathcal{U}'_T$ maps to X dominantly and generically finitely, and we denote this map by $s' \colon \mathcal{U}'_T \to X$. Then by the ramification formula we have

$$\begin{aligned} a(Y, L)(s')^*L|_Y + K_{Y'} &= (a(Y, L)(s')^*L + K_{\mathcal{U}'_T})|_Y \\ &= (a(Y, L)(s')^*L + (s')^*K_X + R)|_Y, \end{aligned}$$

where R is an effective divisor. Then this divisor is adjoint rigid by the assumptions, so $(a(Y, L)(s')^*L + (s')^*K_X)|_Y$ must have the Iitaka dimension 0 for a general Y. Thus our claim follows. Then our assertion follows from the adjoint rigid case as the fibers of π are adjoint rigid. □

Proposition 5.1.8 is particularly easy to prove when subvarieties are curves.

Example 5.1.9 Suppose we have a dominant family of curves C on X such that $a(X, L) = a(C, L|_C)$. Then C is a rational curve and we have

$$a(X, L) = a(C, L|_C) = \frac{2}{L.C}.$$

Since C is movable and $a(X, L)L + K_X$ is pseudo-effective we have

$$(a(X, L)L + K_X).C \geq 0.$$

Thus we conclude that $-K_X.C \leq 2$. On the other hand, since a family is dominant, a general C is free in the sense that if $\beta\colon C' \to C$ is the normalization, then $\beta^* T_X = \oplus \mathcal{O}(a_i)$ with $a_i \geq 0$ for all i. In particular, a component of the moduli space of rational curves $\mathrm{Mor}(\mathbb{P}^1, X)$ containing $C' \to X$ has expected dimension $-K_X.C + \dim X$. (See Chapter 7 for more details about freeness and expected dimension for moduli of rational curves.) Since the family of curves C is dominant, we must have $-K_X.C = 2$, implying that $\dim \mathcal{U} = \dim X$. (Note that the space $\mathrm{Mor}(\mathbb{P}^1, X)$ also counts automorphisms of $\mathbb{P}^1$ which form a 3-dimensional variety.) Thus the evaluation map $s\colon \mathcal{U} \to X$ is dominant and generically finite.

The following theorem is one of the key arithmetic inputs to Theorem 4.3.5 as well as Theorem 5.1.1:

Theorem 5.1.10 (Hilbert's irreducibility theorem) *Let F be a field of characteristic 0. Let $f: Y \to X$ be a generically finite dominant morphism between geometrically integral normal projective varieties defined over F.*

Assume that $\overline{F}(\overline{Y})/\overline{F}(\overline{X})$ is a Galois extension. Let $G := \mathrm{Bir}(\overline{Y}/\overline{X})$ where $\mathrm{Bir}(Y/X)$ is the group of birational X-automorphisms of Y. Then there exists a thin set $Z \subset X(F)$ such that for any $x \in X(F) \setminus Z$, we have

(i) *$f^{-1}(x)$ is irreducible; and*
(ii) *$\mathrm{Gal}(F(f^{-1}(x))/F) \cong G$,*

where $F(f^{-1}(x))$ is the residue field of $f^{-1}(x)$.

Proof The following proof is taken from [Ser92, Proposition 3.3.1]. First, let us assume that $\mathrm{Bir}(\overline{Y}/\overline{X}) = \mathrm{Bir}(Y/X)$. After shrinking X and replacing Y by its preimage if necessary, we may assume that $f: Y \to X$ is étale and finite. In this situation, G coincides with the group of X-automorphisms of Y. Hence G acts on Y, and its quotient Y/G is isomorphic to X. Let Σ be the set of all proper subgroups H of G. We consider the quotient Y/H and its induced projection $\pi_H : Y/H \to X$. We define

$$Z = \bigcup_{H \in \Sigma} \pi_H((Y/H)(F)),$$

which is a thin set because any $H \in \Sigma$ is a proper subgroup of G.

Let $x \in X(F) \setminus Z$. Let $y \in Y$ be a closed point above x and let $\overline{y}$ be a geometric closed point above y. We claim that the set

$$M := \{g \in G \mid \text{there exists } \gamma \in \mathrm{Gal}(\overline{F}/F) \text{ such that } g\overline{y} = \gamma\overline{y}\}$$

is a subgroup of G. The only non-trivial statements here are that (i) M is closed under multiplication and (ii) M is closed under taking the inverse. These follow from the fact that the Galois action on G is trivial so that the G-action and the Galois action on the set of geometric closed points above x commute.

Now we have $x \in \pi_H((Y/M)(F))$ because the image of y in Y/M is a rational point and this means that $M = G$ since $x \notin Z$. Thus $f^{-1}(x)$ is irreducible. Then we have an injective homomorphism $\mathrm{Gal}(F(f^{-1}(x))/F) \hookrightarrow G$ mapping $\gamma \in \mathrm{Gal}(F(f^{-1}(x))/F)$ to $g \in G$ such that $g\overline{y} = \gamma\overline{y}$. Indeed, for any

$$\gamma \in \mathrm{Gal}(F(f^{-1}(x))/F),$$

we have a unique $g \in G$ satisfying $g\overline{y} = \gamma\overline{y}$. Thus this map is well defined. Moreover, $\gamma\overline{y}$ uniquely determines $\gamma \in \mathrm{Gal}(F(f^{-1}(x))/F)$, showing that this map is injective. Finally, this becomes a homomorphism for the same reason that M is a subgroup of G. However, since we have $\#G = [F(f^{-1}(x)) : F]$, this map must be an isomorphism.

In general, let F'/F be a finite extension such that

$$\mathrm{Bir}(\overline{Y}/\overline{X}) = \mathrm{Bir}(Y \otimes F'/X \otimes F').$$

Then there exists a thin set $Z' \subset X(F')$ such that, for any $x \in X(F') \setminus Z'$, we have

- $f^{-1}(x)$ is irreducible; and
- $\mathrm{Gal}(F'(f^{-1}(x))/F') \cong G$,

We set $Z = Z' \cap X(F)$, which is a thin set by Lemma 4.2.8. This Z satisfies the desired property.

□

Here is one of the applications of Hilbert's irreducibility theorem (Theorem 5.1.10):

Proposition 5.1.11 ([LT17, Proposition 5.1]) *Let F be a field of characteristic 0. Suppose that we have a geometrically uniruled smooth projective variety X defined over F, and a big and nef $\mathbb{Q}$-divisor on X such that*

- $a(X, L)L + K_X$ *is rigid; and*
- $N^1(X) = N^1(\overline{X})$.

Let $p\colon X \to W$ be a morphism to a normal projective variety defined over F such that (a) $\overline{p}\colon \overline{X} \to \overline{W}$ is an algebraic fiber space and (b) for a general closed point $w \in W$ the fiber $X_w = p^{-1}(w)$ satisfies $a(X_w, L|_{X_w}) = a(X, L)$ and $(X_w, L|_{X_w})$ is adjoint rigid. Let $W^\circ \subset W$ be a non-empty Zariski open set such that

- $p\colon X^\circ = p^{-1}(W^\circ) \to W^\circ$ *is smooth; and*
- *for any closed point $w \in W^\circ$, X_w is integral and smooth, $a(X_w, L|_{X_w}) = a(X, L)$, and $(X_w, L|_{X_w})$ is adjoint rigid.*

Then

$$\{w \in W^\circ(F) \mid b(F, X, L) \leq b(F, X_w, L|_{X_w})\} \subset W(F)$$

is a thin set of W.

Proof First of all, if W is not geometrically integral, then $W(F)$ is contained in a proper closed subset. Thus we may assume that W is geometrically integral. Since (X, L) is adjoint rigid, there exists a unique effective $\mathbb{Q}$-divisor

$$D = \sum_i a_i D_i \geq 0,$$

where $a_i > 0$ and the D_is are irreducible and distinct such that $a(X, L)L + K_X \sim_{\mathbb{Q}} D$. Let $\mathsf{V} = \langle D_i \mid i \rangle \subset N^1(X)$ where $\langle D_i \mid i \rangle$ is the subspace spanned by D_is. It follows from Corollary 4.1.18 that

$$b(F, X, L) = \dim N^1(X)/\mathsf{V}.$$

Let $\overline{D} \cap \overline{X}_w = \sum_j b_j \overline{D}'_j$ with $b_j > 0$ and $\overline{D}'_j$s irreducible, and let

$$\mathsf{V}' = \langle \overline{D}'_j \mid j \rangle \subset N^1(\overline{X}_w),$$

where X_w is the fiber of p at $w \in W^\circ(F)$. Let G be the monodromy of $\pi_1(\overline{W}^\circ, w)$ on $N^1(\overline{X}_w)$. This is a finite group. Since $\overline{X}$ and $\overline{X}_w$ are rationally connected by Proposition 4.1.9, it follows from the invariant theorem of Deligne ([Del68]) that there is a surjective homomorphism

$$N^1(X)/\mathsf{V} = N^1(\overline{X})/\mathsf{V} \twoheadrightarrow N^1(\overline{X}_w)^G/\mathsf{V}'^G$$

and moreover this homomorphism admits a non-trivial kernel.

Let $\overline{\pi}\colon \overline{T}^{\circ} \to \overline{W}^{\circ}$ be the étale cover killing the monodromy G. Pick a rational point $w \in W^{\circ}(F)$. Pick a point $\overline{t} \in \overline{T}^{\circ}$ above $\overline{w}$. Then the quotient of $\pi_1^{\text{ét}}(W^{\circ}, \overline{w})$ by $\pi_1^{\text{ét}}(\overline{T}^{\circ}, \overline{t})$ is a semi-direct product of G with the Galois group $\mathrm{Gal}(\overline{F}/F)$

$$1 \to G \to \widetilde{G} \to \mathrm{Gal}(\overline{F}/F) \to 1$$

with a splitting $\iota\colon \mathrm{Gal}(\overline{F}/F) \to \widetilde{G}$ coming from a rational point $w \in W^{\circ}(F)$. We consider the Galois cover corresponding to $\iota(\mathrm{Gal}(\overline{F}/F))$ and denote it by $\pi\colon T^{\circ} \to W^{\circ}$. Note that here T° is defined over F. We denote the base change of $p\colon X^{\circ} \to W^{\circ}$ via $T^{\circ} \to W^{\circ}$ by $q^{\circ}\colon Y^{\circ} \to T^{\circ}$. Then we choose a projective compactification $q\colon Y \to T$ of q° such that Y is smooth and Y admits a morphism to X which is compatible with $Y^{\circ} \to X^{\circ}$. Let F'/F be a finite extension such that the Galois action of $\mathrm{Gal}(\overline{F}/F')$ on $N^1(\overline{Y})$ is trivial.

It follows from Hilbert's irreducibility theorem (Theorem 5.1.10) that there exists a thin set $Z' \subset W^{\circ}(F')$ such that for any $w' \in W^{\circ}(F') \setminus Z'$, $\pi^{-1}(w')$ is irreducible and its Galois group is G. Since the action of $\mathrm{Gal}(\overline{F}/F')$ on $N^1(\overline{Y})$ is trivial and the restriction map $N^1(\overline{Y}) \to N^1(\overline{Y}_{\overline{t}})$ is surjective, we conclude that for any $w' \in W^{\circ}(F') \setminus Z'$, $\mathrm{Gal}(\overline{F}/F')$ acts on $N^1(\overline{X}_{w'})$ as G does. Thus for such a w' we have

$$b(F', X_{w'}, L|_{X_{w'}}) = \dim N^1(\overline{X}_{w'})^G/V'^G < \dim N^1(X)/V = b(F, X, L).$$

Let $Z = Z' \cap W^{\circ}(F)$, which is a thin set by Lemma 4.2.8. Then for any $w' \in W^{\circ}(F) \setminus Z$ we have

$$b(F, X_{w'}, L|_{X_{w'}}) \le b(F', X_{w'} \otimes F', (L|_{X_{w'}}) \otimes F') < b(F, X, L).$$

This completes the proof of our assertion. □

Let us discuss Batyrev–Tschinkel's example (Example 4.2.17) in the context of the above proposition:

Example 5.1.12 Let F be a field of characteristic 0, and $X \subset \mathbb{P}^3_x \times \mathbb{P}^3_y$ be a smooth Fano fivefold defined by

$$\sum_{i=0}^{3} x_i y_i^3 = 0,$$

which is considered in Example 4.2.17. Let $L = -K_X$ and $\pi\colon X \to \mathbb{P}^3_x$ be the first projection. Let $U \subset \mathbb{P}^3_x$ be the Zariski open subset defined by $x_0x_1x_2x_3 \neq 0$. Proposition 5.1.11 shows that

$$\{w \in U(F) \mid b(F, X, L) = 2 \le b(F, X_w, L|_{X_w})\}$$

is a thin set of $\mathbb{P}^3_x$. Thus this example is compatible with Manin's conjecture (Conjecture 4.2.9).

Now we will prove Theorem 5.1.1:

Proof of Theorem 5.1.1 We define

$$Z := \bigcup_Y Y(F) \subset X(F),$$

where Y runs over all geometrically integral subvarieties defined over F such that

$$(a(X, L), b(F, X, L)) \leq (a(Y, L|_Y), b(F, Y, L|_Y))$$

in the lexicographic order. Our goal is to show that Z is a thin set of X.

First of all, we may assume that X is geometrically integral; otherwise $X(F)$ is empty. Let V be the union of integral subvarieties $Y \subset X$ such that

$$a(Y, L|_Y) > a(X, L).$$

By Theorem 5.1.6, V is a proper closed subset. Let $\overline{W}$ be the closure in $\mathrm{Hilb}(\overline{X})$ of the locus parametrizing integral subvarieties $\overline{Y}$ defined over $\overline{F}$ such that $\overline{Y} \not\subset \overline{V}$, $a(\overline{Y}, L|_{\overline{Y}}) = a(X, L)$, and $(\overline{Y}, L|_{\overline{Y}})$ is adjoint rigid. It follows from Theorem 5.1.4 that this set is a bounded set. After replacing $\overline{W}$ by the union of its Galois orbit, we may assume that $\overline{W}$ descends to a closed subset W defined over F. Let $W = \cup_i W_i$ be the irreducible decomposition. Let $\pi_i : \mathcal{U}_i \to W_i$ be the universal family with the evaluation map $s_i : \mathcal{U}_i \to X$.

When $s_i : \mathcal{U}_i \to X$ is not dominant, we define a proper closed subset $V_i := s_i(\mathcal{U}_i) \subset X$ and a type I thin set $Z_i := V_i(F) \subset X(F)$.

Suppose that $s_i : \mathcal{U}_i \to X$ is dominant. Then, by Proposition 5.1.8, the map

$$s_i : \mathcal{U}_i \to X$$

is generically finite. Let $\beta_i : \widetilde{\mathcal{U}_i} \to \mathcal{U}_i$ be a smooth resolution with $\widetilde{s}_i = s_i \circ \beta_i$, and $\widetilde{\pi}_i = \pi_i \circ \beta$.

Assume that W_i is not geometrically integral. Then $\mathcal{U}_i(F)$ is not Zariski dense in $\mathcal{U}_i$. Let K_i be a proper closed subset of W_i such that $\widetilde{\pi}_i$ is smooth over $W_i \setminus K_i$. Let R_i be the ramification locus of $\widetilde{s}_i : \widetilde{\mathcal{U}_i} \to X$. We define a proper closed subset $V_i = \overline{s_i(\mathcal{U}_i(F))} \cup \widetilde{s}_i(\widetilde{\pi}_i^{-1}(K_i) \cup R_i)$. We define a type I thin set $Z_i = V_i(F)$.

Assume that W_i is geometrically integral. If $s_i : \mathcal{U}_i \to X$ has degree ≥ 2, then we let R_i be the ramification locus of $\widetilde{s}_i : \widetilde{\mathcal{U}_i} \to X$. Let K_i be a proper closed subset of W_i such that $\widetilde{\pi}_i$ is smooth over $W_i \setminus K_i$. We set a proper closed subset $V_i = \widetilde{s}_i(R_i \cup \widetilde{\pi}_i^{-1}(K_i))$. We define a thin set Z_i by

$$Z_i = s_i(\mathcal{U}_i(F)) \cup V_i(F).$$

Finally, assume that $s_i : \mathcal{U}_i \to X$ is birational. Let R_i be the ramification locus of $\widetilde{s}_i : \widetilde{\mathcal{U}}_i \to X$. Let K_i be a proper closed subset of W_i such that $\widetilde{\pi}_i$ is smooth over $W_i \setminus K_i$, and for any $w \in W_i \setminus K_i$, $Y_w := \widetilde{\pi}_i^{-1}(w)$ is integral, $a(Y_w, \widetilde{s}_i^* L|_{Y_w}) = a(X, L)$, and $(Y_w, \widetilde{s}_i^* L|_{Y_w}))$ is adjoint rigid. We define a proper closed subset $V_i := \widetilde{s}_i(R_i \cup \widetilde{\pi}_i^{-1}(K_i))$. Consider

$$Z_i = \bigcup_w Y_w(F) \cup V_i(F),$$

where $w \in (W_i \setminus K_i)(F)$ runs over all rational points such that

$$(a(X, L), b(F, X, L)) \leq (a(Y_w, \widetilde{s}_i^* L|_{Y_w}), b(F, Y_w, \widetilde{s}_i^* L|_{Y_w}))$$

in the lexicographic order. It follows from Proposition 5.1.11 that this is a thin set.

Then we define a thin set Z' by

$$Z' = \bigcup_i Z_i \cup V(F) \subset X(F).$$

We claim that $Z \subset Z'$. Let Y be a geometrically integral subvariety such that

$$(a(X, L), b(F, X, L)) \leq (a(Y, L|_Y), b(F, Y, L|_Y)).$$

We would like to show that $Y(F) \subset Z'$. This is clear when $Y \subset V$. Thus we may assume that $Y \not\subset V$. This means that $a(Y, L|_Y) = a(X, L)$. Lemma 5.1.3 implies that there exists i and a subvariety $S \subset W_i$ such that $Y = s_i(\widetilde{\pi}_i^{-1}(S))$ and $\widetilde{s}_i : \widetilde{S} \to Y$ is birational where $\widetilde{S}$ is the main component of $\widetilde{\pi}_i^{-1}(S)$. We may assume that $s_i : \mathcal{U}_i \to X$ is dominant and W_i is geometrically integral; otherwise it is clear from the construction that $Y(F) \subset Z_i$. Assume that $s_i : \mathcal{U}_i \to X$ has degree ≥ 2. If $S \subset K_i$ then our assertion is clear. So we may assume the contrary. In this setting, the morphism $\widetilde{s}_i : \widetilde{\pi}_i^{-1}(S) \to Y$ is bijective outside of $\widetilde{s}_i(R_i)$, thus we conclude that every point in $Y(F)$ is contained in $\widetilde{s}_i(R_i)$ or the s_i-image of a rational point in $\mathcal{U}_i(F)$. We conclude that $Y(F) \subset Z_i$. Finally, assume that $s_i : \mathcal{U}_i \to X$ is birational. Again if $S \subset K_i$, then our assertion is clear. Otherwise again the morphism $\widetilde{s}_i : \widetilde{\pi}_i^{-1}(S) \to Y$ is bijective outside of $\widetilde{s}_i(R_i)$, thus Lemma 5.1.3 implies that $Y(F) \subset Z_i$. Hence the claim follows. □

5.2 Covers

In this section, we discuss the main results of [Sen21]. First, let us introduce the following definition:

Definition 5.2.1 Let F be a field of characteristic 0. Let X be a geometrically uniruled projective variety defined over F, and L be a big and nef $\mathbb{Q}$-Cartier divisor on X. We say a dominant generically finite morphism $f: Y \to X$ from a projective variety Y defined over F is an *a-cover* if $a(Y, f^*L) = a(X, L)$. An a-cover $f: Y \to X$ is adjoint rigid if (Y, f^*L) is an adjoint rigid pair in the sense of Definition 4.1.8.

Next, let us observe the behavior of the a-invariant under the pullback of a dominant generically finite morphism:

Lemma 5.2.2 *Let F be a field of characteristic 0. Let X be a geometrically uniruled projective variety defined over F, and L be a big and nef $\mathbb{Q}$-Cartier divisor on X. Let $f: Y \to X$ be a dominant generically finite morphism from a projective variety Y defined over F. Then we have*

$$a(Y, f^*L) \leq a(X, L).$$

Furthermore, if $f: Y \to X$ is an adjoint rigid a-cover, then (X, L) is adjoint rigid.

Proof After taking resolutions, we may assume that X and Y are smooth. Then the ramification formula shows that there exists an effective divisor $R \geq 0$ on Y such that

$$K_Y \sim f^*K_X + R.$$

Thus

$$a(X, L)f^*L + K_Y \sim_{\mathbb{Q}} f^*(a(X, L)L + K_X) + R$$

is pseudo-effective. This implies that $a(Y, f^*L) \leq a(X, L)$. Furthermore if (X, L) is not adjoint rigid, then the above formula implies that either

$$a(Y, f^*L) < a(X, L)$$

or $a(Y, f^*L) = a(X, L)$ but (Y, f^*L) is not adjoint rigid. Thus the last statement follows. □

The main result of this section is the following theorem:

Theorem 5.2.3 ([Sen21, Corollary 1.4]) *Let F be a field of characteristic 0. Suppose that we have a smooth geometrically uniruled projective variety X defined over F and a big and nef $\mathbb{Q}$-divisor L on X. We assume that*

$$a(X, L)L + K_X$$

is rigid.

As we let $f: Y \to X$ run over all adjoint rigid a-covers with Y smooth such that

(i) $(a(X,L),b(F,X,L)) \leq (a(Y,f^*L),b(F,Y,f^*L))$; *and*
(ii) $f\colon Y \to X$ *is face-contracting,*

the union

$$\bigcup_f f(Y(F)) \subset X(F)$$

is a thin set of X.

To this end, we first prove the finiteness of adjoint rigid a-covers up to birational equivalence over an algebraic closure using the minimal model program and the boundedness of singular Fano varieties. The main technical input of Theorem 5.2.3 from the minimal model program is the following proposition:

Proposition 5.2.4 ([Sen21, Proposition 2.18]) *We assume that* $\overline{F} = F$. *Let* X *be a smooth projective variety defined over* F, *and* $\Delta \geq 0$ *be an effective* $\mathbb{Q}$-*divisor on* X *which is big and nef. Let* $\psi\colon X \dashrightarrow X'$ *be a minimal model for* (X,Δ) *obtained by Theorem 3.4.1. Let* E *be an exceptional prime divisor for* ψ, *and* $\pi\colon \widetilde{E} \to E$ *be a resolution. Then*

$$K_{\widetilde{E}} + \pi^*(\Delta|_E)$$

is not pseudo-effective.

Proof The following proof is taken from [Sen21, Proposition 2.18]. First, we reduce our problem to the case when E is smooth and Δ is ample. Let $\pi\colon \widetilde{X} \to X$ be a log resolution of $(X, E+\Delta)$, that is, $\pi\colon \widetilde{X} \to X$ is a smooth resolution such that the union of the support of the strict transform of $E+\Delta$ and the exceptional divisors is a divisor with strict normal crossings. We further assume that $\phi = \psi \circ \pi\colon \widetilde{X} \to X'$ is a morphism. Let $\widetilde{E}$ be the strict transform of E. Our goal is to show that $K_{\widetilde{E}} + \pi^*(\Delta|_E)$ is not pseudo-effective. By Remark 3.4.3, we can find $\widetilde{\Delta} \sim_{\mathbb{Q}} \pi^*\Delta$ such that $(\widetilde{X},\widetilde{\Delta})$ is a terminal pair and $\phi\colon \widetilde{X} \to X'$ is a $(K_{\widetilde{X}} + \widetilde{\Delta})$-minimal model such that $(X', \phi_*\widetilde{\Delta})$ is terminal. Let H be an ample divisor $\widetilde{X}$. For a sufficiently small $\epsilon > 0$, $(\widetilde{X}, \widetilde{\Delta} + \epsilon H)$ and $(X', \phi_*\widetilde{\Delta} + \epsilon\phi_*H)$ are still terminal pairs. Applying $(K_{X'} + \phi_*\widetilde{\Delta} + \epsilon\phi_*H)$-MMP ([BCHM10, Corollary 1.4.2]) to X' if necessary, we obtain a $(K_{\widetilde{X}} + \widetilde{\Delta} + \epsilon H)$-minimal model $\phi'\colon \widetilde{X} \dashrightarrow X''$ such that ϕ' contracts $\widetilde{E}$. Then since minimal models are isomorphic in codimension 1 ([BCHM10, Corollary 1.1.3]), any outcome of $(K_{\widetilde{X}} + \widetilde{\Delta} + \epsilon H)$-MMP ([BCHM10, Corollary 1.4.2]) will contract $\widetilde{E}$. Thus by replacing X by $\widetilde{X}$, E by $\widetilde{E}$, and Δ by $\widetilde{\Delta} + \epsilon H$, we may assume that E is smooth and Δ is ample.

Using ampleness of Δ, we may find a $\mathbb{Q}$-divisor $\Delta_0 \sim_{\mathbb{Q}} \Delta$ such that (X, Δ_0) is klt, $(E, \Delta_0|_E)$ is canonical, and $(X, E + \Delta_0)$ is divisorial log terminal. Indeed, these follow from the Bertini theorem combined with [KM98, Corollary 2.31].

Then since $(X, E + \Delta_0)$ is dlt and Δ_0 is ample, it follows from [BCHM10] that we may run a $(K_X + E + \Delta_0)$-MMP. Since the ascending chain condition for log canonical thresholds and special termination for dlt flips are known ([HMX14] and [BCHM10, Lemma 5.1] respectively), this $(K_X + E + \Delta_0)$-MMP terminates with a minimal model $\rho\colon X \dashrightarrow X_m$ by [Bir07, Theorem 1.2]. Then since any $(K_X + \Delta_0)$-MMP contracts E, ρ must contract E as well. Let $\rho_k : X_k \longrightarrow X_{k+1}$ be the divisorial contraction that is the step of the $(K_X + E + \Delta_0)$-MMP contracting the strict transform E_k of E. Let $\Phi_k : X \dashrightarrow X_k$ be the composition of the steps of $(K_X + E + \Delta_0)$-MMP, and $\Delta_k = (\Phi_k)_*\Delta_0$. Then since $(X_k, E_k + \Delta_k)$ is dlt, E_k is normal by [KM98, Proposition 5.51]. It follows from [AK19, Theorem 7] that the restriction

$$\Phi_k|_E : (E, \mathrm{Diff}_E \Delta_0) \dashrightarrow (E_k, \mathrm{Diff}_E \Delta_k)$$

is a composition of steps of a $(K_E + \mathrm{Diff}_E \Delta_0)$-MMP. (Here Diff_E denotes "different." See [Kol13, Definition 4.2] for the definition and its basic properties.) Since $\mathrm{Diff}_E \Delta_0 = \Delta_0|_E$, it suffices to show that $K_{E_k} + \mathrm{Diff}_E \Delta_k$ is not pseudo-effective. Since E_k is covered by curve C such that

$$(K_{E_k} + \mathrm{Diff}_E \Delta_k).C = (K_{E_k} + E_k + \Delta_k).C < 0,$$

our assertion follows. □

As a corollary, we have the following restrictions on branch divisors of adjoint rigid a-covers:

Corollary 5.2.5 ([Sen21, Corollary 2.20]) *Let F be a field of characteristic 0. Let X be a smooth geometrically uniruled and geometrically integral projective variety defined over F, and L be a big and nef $\mathbb{Q}$-divisor on X. Suppose that we have an adjoint rigid a-cover $f : Y \longrightarrow X$ from a normal geometrically integral projective variety. Let R_0 be a component of the ramification locus of f such that $B_0 = f(R_0)$ is a branch divisor. Then we have*

$$a(B_0, L|_{B_0}) > a(X, L).$$

Proof In view of Lemma 4.1.5, we may assume that $\overline{F} = F$. After applying a smooth resolution we may assume that Y is smooth. Then there exists the ramification divisor $R \geq 0$ such that

$$K_Y \sim f^*K_X + R,$$

and R contains R_0 in its support. Since $a(X, L)L + K_X$ is $\mathbb{Q}$-linearly equivalent to an effective $\mathbb{Q}$-divisor, $a(X, L)f^*L + K_Y \sim_{\mathbb{Q}} f^*(a(X, L)L + K_X) + R$ is also $\mathbb{Q}$-linearly equivalent to an effective $\mathbb{Q}$-divisor containing R_0 in its support. By the assumption of adjoint rigid a-cover, this divisor has the Iitaka dimension 0. In particular, any outcome of $(a(X, L)f^*L + K_Y)$-MMP (Theorem 3.4.1) contracts R_0. Thus it follows from Proposition 5.2.4 that if we denote a resolution of R_0 by $\pi\colon \widetilde{R}_0 \to R_0$, then $a(X, L)\pi^*((f^*L)|_{R_0}) + K_{\widetilde{R}_0}$ is not pseudo-effective. This implies that

$$a(R_0, (f^*L)|_{R_0}) > a(X, L).$$

Finally, by Lemma 5.2.2, we have $a(B_0, L|_{B_0}) \geq a(R_0, (f^*L)|_{R_0}) > a(X, L)$. Hence the claim. □

We also need the following input from the weak BAB conjecture (Corollary 3.5.6):

Proposition 5.2.6 ([Sen21, Proposition 2.17]) *Let F be a field of characteristic 0. Let X be a smooth geometrically uniruled and geometrically integral projective variety defined over F, and L be a big and nef $\mathbb{Q}$-divisor on X. Then there exists a constant $C > 0$ depending only on* $\dim X$, $\mathrm{Vol}(L)$, *and* $a(X, L)$ *such that, for any adjoint rigid a-cover $f: Y \to X$ from a geometrically integral projective variety Y, we have*

$$\deg f \leq C.$$

Proof Again we may assume that $\overline{F} = F$. Suppose that we have an adjoint rigid a-cover $f\colon Y \to X$. After applying a resolution, we may assume that Y is smooth. Then it follows from Proposition 4.1.9 that there exists a birational contraction $\phi\colon Y \dashrightarrow Y'$ such that

- Y' is a weak Fano variety with terminal singularities; and
- $a(X, L)\phi_*L + K_{Y'} \sim_{\mathbb{Q}} 0$.

Then it follows from the negativity lemma (Lemma 3.2.16) that

$$(\deg f) \cdot L^{\dim X} = (f^*L)^{\dim X} = \mathrm{Vol}(f^*L) \leq \mathrm{Vol}(\phi_*f^*L) = \frac{\mathrm{Vol}(-K_{Y'})}{a(X, L)^{\dim X}}.$$

Thus our assertion follows from Corollary 3.5.6. □

The following theorem states the finiteness of adjoint rigid a-covers up to birational equivalence over an algebraically closed field, which was conjectured in [LT17, Conjecture 1.7] by Lehmann and the author. This was proved by Sengupta in [Sen21, Theorem 1.1]:

Theorem 5.2.7 ([Sen21, Theorem 1.1]) *Assume that $\overline{F} = F$. Let X be a smooth projective uniruled variety defined over F, and L be a big and nef $\mathbb{Q}$-divisor on X. Then there are only finitely many adjoint rigid a-covers $f: Y \to X$ up to birational equivalence.*

Proof Let V be a proper closed subset as defined in Theorem 5.1.6. It follows from Corollary 5.2.5 that the cover

$$f: Y \setminus f^{-1}(V) \to X \setminus V$$

is étale in codimension 1. Moreover, by Proposition 5.2.6 we have

$$\deg f \leq C.$$

It follows from [Gea72, Théoréme II.2.3.1] that $\pi_1^{\text{ét}}(X \setminus V, x)$ is topologically finitely generated so that it contains only finitely many subgroups of given index. This completes the proof. □

Although we prove the finiteness of adjoint rigid a-covers over an algebraic closure, the corresponding statement over a number field is not owing due to the existence of twists. To overcome this issue, we again use Hilbert's irreducibility theorem (Theorem 5.1.10) in a systematic way. First, let us record the following lemma:

Lemma 5.2.8 ([Che04]) *Let $f: Y \dashrightarrow X$ be a dominant generically finite rational map between normal projective varieties over a field F of characteristic 0. Then there exists a birational modification $f': Y' \to X$ of f such that Y' is smooth and projective and* $\mathrm{Bir}(\overline{Y}'/\overline{X}) = \mathrm{Aut}(\overline{Y}'/\overline{X})$.

Thus any twist of $f: Y \dashrightarrow X$ is birational to a twist of $f': Y' \to X$.

Proof The following proof is taken from [LST22, Lemma 6.1]. After applying a resolution, we may assume that f is a morphism. Then we replace Y by its Stein factorization so that we have $\mathrm{Bir}(\overline{Y}/\overline{X}) = \mathrm{Aut}(\overline{Y}/\overline{X})$. Let F'/F be a finite Galois extension such that $G = \mathrm{Aut}(\overline{Y}/\overline{X}) = \mathrm{Aut}(Y \otimes F'/X \otimes F')$. Then $G \rtimes \mathrm{Gal}(F'/F)$ acts on $Y \otimes F'$. By [AW97, Theorem 0.1] we resolve singularities equivariantly with respect to $G \rtimes \mathrm{Gal}(F'/F)$ to obtain a smooth projective variety Y'' defined over F' and define Y' as the quotient of Y'' by $\mathrm{Gal}(F'/F)$. This satisfies the desired property. □

The following two propositions are applications of Hilbert's irreducibility theorem (Theorem 5.1.10):

Proposition 5.2.9 ([LT17, Proposition 8.2]) *Let F be a field of characteristic 0. Suppose that we have a dominant generically finite morphism $f\colon Y \to X$ between geometrically integral normal projective varieties defined over F such that the extension $\overline{F}(\overline{Y})/\overline{F}(\overline{X})$ is not Galois. Then the union*

$$\bigcup_{\sigma \in H^1(F, \mathrm{Aut}(\overline{Y}/\overline{X}))} f^\sigma(Y^\sigma(F)) \subset X(F)$$

is a thin set.

Proof After replacing Y by its Stein factorization, we may assume that f is a finite morphism. Let X° be a Zariski open subset such that

$$f\colon Y^\circ = f^{-1}(X^\circ) \to X^\circ$$

is a proper étale covering. Then after replacing Y by its twist if necessary, we may assume that $Y^\circ(F) \neq \emptyset$; otherwise our statement is obvious. We claim that there exists a finite cover $g\colon W \to Y$ defined over F such that $\overline{F}(\overline{W})$ is the Galois closure of $\overline{F}(\overline{Y})$ over $\overline{F}(\overline{X})$. Pick a rational point $y \in Y^\circ(F)$ and let $x = f(y)$. Then we have

$$H := \pi_1^{\text{ét}}(\overline{Y}^\circ, y) \subset \pi_1^{\text{ét}}(\overline{X}^\circ, x) =: G.$$

Since $\overline{F}(\overline{Y})/\overline{F}(\overline{X})$ is not Galois, H is not normal in G. Let

$$H' = \cap_{g \in G} gHg^{-1}.$$

Then H' is a finite index subgroup of H and this corresponds to the Galois closure $\overline{W}^\circ$. The fundamental group exact sequence is

$$1 \to \pi_1^{\text{ét}}(\overline{Y}^\circ, y) \to \pi_1^{\text{ét}}(Y^\circ, y) \to \mathrm{Gal}(\overline{F}/F) \to 1.$$

Then our rational point y induces a section ι of $\mathrm{Gal}(\overline{F}/F)$ into $\pi_1^{\text{ét}}(Y^\circ, y)$ and the conjugate action of this section fixes H and G, and therefore, it fixes H' as well. The cover W° corresponding to $H' \cdot \iota(\mathrm{Gal}(\overline{F}/F)) \subset \pi_1^{\text{ét}}(Y^\circ, y)$ satisfies the desired property.

In particular, for any twist Y^σ with $y \in (Y^\circ)^\sigma(F)$, we may find a twist $W^{\sigma'} \to X$ such that $W^{\sigma'}$ factors through Y^σ. Moreover, by the construction, $W^{\sigma'}$ comes with a rational point w mapping to y. Let F_1/F be a finite extension such that $\mathrm{Aut}(\overline{W}/\overline{X}) = \mathrm{Aut}(W \otimes F_1/X \otimes F_1)$. By Hilbert's irreducibility theorem (Theorem 5.1.10), there exists a thin set $Z' \subset X(F_1)$ such that for any

$$x \in X(F_1) \setminus Z',$$

$(f \circ g)^{-1}(x)$ is irreducible and $\mathrm{Gal}(F_1((f \circ g)^{-1}(x))/F_1)$ is isomorphic to

$$G = \mathrm{Aut}(\overline{W}/\overline{X}).$$

Let $Z = X(F) \cap Z'$. This is a thin set by Lemma 4.2.8.

We claim that

$$\bigcup_{\sigma \in H^1(F, \mathrm{Aut}(\overline{Y}/\overline{X}))} f^\sigma(Y^\sigma(F)) \subset Z \cup (X \setminus X^\circ)(F).$$

Indeed, for any $x = f(y) \in f^\sigma((Y^\circ)^\sigma(F))$, there exists $\sigma' \in H^1(F, \mathrm{Aut}(\overline{W}/\overline{X}))$ such that $g^{\sigma'} : W^{\sigma'} \to X$ factors through Y^σ with a rational point w mapping to y. This implies that $\sigma' : \mathrm{Gal}(\overline{F}/F) \to \mathrm{Aut}(\overline{W}/\overline{X})$ factors through the normalizer of $\mathrm{Aut}(\overline{W}/\overline{Y})$ in G, which is a proper subgroup. Indeed, let F'/F_1 be a finite extension such that we have compatible isomorphisms

$$\begin{array}{ccc} W \otimes F' & \xrightarrow{\psi} & W^{\sigma'} \otimes F' \\ \downarrow & & \downarrow \\ Y \otimes F' & \xrightarrow{\phi} & Y^{\sigma} \otimes F' \end{array}$$

Then for any $s \in \mathrm{Gal}(F'/F)$, $\sigma'_s = \psi^{-1} \circ \psi^s$ is compatible with $\sigma_s = \phi^{-1} \circ \phi^s$. This implies that σ'_s is contained in the normalizer of $\mathrm{Aut}(\overline{W}/\overline{Y})$ in G.

Then the existence of w implies that $\mathrm{Gal}(F_1((f \circ g)^{-1}(x))/F_1)$ is in the normalizer of $\mathrm{Aut}(\overline{W}/\overline{Y})$ in G. Thus we conclude that $x \in Z$. □

Next, we discuss the case when $f : Y \to X$ is geometrically Galois:

Proposition 5.2.10 ([LT17, Proposition 8.4] and [LST22, Theorem 6.3]) *Let F be a field of characteristic 0. Suppose that we have a dominant generically finite morphism $f : Y \to X$ between geometrically integral normal projective varieties defined over F such that the extension $\overline{F}(\overline{Y})/\overline{F}(\overline{X})$ is Galois. Let L be a big and nef $\mathbb{Q}$-Cartier divisor on X. As $\sigma \in H^1(F, \mathrm{Aut}(\overline{Y}/\overline{X}))$ varies over all 1-cocycles such that the twist $f^\sigma : Y^\sigma \to X$ satisfies*

(i) *$(a(X, L), b(F, X, L)) \leq (a(Y, f^*L), b(F, Y, f^*L))$ in the lexicographic order; and*
(ii) *if the equality is achieved, then $f^\sigma : Y^\sigma \to X$ is face-contracting,*

the union

$$\bigcup_\sigma f^\sigma(Y^\sigma(F)) \subset X(F)$$

is a thin set.

Proof We need to address the statement only when $a(Y, f^*L) = a(X, L)$. Applying Lemma 5.2.8, we may assume that X and Y are smooth and

$$G =: \mathrm{Bir}(\overline{Y}/\overline{X}) = \mathrm{Aut}(\overline{Y}/\overline{X}).$$

Let F_1/F be a finite extension such that $N^1(Y \otimes F_1) = N^1(\overline{Y})$ and

$$\mathrm{Aut}(Y \otimes F_1/X \otimes F_1) = \mathrm{Aut}(\overline{Y}/\overline{X}).$$

By Theorem 5.1.10, there exists a thin set $Z' \subset X(F_1)$ such that for any $x \in X(F_1) \setminus Z'$, $f^{-1}(x)$ is irreducible and $\mathrm{Gal}(F_1(f^{-1}(x))/F_1)$ is isomorphic to G. We let $Z = X(F) \cap Z'$, which is a thin set of $X(F)$ by Lemma 4.2.8.

We will prove that if a twist $\sigma \in H^1(F, \mathrm{Aut}(\overline{Y}/\overline{X}))$ satisfies $f^\sigma(Y^\sigma(F)) \not\subset Z$, then we have $b(F, Y^\sigma, (f^\sigma)^*L) \leq b(F, X, L)$ and if the equality is achieved then f^σ is face-contracting. We claim that $N^1(\overline{Y})^G$ is spanned by $N^1(\overline{X})$ and G-invariant $\overline{f}$-exceptional divisors. Let $\overline{g}\colon \overline{Y} \to \overline{W}$ and $\overline{h}\colon \overline{W} \to \overline{X}$ be the Stein factorization of $\overline{f}\colon \overline{Y} \to \overline{X}$. Let $\overline{D}_{\overline{Y}}$ be a G-invariant $\mathbb{Q}$-divisor on $\overline{Y}$. Let $\overline{D}_{\overline{X}} = \overline{f}_*\overline{D}_{\overline{Y}}$. Then we have $|G|\overline{g}_*\overline{D}_{\overline{Y}} = \overline{h}^*\overline{D}_{\overline{X}}$ as $\mathbb{Q}$-divisors. Thus $\overline{D}_{\overline{Y}} - \frac{1}{|G|}\overline{f}^*\overline{D}_{\overline{X}}$ is $\overline{g}$-exceptional, hence it is $\overline{f}$-exceptional.

Let $\mathcal{F}^{\overline{X}}$ be the minimal face of $\overline{\mathrm{Eff}}^1(\overline{X})$ containing $a(X, L)L + K_X$; in other words it is the dual face of $\mathcal{F}(\overline{X}, L)$. Similarly let $\mathcal{F}^{\overline{Y}}$ be the minimal face of $\overline{\mathrm{Eff}}^1(\overline{Y})$ containing $a(X, L)f^*L + K_Y$. By the ramification formula, $\mathcal{F}^{\overline{Y}}$ contains all $\overline{f}$-exceptional divisors. This implies that a natural map

$$N^1(\overline{X})/\langle \mathcal{F}^{\overline{X}} \rangle \to N^1(\overline{Y})^G/\langle \mathcal{F}^{\overline{Y}} \rangle^G$$

is surjective, where $\langle \mathcal{F}^{\overline{X}} \rangle$ and $\langle \mathcal{F}^{\overline{Y}} \rangle$ are subspaces generated by $\mathcal{F}^{\overline{X}}$ and $\mathcal{F}^{\overline{Y}}$ respectively.

We now suppose that we have $x \in X(F) \setminus Z$ with $y \in Y^\sigma(F)$ such that $x = f^\sigma(y)$. Since $f^{-1}(x)$ is irreducible even after taking the base change to F_1, the existence of y implies that a cocycle σ induces a surjection from $\mathrm{Gal}(\overline{F}/F_1)$ to G. Since the action of $\mathrm{Gal}(\overline{F}/F_1)$ is trivial on $N^1(\overline{Y})$, one can conclude that the action of $\mathrm{Gal}(\overline{F}/F_1)$ on $N^1(\overline{Y}^\sigma)$ acts as G does. Thus after taking a finite quotient we have a surjection

$$\begin{aligned} N^1(X)/\langle \mathcal{F}^X \rangle &\to (N^1(\overline{Y}^\sigma)^G/\langle \mathcal{F}^{\overline{Y}^\sigma} \rangle^G)^{\mathrm{Gal}(\overline{F}/F)}, \\ &= (N^1(\overline{Y}^\sigma)/\langle \mathcal{F}^{\overline{Y}^\sigma} \rangle)^{\mathrm{Gal}(\overline{F}/F)} \\ &= (N^1(Y^\sigma)/\langle \mathcal{F}^{Y^\sigma} \rangle). \end{aligned}$$

Our assertion follows. □

As a corollary we have the following statement:

Corollary 5.2.11 ([LST22, Theorem 6.3]) *Let F be a field of characteristic 0. Let X be a smooth geometrically uniruled projective variety defined over F, and L be a big and nef $\mathbb{Q}$-divisor on X. Suppose that $f\colon Y \to X$ is a dominant generically finite morphism from a normal projective variety Y defined over F. Let σ vary over all $\sigma \in H^1(F, \mathrm{Aut}(\overline{Y}/\overline{X}))$ such that*

(i) Y^σ *is irreducible;*

(ii) $(a(X, L), b(F, X, L)) \leq (a(Y^\sigma, (f^\sigma)^*L), b(F, Y^\sigma, (f^\sigma)^*L))$ *in the lexicographic order; and*

(iii) *if the above inequality is equal, then* $f^\sigma : Y^\sigma \to X$ *is face-contracting.*

Then the union

$$\bigcup_\sigma f^\sigma(Y^\sigma(F)) \subset X(F)$$

is a thin set.

Proof When X is not geometrically integral, $X(F)$ is empty and the statement is vacuous. So we may assume that X is geometrically integral. When Y is not geometrically integral, if there is a rational point on Y^σ that is a smooth point of Y^σ, then Y^σ is reducible. Hence the union in the statement is contained in $f(\mathrm{Sing}(Y))(F)$, which is a type I thin set. Hence we may assume that Y is geometrically integral. If $\overline{F}(\overline{Y})/\overline{F}(\overline{X})$ is not Galois, then our assertion follows from Proposition 5.2.9. When $\overline{F}(\overline{Y})/\overline{F}(\overline{X})$ is Galois, our assertion follows from Proposition 5.2.10. □

Finally, we prove Theorem 5.2.3. Before the proof let us record the following lemma which is well known to the experts:

Lemma 5.2.12 (Lang–Nishimura) *Let F be a field of characteristic* 0. *Let X be a smooth projective variety defined over F, and $\beta\colon X' \to X$ be a birational morphism from another smooth projective variety defined over F. Then β induces a surjective map $X'(F) \to X(F)$.*

Proof Let $x \in X(F)$. We consider the blowup of X at x and denote it by $\phi\colon \mathrm{Bl}_x(X) \to X$ with the exceptional divisor E, which is isomorphic to the projective space $\mathbb{P}^{\dim X - 1}$. Then $\beta^{-1} \circ \phi$ is well defined at the generic point of E. We choose a general rational point on E; its image on X' will map to x. Hence the claim. □

Proof of Theorem 5.2.3 We may assume that X is geometrically integral; otherwise $X(F)$ is empty. It follows from Theorem 5.2.7 that, over an algebraic closure, there are only finitely many adjoint rigid a-covers for (X, L) up to birational equivalence. Let $\overline{f}_i\colon \overline{Y}_i \to \overline{X}$ denote these a-covers. Among them, we are concerned only with a-covers that descend to F after possible birational modifications, so we may assume that $f_i\colon Y_i \to X$s are defined over F. After applying Lemma 5.2.8, we may assume that each Y_i is smooth and we have $\mathrm{Bir}(\overline{Y}_i/\overline{X}) = \mathrm{Aut}(\overline{Y}_i/\overline{X})$. Let V_i be the image of the ramification divisor of $f_i\colon Y_i \to X$. For each i we let σ vary over all $\sigma \in H^1(F, \mathrm{Aut}(\overline{Y}_i/\overline{X}))$

such that $(a(X, L), b(F, X, L)) \leq (a(Y_i^\sigma, (f_i^\sigma)^* L), b(F, Y_i^\sigma, (f_i^\sigma)^* L))$ and f_i^σ is face-contracting. We define Z_i as

$$Z_i = \bigcup_\sigma f_i^\sigma(Y_i^\sigma(F)) \subset X(F).$$

This is a thin set by Corollary 5.2.11. Then we define

$$Z' = \bigcup_i (Z_i \cup V_i(F)).$$

By construction, this is a thin set of X.

Now let $f\colon Y \to X$ be an adjoint rigid a-cover from a smooth projective variety such that $(a(X, L), b(F, X, L)) \leq (a(Y^\sigma, (f^\sigma)^* L), b(F, Y^\sigma, (f^\sigma)^* L))$ and f is face-contracting. Then $f\colon Y \to X$ is birational to a twist $f_i^\sigma : Y_i^\sigma \to X$ for some i. Hence we conclude by Lemma 5.2.12 that

$$f(Y(F)) \subset Z_i \cup V_i(F).$$

Therefore the claim holds. □

5.3 Generically Finite Maps in Families

Finally, let us explain the ideas for the proof of Theorem 4.3.5. First, we will prove that Z_3 constructed in Section 4.3.2 is contained in a proper closed subset:

Lemma 5.3.1 ([LST22, Lemma 4.23]) *Let F be a field of characteristic 0. Let X be a geometrically uniruled smooth projective variety defined over F, and L be a big and nef $\mathbb{Q}$-divisor on X. Then there exists a proper closed subset $V \subset X$ such that if $f\colon Y \to X$ is a thin map satisfying $a(Y, f^* L) \geq a(X, L)$ and $d(Y, f^* L) > d(X, L)$, then $f(Y) \subset V$.*

Proof Without loss of generality we may assume that our ground field F is algebraically closed. Theorem 5.1.4 implies that subvarieties S such that $S \not\subset \mathbf{B}_+(L)$, $a(S, L|_S) \geq a(X, L)$, and $(S, L|_S)$ is adjoint rigid are parametrized by finitely many families. Among them let us denote non-dominant families by $\pi_i\colon \mathcal{U}_i \to W_i$ with the evaluation map $s_i\colon \mathcal{U}_i \to X$. We define a proper closed subset V by

$$V := \mathbf{B}_+(L) \cup \bigcup_i \overline{s_i(\mathcal{U}_i)}.$$

Let $f\colon Y \to X$ be a thin map such that

$$a(Y, f^* L) \geq a(X, L), \text{ and } d(Y, f^* L) > d(X, L).$$

We claim that $f(Y) \subset V$. Assume for a contradiction that $f(Y) \not\subset V$.

Let Γ be a general fiber of the canonical map for (Y, f^*L) and denote its image on X by S. It follows from Lemma 5.2.2 that

$$a(S, L|_S) \geq a(\Gamma, f^*L|_\Gamma) = a(Y, f^*L) \geq a(X, L).$$

If $a(S, L|_S) > a(X, L)$, then Lemma 5.1.2 implies $S \subset V$, therefore $f(Y) \subset V$, a contradiction. Thus we conclude that $a(S, L|_S) = a(X, L)$ and, by Lemma 5.2.2, $(S, L|_S)$ is adjoint rigid. Since $S \not\subset V$, S must be a member of a dominant family. Hence it suffices to show that there is no dominant family of adjoint rigid subvarieties S such that $\dim S > d(X, L)$.

After applying a resolution, we may assume that the canonical map for (X, L) is a morphism $\pi\colon X \to W$. Using Theorem 3.4.5, one can conclude that there exists an ample $\mathbb{Q}$-Cartier divisor H on W and an effective $\mathbb{Q}$-divisor E on X such that $a(X, L)L + K_X$ is $\mathbb{Q}$-linearly equivalent to $\pi^*H + E$. Now suppose that we have a family $q\colon \mathcal{S} \to T$ with the dominant evaluation map $g\colon \mathcal{S} \to X$ such that q generically parametrizes adjoint rigid subvarieties S with $a(S, g^*L|_S) = a(X, L)$ and $\dim S > d(X, L)$, and $\mathcal{S}$ is smooth and projective. It follows from Proposition 5.1.8 that g is generically finite. Then by the ramification formula, we have an effective $\mathbb{Q}$-divisor R such that

$$a(X, L)g^*L + K_{\mathcal{S}} \sim_{\mathbb{Q}} g^*(a(X, L) + K_X) + R \sim_{\mathbb{Q}} g^*\pi^*H + g^*E + R.$$

However, the restriction of $g^*\pi^*H + g^*E + R$ to S has Iitaka dimension at least 1 so this contradicts the fact that $(S, g^*L|_S)$ is adjoint rigid. Hence the claim. □

Thus to prove Theorem 4.3.5, we only need to show that Z_1 and Z_2 constructed in Section 4.3.2 are thin sets. To this end, we need to state a very technical lemma (Lemma 5.3.5); however its proof is beyond the scope of this book. To state this lemma, we first need to introduce some terminology:

Definition 5.3.2 ([LST22, Definition 7.1]) We assume that F is algebraically closed. A *good family of adjoint rigid varieties* is a morphism $p\colon \mathcal{U} \to W$ between smooth quasi-projective varieties with a relatively p-big and p-nef $\mathbb{Q}$-divisor L on $\mathcal{U}$ satisfying the following conditions:

(i) The map p is projective, surjective, and smooth with irreducible fibers.
(ii) For the fibers $\mathcal{U}_w$ over closed points $w \in W$, the a-invariant $a(\mathcal{U}_w, L|_{\mathcal{U}_w})$ is constant and positive, and $(\mathcal{U}_w, L|_{\mathcal{U}_w})$ is adjoint rigid.
(iii) The b-value $b(F, \mathcal{U}_w, L|_{\mathcal{U}_w})$ is also constant.
(iv) Let Q denote the union of all prime divisors D in fibers $\mathcal{U}_w$ such that $a(D, L|_D) > a(\mathcal{U}_w, L|_{\mathcal{U}_w})$. Then Q is closed and flat over W. Moreover, let $\mathcal{V} = \mathcal{U} \setminus Q$, then there is a projective birational morphism $\phi\colon \mathcal{U}' \to \mathcal{U}$

which is an isomorphism over $\mathcal{V}$ such that $\mathcal{U}'$ is smooth over W and $\mathcal{U}'\setminus\mathcal{V}$ is a strict normal crossings divisor relative to W.

A base change of a good family is the good family induced via a base change by a map $g: T \to W$.

A *good morphism of good families* is a diagram

$$\begin{array}{ccc} \mathcal{Y} & \xrightarrow{f} & \mathcal{U} \\ {\scriptstyle q}\downarrow & & \downarrow{\scriptstyle p} \\ T & \xrightarrow[g]{} & W \end{array}$$

such that f and g are generically finite, p and q are good families of adjoint rigid families with respect to L and f^*L respectively, the dimensions of fibers of p and q are the same, and $a(\mathcal{Y}_t, f^*L|_{\mathcal{Y}_t}) = a(\mathcal{U}_{g(t)}, L|_{\mathcal{U}_{g(t)}})$ for all closed points $t \in T$.

Definition 5.3.3 ([LST22, Definition 8.2]) Let F be a field of characteristic 0. A good family of adjoint rigid varieties over F is a morphism $p: \mathcal{U} \to W$ between smooth quasi-projective varieties over F and p-big and p-nef divisor L on $\mathcal{U}$ such that its base change to the algebraic closure $\overline{p}: \overline{\mathcal{U}} \to \overline{W}$ is a good family of adjoint rigid varieties over $\overline{F}$ with respect to L. We similarly define a good morphism of good families defined over F.

If one has a family of adjoint rigid varieties, then one can obtain a good family by shrinking the base. Since we do not give full proofs for some statements in this section, we omit a proof of the following lemma:

Lemma 5.3.4 ([LST22, Lemma 7.3]) *Let F be a field of characteristic* 0. *Let $p: X \to W$ be a surjective projective morphism of varieties defined over F such that X is smooth and p has geometrically connected fibers. Assume that L is a relatively p-big $\mathbb{Q}$-divisor and it is the restriction of a nef $\mathbb{Q}$-Cartier $\mathbb{Q}$-divisor on some projective compactification of X. Suppose that a general fiber X_w is adjoint rigid with respect to L. Then there exists a non-empty open subset $W^\circ \subset W$ such that $p: X^\circ = p^{-1}(W^\circ) \to W^\circ$ is a good family of adjoint rigid varieties.*

The following lemma is one of the crucial ingredients for the proof of Theorem 4.3.5, and we call the families constructed in this lemma *universal families up to twisting*. Its proof is beyond the scope of this book, but it is based on intricate arguments involving arithmetic fundamental groups and homotopy lifting properties of certain covering spaces with rational points.

Readers interested in its proof should consult the original paper [LST22, Lemma 8.3]:

Lemma 5.3.5 ([LST22, Lemma 8.3]) *Let F be a field of characteristic 0. Suppose that X is a geometrically uniruled geometrically integral smooth projective variety defined over F, and L is a big and nef $\mathbb{Q}$-divisor on X. Let $p\colon \mathcal{U} \to W$ be a surjective morphism between projective varieties with a morphism $s\colon \mathcal{U} \to X$ which is birational. We assume that there exists a non-empty Zariski open subset $W^\circ \subset W$ such that $p\colon \mathcal{U}^\circ = p^{-1}(W^\circ) \to W^\circ$ is a good family of adjoint rigid subvarieties over F and any fiber over W° has the same a-invariant with respect to s^*L as X does with respect to L.*

Then there exists a proper closed subset $C \subset X$ and a finite set of dominant generically finite morphisms $\{f_j\colon \mathcal{Y}_j \to \mathcal{U}\}$ defined over F to fit into a commutative diagram

$$\begin{array}{ccc} \mathcal{Y}_j & \xrightarrow{f_j} & \mathcal{U} \\ {\scriptstyle q_j}\downarrow & & \downarrow{\scriptstyle p} \\ T_j & \xrightarrow[g_j]{} & W \end{array}$$

for which the following properties are satisfied:

(i) *Both $\mathcal{Y}_j$ and T_j are projective and geometrically integral, $\mathcal{Y}_j$ is smooth, T_j is normal, and $q_j\colon \mathcal{Y}_j \to T_j$ is a generically good family of adjoint rigid subvarieties.*

(ii) *The canonical map for $a(X, L)f_j^*L + K_{\mathcal{Y}_j}$ is a morphism and coincides with q_j over some non-empty open subset of T_j.*

(iii) *The morphism $g_j\colon T_j \to W$ is dominant, finite, and Galois.*

(iv) *We have* $\mathrm{Bir}(\overline{\mathcal{Y}_j}/\overline{X}) = \mathrm{Aut}(\overline{\mathcal{Y}_j}/\overline{X})$.

(v) *There is a non-empty Zariski open subset $W' \subset W^\circ$ such that for any j, any twist $\mathcal{Y}_j^\sigma$ over X, and any closed point t on the preimage $(T_j')^\sigma$ of W' we have an isomorphism $(s_j^\sigma)_*\colon \mathcal{F}(\mathcal{Y}_{j,t}^\sigma, s_j^*L|_{\mathcal{Y}_{j,t}^\sigma}) \to \mathcal{F}(\mathcal{Y}_j, s_j^*L)$. Here $s_j\colon \mathcal{Y}_j \to X$ is the composition of s and f_j, and $\mathcal{Y}_{j,t}^\sigma$ denotes a fiber of $q_j^\sigma\colon \mathcal{Y}_j^\sigma \to T_j^\sigma$ at t.*

(vi) *Assume that $q\colon \mathcal{Y} \to T$ is a projective surjective morphism of varieties over F such that $\mathcal{Y}$ is smooth and geometrically integral and we have a* *diagram*

$$\begin{array}{ccc} \mathcal{Y} & \xrightarrow{f} & \mathcal{U} \\ {\scriptstyle q}\downarrow & & \downarrow{\scriptstyle p} \\ T & \xrightarrow[g]{} & W \end{array}$$

satisfying the following conditions:

(a) *there exists a non-empty Zariski open subset $T' \subset T$ such that $\mathcal{Y}$ is a good family of adjoint rigid varieties over T' and the map $f : q^{-1}(T') \to \mathcal{U}$ has image in $\mathcal{U}^\circ$ and is a good morphism of good families; and*

(b) *there is a rational point $y \in \mathcal{Y}(F)$ such that $s \circ f(y) \notin C$.*

Then there exists j and a twist $f_j^\sigma : \mathcal{Y}_j^\sigma \to \mathcal{U}$ such that $f(y) \in f_j^\sigma(\mathcal{Y}_j^\sigma(F))$. Moreover, there exists a dominant generically finite morphism $\widetilde{T} \to T$ such that the main component $\widetilde{q}\colon \widetilde{\mathcal{Y}} \to \widetilde{T}$ of the base change of q by $\widetilde{T} \to T$ satisfies the properties that the induced map $\widetilde{f}\colon \widetilde{\mathcal{Y}} \to \mathcal{U}$ factors rationally through f_j^σ and a general geometric fiber of $\widetilde{q}\colon \widetilde{\mathcal{Y}} \to \widetilde{T}$ maps birationally to a geometric fiber of $q_j^\sigma : \mathcal{Y}_j^\sigma \to T_j^\sigma$.

Using Lemma 5.3.5, we will prove Theorem 4.3.5.

Proof of Theorem 4.3.5 We only need to show that Z_1 and Z_2 are thin sets. Let V be a proper closed subset of X as defined in Theorem 5.1.6. We consider subvarieties $\overline{Y}$ of $\overline{X}$ defined over $\overline{F}$ and not contained in V such that

$$a(Y, L|_Y) = a(X, L)$$

and $(Y, L|_Y)$ is adjoint rigid. By Theorem 5.1.4, such subvarieties are parametrized by a bounded set $\overline{W} \subset \mathrm{Hilb}(\overline{X})$. Replacing $\overline{W}$ by its closure we may assume that $\overline{W}$ is projective. Next, replacing $\overline{W}$ by the union of its Galois orbit, we may assume that $\overline{W}$ descends to W defined over F. Let the W_is be the irreducible components of W, and $p_i : \mathcal{U}_i \to W_i$ be the pullback of the universal family with the evaluation map $s_i : \mathcal{U}_i \to X$. Here we endow the reduced structures with those schemes. If s_i is not dominant, then we add $s_i(\mathcal{U}_i)$ to V and ignore this family. Therefore, we may assume that, for any i, the evaluation map $s_i : \mathcal{U}_i \to X$ is dominant. Then it follows from Proposition 5.1.8 that $s_i : \mathcal{U}_i \to X$ is generically finite. If $\mathcal{U}_i$ is not geometrically integral, then $\overline{s_i(\mathcal{U}_i(F))}$ is a proper closed subset; we enlarge V by adding this proper closed subset.

Suppose that $\mathcal{U}_i$ is geometrically integral. After applying a resolution, we may assume that $\mathcal{U}_i$ is smooth. By Lemma 5.3.4, there exists a non-empty Zariski open subset W_i° such that $p_i : p_i^{-1}(W_i^\circ) \to W_i^\circ$ is a good family of adjoint rigid subvarieties with a flat and proper divisor $Q_i \subset p_i^{-1}(W_i^\circ)$. We set $\mathcal{V}_i = p_i^{-1}(W_i^\circ) \setminus Q_i$. We enlarge V by adding $s_i(\mathcal{U}_i \setminus \mathcal{V}_i) \cup S_i(E_i)$ where E_i is the ramification divisor of $s_i \colon \mathcal{U}_i \to X$.

Next, we may further assume that $s_i \colon \mathcal{U}_i \to X$ is birational. We apply Lemma 5.3.5 and obtain the universal families up to twisting $q_{i,j} \colon \mathcal{Y}_{i,j} \to T_{i,j}$ with the evaluation map $s_{i,j} \colon \mathcal{Y}_{i,j} \to X$. We shrink W_i° if necessary so that Lemma 5.3.5 (v) is valid for all closed points on $T_{i,j}^{\circ\sigma}$ where $T_{i,j}^\circ$ is the preimage of $W_{i,j}^\circ$. We add $s_i(p_i^{-1}(W_i \setminus W_i^\circ))$ and C_i to V where C_i is a proper closed subset from Lemma 5.3.5.

We now construct a thin set Z' in several steps. First, we set $Z' = V(F)$. This is a type I thin set. Next, for each i, if $\mathcal{U}_i$ is geometrically integral and $s_i \colon \mathcal{U}_i \to X$ has degree > 1, then we add $s_i(\mathcal{U}_i(F))$ to Z'.

If $\mathcal{U}_i$ is geometrically integral and $s_i \colon \mathcal{U}_i \to X$ is birational, then for each j we let σ run over all $\sigma \in H^1(F, \mathrm{Aut}(\overline{\mathcal{Y}}_{i,j}/\overline{X}))$ such that

$$(a(X, L), b(F, X, L)) \leq (a(\mathcal{Y}_{i,j}^\sigma, (s_{i,j}^\sigma)^* L), b(F, \mathcal{Y}_{i,j}^\sigma, (s_{i,j}^\sigma)^* L))$$

in the lexicographic order and $s_{i,j}^\sigma \colon \mathcal{Y}_{i,j}^\sigma \to X$ is face-contracting. We add the set

$$\bigcup_\sigma s_{i,j}^\sigma(\mathcal{Y}_{i,j}^\sigma(F)) \subset X(F)$$

to Z'. It follows from Corollary 5.2.11 that this is a thin set.

We claim that $\mathsf{Z}_1 \cup \mathsf{Z}_2 \subset Z'$. Let us prove $\mathsf{Z}_1 \subset Z'$. Let $f \colon Y \to X$ be a thin map such that Y is projective and smooth, $d(Y, f^*L) < d(X, L)$, and

$$(a(X, L), b(F, X, L)) \leq (a(Y, f^*L), b(F, Y, f^*L))$$

in the lexicographic order. Our goal is to show that $f(Y(F)) \subset Z'$. Pick a rational point $y \in Y(F)$; we will prove that $f(y) \in Z'$. We may assume that $f(y) \notin V$; otherwise our claim is immediate. This assumption forces $a(Y, f^*L)$ to be equal to $a(X, L)$. Moreover, by Lemma 5.2.12, we are free to replace Y by its resolution and y by its preimage. Let $\phi \colon Y \dashrightarrow B$ be the canonical map for $a(Y, f^*L)f^*L + K_Y$. After applying a resolution, we may assume that this is a morphism. By Lemma 5.2.2, the image of a general fiber of ϕ has the same a-invariant as X does. This implies that for some i there is a rational map $g : B \dashrightarrow W_i$ and after some birational modifications of Y and B, we may assume that g is a morphism. Then $f \colon Y \to X$ rationally factors through $f' \colon Y \dashrightarrow B \times_{W_i} \mathcal{U}_i \to \mathcal{U}_i$. After applying a birational modification again, we may assume that f' is a morphism and is generically a good morphism of good families. If $\mathcal{U}_i$ is not geometrically integral or $s_i \colon \mathcal{U}_i \to X$ is not birational, then it follows from the construction that $f(Y(F)) \subset Z'$. Thus we may assume that $\mathcal{U}_i$ is geometrically integral and $s_i : \mathcal{U}_i \to X$ is birational.

Now we apply Lemma 5.3.5 (vi) and conclude that there exists j and a twist σ such that $f(y) \in s^\sigma_{i,j}(\mathcal{Y}_{i,j}(F))$. Moreover, there is a generically finite dominant morphism $\psi : \widetilde{B} \to B$ from a normal projective variety such that for the main component $\widetilde{\phi} : \widetilde{Y} \to \widetilde{B}$ of the base change of ϕ by ψ the induced map $\widetilde{f} : \widetilde{Y} \to X$ rationally factors through $s^\sigma_{i,j} : \mathcal{Y}^\sigma_{i,j} \to X$.

It only remains to prove that

$$(a(X, L), b(F, X, L)) \leq (a(\mathcal{Y}^\sigma_{i,j}, (s^\sigma_{i,j})^* L), b(F, \mathcal{Y}^\sigma_{i,j}, (s^\sigma_{i,j})^* L))$$

and if the equality is achieved, then $s^\sigma_{i,j}$ is face-contracting. We know that the a-invariants are the same; let us show that $b(F, X, L) \leq b(F, \mathcal{Y}^\sigma_{i,j}, (s^\sigma_{i,j})^* L)$. Let $c \in B$ be a general closed point. (This point may not be a rational point.) Lemma 5.1.3 shows that

$$b(F, Y, f^* L) \leq b(F, Y_c, f^* L).$$

Since $f(y) \notin V$, Y_c maps to a fiber $\mathcal{Y}^\sigma_{i,j,t}$ where $t \in (T^\circ_{i,j})^\sigma$. Then, since every geometric component of Y_c maps birationally to a geometric component of $\mathcal{Y}^\sigma_{i,j,t}$, we may conclude that

$$b(F, Y_c, f^* L) = b(F, \mathcal{Y}^\sigma_{i,j,t}, (s^\sigma_{i,j})^* L|_{\mathcal{Y}^\sigma_{i,j,t}}).$$

See [LST22, Lemma 4.12] for more details. On the other hand, by Lemma 5.3.5 (v), we have

$$b(F, \mathcal{Y}^\sigma_{i,j,t}, (s^\sigma_{i,j})^* L|_{\mathcal{Y}^\sigma_{i,j,t}}) = b(F, \mathcal{Y}^\sigma_{i,j}, (s^\sigma_{i,j})^* L).$$

Thus our claim follows. Finally, when the equality

$$(a(X, L), b(F, X, L)) = (a(\mathcal{Y}^\sigma_{i,j}, (s^\sigma_{i,j})^* L), b(F, \mathcal{Y}^\sigma_{i,j}, (s^\sigma_{i,j})^* L))$$

is achieved, using the fact that $d(\mathcal{Y}^\sigma_{i,j}, (s^\sigma_{i,j})^* L) = d(Y, f^* L) < d(X, L)$, one can show that $s^\sigma_{i,j} : \mathcal{Y}^\sigma_{i,j} \to X$ is face-contracting. See [LST22, Lemma 4.27] for more details. □

6
Further Examples

In this chapter we compute our conjectural exceptional sets constructed in Section 4.3.2 for several examples. Since the uniform bounds in Corollary 3.5.6 are not effective in general, our general proofs of Theorem 4.3.5 do not provide effective algorithms to compute these exceptional sets. Instead, we appeal to adjunction theory, which provides us with powerful and practical tools to compute them. The examples that are discussed in this chapter are (1) projective spaces and smooth quadric hypersurfaces, (2) Le Rudulier's example, (3) certain quadric bundles, (4) surfaces, (5) smooth prime Fano threefolds, and (6) Fano hypersurfaces.

6.1 Projective Spaces and Smooth Quadric Hypersurfaces

Although our exceptional sets are computed for generalized flag varieties in Example 4.3.9, here we discuss a more geometric approach to this computation for projective spaces and smooth quadric hypersurfaces.

First, we discuss a result from adjunction theory which gives an upper bound for the a-invariant for big and nef Cartier divisors:

Proposition 6.1.1 (Siu) *Let F be a field of characteristic 0. Let X be a smooth projective variety of dimension n defined over F, and L be a big and nef divisor on X. Then $K_X + (n+1)L$ is $\mathbb{Q}$-linearly equivalent to an effective $\mathbb{Q}$-divisor.*

Fujita's conjecture claims that the adjoint divisor in the statement is base-point free when L is ample. The proof of the above proposition is straightforward; we include it here for completeness:

Proof The following argument is due to Siu. Let

$$P(m) = \chi(X, \mathcal{O}_X(K_X + mL)).$$

This is a polynomial of degree at most n by the Hirzebruch–Riemann–Roch theorem. It follows from the Kawamata–Viehweg vanishing theorem ([KM98, Theorem 2.64]) that for any positive integer $m > 0$ we have

$$P(m) = h^0(X, \mathcal{O}_X(K_X + mL)).$$

On the other hand, since $L^n > 0$, the asymptotic Riemann–Roch (see, e.g., [KM98, Theorem 1.36]) implies that, for $m \gg 0$, $P(m) > 0$. Hence $P(m)$ is not identically zero. Since $P(m)$ has degree at most n, it cannot have $n + 1$ distinct roots. Thus there exists $1 \leq m \leq n + 1$ such that

$$P(m) = h^0(X, \mathcal{O}_X(K_X + mL)) > 0.$$

Since L is $\mathbb{Q}$-linearly equivalent to an effective $\mathbb{Q}$-divisor, our assertion follows. □

Let us state a result for projective spaces:

Proposition 6.1.2 ([LT17, Corollary 9.3]) *Let F be a field of characteristic* 0. *Let $X = \mathbb{P}^n$ be the n-dimensional projective space defined over F, and H be the hyperplane class on X. Let $f : Y \to X$ be a thin map from a smooth projective variety Y defined over F. Then we have $a(Y, f^*H) < a(X, H)$.*

Proof We may assume that $\overline{F} = F$. First, note that $a(X, H) = n + 1$. Thus, when $\dim Y < \dim X = n$, it follows from Proposition 6.1.1 that

$$a(Y, f^*H) \leq \dim Y + 1 < n + 1.$$

Hence we may assume that $\dim Y = n$. If $a(Y, f^*H)f^*H + K_Y$ has positive Iitaka dimension, then it follows from Lemma 5.1.3 that Y is covered by lower dimensional subvarieties with the same a-invariant as Y. Thus

$$a(Y, f^*H) < n + 1$$

follows. Finally, assume that (Y, f^*H) is adjoint rigid. Since there is no subvariety S of X such that $a(S, H|_S) \geq a(X, H)$, Corollary 5.2.5 implies that $f : Y \to X$ is étale in codimension 1. Since $\pi_1^{\text{ét}}(X)$ is trivial, f must be birational. But this contradicts the fact that f is a thin map. Thus our assertion follows. □

Proposition 6.1.2 shows that the exceptional set Z constructed in Section 4.3.2 is the empty set for projective space.

Next, we discuss the exceptional set for smooth quadric hypersurfaces. We now state the following lemma:

Lemma 6.1.3 *Let F be a field of characteristic* 0. *Let X be a smooth projective variety of dimension n, and H be a big and nef divisor on X. Assume that $a(X,H) > n$. Then $H^n = 1$.*

Proof The following proof is taken from [LT17, Proposition 9.1]. Under our assumption, we must have $\kappa(K_X + a(X,H)H) = 0$. Indeed, if the Iitaka dimension of this adjoint divisor is positive, then X is covered by subvarieties with the same a-invariant as X by Lemma 5.1.3. However, by Proposition 6.1.1 these subvarieties must have a-invariants less than or equal to n, and this is a contradiction.

Let $P(m) = \chi(X, \mathcal{O}_X(K_X + mH))$, which is a polynomial of degree n by the Hirzebruch–Riemann–Roch theorem. Again by Kawamata–Viehweg vanishing theorem ([KM98, Theorem 2.64]), for any positive integer m we have

$$P(m) = h^0(X, \mathcal{O}_X(K_X + mH)).$$

Since $a(X,H) > n$, $P(m) = 0$ holds for $m = 1, \ldots, n$. Thus we must have $P(n+1) > 0$. Since $a(X,H)H + K_X$ is rigid, we conclude that $P(n+1) = 1$. This determines the polynomial $P(m)$ as

$$P(m) = \binom{m-1}{n}.$$

It follows from [KM98, Theorem 1.36] that $H^n = 1$. □

Here is a geometric approach to smooth quadric hypersurfaces:

Proposition 6.1.4 ([LT17, Proposition 9.4]) *Let F be a field of characteristic* 0. *Let $Q \subset \mathbb{P}^{n+1}$ be a smooth quadric hypersurface of dimension $n \geq 3$ defined over F, and H be the hyperplane class on Q. Then for any thin map $f : Y \to Q$ from a smooth projective variety, we have $a(Y, f^*H) < a(Q,H)$.*

Proof We may assume that $\overline{F} = F$. First note that $a(Q,H) = n$. Thus our assertion follows from Proposition 6.1.1 when $\dim Y < n-1$.

Next, we consider the case when $\dim Y = n-1$. Suppose that $a(Y, f^*H) = n$. Then by Lemma 6.1.3, we have $(f^*H)^{n-1}.Y = 1$. This implies that f is birational to the image Y', and the image Y' satisfies $H^{n-1}.Y' = 1$. This means that Y' is a linear subspace of codimension 1, which yields a contradiction in our dimension range $n \geq 3$.

Finally, when $\dim Y = n$, one can argue as in Proposition 6.1.2 to conclude our assertion. □

Again it follows from Proposition 6.1.4 that the exceptional set Z constructed in Section 4.3.2 is the empty set for any smooth quadric hypersurface of dimension $n \geq 3$.

6.2 Le Rudulier's Example

In this section, we study Example 4.2.18 in greater detail. Let F be a field of characteristic 0 and S be $\mathbb{P}^1 \times \mathbb{P}^1$ defined over F. We consider $X = \mathrm{Hilb}^{[2]}(S)$ with the polarization $L = -K_X$. In [LR19], Le Rudulier proved Manin's conjecture (Conjecture 4.2.9) for this example when $F = \mathbb{Q}$ after removing the contribution of a Zariski dense thin set. In this section, we compute our conjectural exceptional set Z and confirm that it coincides with the exceptional set used by Le Rudulier.

6.2.1 Fujita-type Statements for Low Dimensional Varieties

Here, we collect several results from adjunction theory that resemble Reider's theorem for smooth projective surfaces in [Rei88]. We assume that our ground field is an algebraically closed field of characteristic 0.

Proposition 6.2.1 ([LT17, Proposition 3.6]) *Let X be a smooth projective variety and L be a big and nef $\mathbb{Q}$-divisor on X.*

(i) *Suppose that* $\dim X = 1$. *If* $\mathrm{Vol}(L) > 2$, *then $K_X + L$ is big.*
(ii) *Suppose that* $\dim X = 2$. *If* $\mathrm{Vol}(L) > 9$ *and $L.C > 2$ for every curve $C \subset X$ through a general point of X, then $K_X + L$ is big. Moreover, if X is uniruled and $a(X, L)L + K_X$ is rigid, then* $\mathrm{Vol}(L) \leq 9/a(X, L)^2$.
(iii) *Suppose that* $\dim X = 3$. *If* $\mathrm{Vol}(L) > 64$, *$L^2.S > 9$ for every surface $S \subset X$ through a general point of X, and $L.C > 2$ for every curve $C \subset X$ through a general point of X, then $K_X + L$ is big.*

Proof When X is not uniruled, our assertion is obvious because K_X is pseudo-effective. Thus we assume that X is uniruled. We prove that under the assumptions of the statement, we have $a(X, L) < 1$ by induction on the dimension. We discuss only the cases of dimension 2 and 3 as the case of curves is straightforward.

Suppose that X is a surface. When $a(X, L)L + K_X$ is not adjoint rigid, we consider the canonical map $\pi\colon X \dashrightarrow W$ and we denote its general fiber by Y. Note that this is a morphism because X is a surface and any outcome of the MMP is a birational morphism. Then it follows from Lemma 5.1.3 that

$$a(X, L) = a(Y, L|_Y) = \frac{2}{L.Y} < 1.$$

Note that Y is rational by Proposition 4.1.9. Thus our assertion follows.

Assume that $a(X, L)L + K_X$ is adjoint rigid. We apply a $(K_X + a(X, L)L)$-MMP and obtain a minimal model $\phi\colon X \dashrightarrow X'$. Again this is a morphism because X is a surface. Then X' is a weak del Pezzo surface, so that we have

$$\mathrm{Vol}(L) \leq \mathrm{Vol}(\phi^*\phi_*L) = \mathrm{Vol}(\phi_*L) = \frac{\mathrm{Vol}(-K_{X'})}{a(X,L)^2}.$$

Since we have $\mathrm{Vol}(-K_{X'}) \leq 9$, our assertion follows from the assumption $\mathrm{Vol}(L) > 9$.

Next, we assume that X is a threefold. Suppose that $a(X,L)L + K_X$ is not rigid. We consider the canonical map $\pi\colon X \dashrightarrow W$ for $(X, a(X,L)L)$ and denote its general fiber by Y. Then it follows from Lemma 5.1.3 that

$$a(Y, L|_Y) = a(X, L).$$

Thus our assertion follows from the statement for lower dimensional subvarieties.

Next, we may assume that $a(X,L)L+K_X$ is rigid. We apply a $(K_X+a(X,L)L)$-MMP and obtain a minimal model $\phi\colon X \dashrightarrow X'$. Then we have

$$a(X,L)\phi_*L + K_{X'} \sim_{\mathbb{Q}} 0.$$

For a sufficiently small $\epsilon > 0$, $\phi\colon X \dashrightarrow X'$ is a $(K_X + (a(X,L) - \epsilon)L$-MMP as well. We continue a $(K_X + (a(X,L) - \epsilon)L)$-MMP further, and the outcome is a Mori fiber space $\psi\colon X \dashrightarrow X''$ with $\rho\colon X'' \to B$. After applying a resolution, we may assume that $\psi\colon X \to X''$ is a morphism. Note that we have $a(X,L)\psi_*L + K_{X''} \sim_{\mathbb{Q}} 0$. We consider the composition $\pi = \rho \circ \psi\colon X \to B$. Then for a general fiber Y of π, we have $a(Y, L|_Y) = a(X,L)$.

If the dimension of Y is 1 or 2, then our assertion follows from the statements for lower dimensional varieties. Thus we may assume that B is a point, that is, X'' is a $\mathbb{Q}$-factorial terminal Fano threefold of Picard rank 1. When X'' is Gorenstein, [Nam97] shows that X'' is smoothable, that is, there is a flat deformation of X'' whose general members are smooth Fano threefolds. In particular, it follows from invariance of log plurigenera [HMX13, Theorem 1.8] (see also Theorem 3.4.7) and the classification of Fano threefolds [IP99] that we have $\mathrm{Vol}(-K_{X''}) \leq 64$. Even when X'' is not Gorenstein, it follows from [Pro07] that $\mathrm{Vol}(-K_{X''}) \leq 125/2 \leq 64$. Now the negativity lemma (Lemma 3.2.16) implies that

$$\mathrm{Vol}(L) \leq \mathrm{Vol}(\psi^*\psi_*L) = \mathrm{Vol}(\psi_*L) = \mathrm{Vol}(-K_{X''})/a(X,L)^3.$$

Thus the claim follows from our assumption. □

By slightly modifying the arguments of the above proposition, one can establish a result which looks more like Reider's theorem:

Proposition 6.2.2 ([LT17, Theorem 3.7]) *Let X be a smooth projective surface* *and L be a big and nef $\mathbb{Q}$-divisor on X. Suppose that $L.C > 3$ for every rational curve $C \subset X$ passing through a general point. Then $K_X + L$ is big.*

Proof Again, let us assume that X is uniruled. We first assume that (X, L) is not adjoint rigid. Let $\pi\colon X \to W$ be the canonical map for $a(X, L)L + K_X$. Let C be a general fiber. This is a smooth rational curve by Proposition 4.1.9. Since we have $(a(X, L)L + K_X).C = 0$, we conclude that

$$a(X, L) = \frac{2}{L.C}.$$

However, since we have $L.C > 3$ we conclude that $a(X, L) < \frac{2}{3}$.

Next, assume that (X, L) is adjoint rigid. Let $\phi\colon X \dashrightarrow X'$ be a minimal model for $K_X + a(X, L)L$. For a sufficiently small $\epsilon > 0$, we continue a $(K_X + (a(X, L) - \epsilon)L)$-MMP and obtain a Mori fiber space $\psi\colon X \to X''$ with $\rho\colon X'' \to B$. If $\dim B = 1$, we may conclude as in the case of non-adjoint rigid. Thus we may assume that B is a point and this means that X'' is a smooth del Pezzo surface of Picard rank 1, that is, $\mathbb{P}^2$. Let C be the strict transform of a general line on $\mathbb{P}^2$. Then we have $(a(X, L)L + K_X).C = a(X, L)L.C - 3 = 0$, that is,

$$a(X, L) = \frac{3}{L.C}.$$

Thus it follows from our assumption that $a(X, L) < 1$. □

Combining the arguments of Propositions 6.2.1 and 6.2.2, we have the following corollary:

Corollary 6.2.3 ([LT17, Corollary 3.8]) *Let X be a smooth projective threefold and L be a big and nef $\mathbb{Q}$-divisor on X. If $\mathrm{Vol}(L) > 64$ and $L.C > 3$ for every rational curve through a general point, then $K_X + L$ is big.*

6.2.2 Subvarieties of $\mathrm{Hilb}^{[2]}(S)$

Let S be $\mathbb{P}^1 \times \mathbb{P}^1$ and X be the Hilbert scheme $\mathrm{Hilb}^{[2]}(S)$. We denote the pullback of the ample generator via the first projection $S \to \mathbb{P}^1$ by H_1 and the second projection $S \to \mathbb{P}^1$ by H_2.

In this section, we review a numerical description of effective cycles on X developed in [BC13, Example 18] and [LT17, Section 9.3]. We assume that our ground field is an algebraically closed field of characteristic 0. We freely use the notations set up in Example 4.2.18.

First, we describe the theory of divisors and curves on X following [BC13, Example 18]. See also [LT17, Section 9.3].

The Néron–Severi space $N^1(X)$ is 3-dimensional. A basis consists of the following divisors:

- $H_1[2]$, the divisor parametrizing all subschemes where at least one point is contained in a fixed fiber of the first projection;
- $H_2[2]$, the divisor parametrizing all subschemes where at least one point is contained in a fixed fiber of the second projection; and
- G, the divisor that is a half of the divisor E parametrizing non-reduced schemes.

The pseudo-effective cone $\overline{\mathrm{Eff}}^1(X)$ is a simplicial cone generated by G and

- $D_1 = H_1[2] - G$, the divisor parametrizing all subschemes supported on some (non-fixed) fiber of the first projection; and
- $D_2 = H_2[2] - G$, the divisor parametrizing all subschemes supported on some (non-fixed) fiber of the second projection.

The nef cone $\mathrm{Nef}^1(X)$ coincides with the movable cone $\overline{\mathrm{Mov}}^1(X)$; it is a simplicial cone generated by $H_1[2], H_2[2]$, and

- $X_{1,1} = H_1[2] + H_2[2] - G$.

Here is the description of $X_{1,1}$: any length-2 subscheme ℓ on S determines a 2-dimensional subspace of $H^0(S, \mathcal{O}_S(1,1))$, which is the space of sections vanishing along ℓ, and this correspondence defines a birational morphism

$$\Phi\colon X \to \mathrm{Gr}(2,4).$$

Then $X_{1,1}$ is the pullback of the ample generator on $\mathrm{Gr}(2,4)$.

Next, we consider curves on X. The space $N_1(X)$ is 3-dimensional and the pseudo-effective cone and the nef cone are simplicial. The pseudo-effective cone $\overline{\mathrm{Eff}}_1(X)$ is generated by

- J_1, the curve parametrizing non-reduced schemes that are scheme theoretically contained in a fixed fiber of the first projection;
- J_2, the curve parametrizing non-reduced schemes that are scheme theoretically contained in a fixed fiber of the second projection; and
- C, the curve parametrizing non-reduced structures on a fixed point.

The nef cone $\mathrm{Nef}_1(X)$ is generated by the following curves:

- $F_1(1,2)$, the curve parametrizing subschemes containing a fixed point and letting the other point vary on a fixed fiber of the first projection;
- $F_2(1,2)$, the curve parametrizing subschemes containing a fixed point and letting the other point vary on a fixed fiber of the second projection; and

- $R(1,1)$, the curve parametrizing subschemes found by fixing a general element of $\mathcal{O}(1,1)$ which is a smooth conic, fixing a $2:1$ map from a conic to $\mathbb{P}^1$, and taking fibers.

For the record, we include the intersection matrix of divisors and curves. Table 6.1 can be found in [LT17, Section 9.3]:

Table 6.1 *The intersection matrix of divisors and curves*

	$X_{1,1}$	$H_1[2]$	$H_2[2]$	G	D_1	D_2
$F_1(1,2)$	1	0	1	0	0	1
$F_2(1,2)$	1	1	0	0	1	0
$R(1,1)$	1	1	1	1	0	0
C	1	0	0	−1	1	1
J_1	0	0	1	1	−1	0
J_2	0	1	0	1	0	−1
P_1	2	0	2	0	0	2
P_2	2	2	0	0	2	0

Finally, we discuss the theory of surfaces on X developed in [LT17, Section 9.3]. Let us introduce some notations: let α and α' be two codimension-2 real cycles on X. Two cycles are numerically equivalent, denoted by $\alpha \equiv \alpha'$, if for any surface T, we have $T.\alpha = T.\alpha'$. We denote the following vector space

$$\{\text{codimension 2 real cycles}\}/\equiv$$

by $N_2(X) = N^2(X)$. This is the *space of 2-cycles up to numerical equivalence.* Note that this space is self dual with respect to the intersection pairing.

The closure of the cone of effective cycles is called the *pseudo-effective cone of surfaces* and is denoted by $\overline{\mathrm{Eff}}_2(X) = \overline{\mathrm{Eff}}^2(X)$.

Any real 2-cycle α is *nef* if for any surface T, $T.\alpha \geq 0$. We denote the *nef cone of surfaces* by $\mathrm{Nef}_2(X) = \mathrm{Nef}^2(X)$.

Let $\alpha = \sum a_i T_i$ be a real effective 2-cycle with $a_i > 0$ and T_i is an irreducible surface. We say α is *movable* if every component T_i is a member of a dominant flat family of surfaces on X. We denote the closure of movable classes as $\overline{\mathrm{Mov}}_2(X) = \overline{\mathrm{Mov}}^2(X)$ and call it the *movable cone of surfaces.*

Let us summarize the descriptions of these cones. First of all $N_2(X)$ is 6-dimensional and $\mathrm{Sym}^2(N^1(X))$ is a 5-dimensional subspace with a relation $H_1[2]H_2[2] + G^2 = H_1[2]G + H_2[2]G$. Let FP be the class of the surface parametrizing all degree-2 subschemes whose support contains a fixed point.

Then classes

$$H_1[2]^2, H_1[2]H_2[2], H_2[2]^2, H_1[2]G, H_2[2]G, FP$$

form a basis for $N_2(X)$, and the intersection matrix with respect to this basis is given by Table 6.2. One can find Table 6.2 in [LT17, Section 9.3].

Table 6.2 *The intersection matrix of surfaces*

	$H_1[2]^2$	$H_1[2]H_2[2]$	$H_2[2]^2$	$H_1[2]G$	$H_2[2]G$	FP
$H_1[2]^2$	0	0	2	0	0	0
$H_1[2]H_2[2]$	0	2	0	0	0	1
$H_2[2]^2$	2	0	0	0	0	0
$H_1[2]G$	0	0	0	0	−2	0
$H_2[2]G$	0	0	0	−2	0	0
FP	0	1	0	0	0	1

To describe various cones in this vector space, we recall several classes of surfaces in terms of the basis $H_1[2]^2, H_1[2]H_2[2], H_2[2]^2, H_1[2]E, H_2[2]E, FP$ in Table 6.3. Table 6.3 is also available in [LT17, Section 9.3]. The rightmost column is a description of a general member of a given surface.

It is shown in [LT17, Section 9.3] that the pseudo-effective cone of surfaces $\overline{\mathrm{Eff}}_2(X)$ is generated by

$$S_1, S_2, V_1, V_2, T_1, T_2, FP, L_{1,2}.$$

It is also shown in [LT17, Section 9.3] that the nef cone of surfaces $\mathrm{Nef}_2(X)$ coincides with the movable cone of surfaces $\overline{\mathrm{Mov}}_2(X)$, and they are generated by

$$FP, L_{1,2}, L_{1,1}, L_{2,2}, M_{1,1}, M_{2,2}, \alpha_1, \alpha_2, \alpha_3, \alpha_4.$$

Below we will use these descriptions of the cones of subvarieties without proofs. For details, see [BC13, Example 18] and [LT17, Section 9.3].

6.2.3 Computation of the Exceptional Set

Let F be a field of characteristic 0. Recall that S is $\mathbb{P}^1 \times \mathbb{P}^1$ and X is $\mathrm{Hilb}^{[2]}(S)$ with $L = -K_X$. We freely use the notation set up in Example 4.2.18 as well as Section 6.2.2.

When $F = \mathbb{Q}$, [LR19] proved Manin's conjecture (Conjecture 4.2.9) after removing a thin set Z' that is given by

$$Z' = (E \cup D_1 \cup D_2)(\mathbb{Q}) \cup f(W(\mathbb{Q})),$$

Table 6.3 *The description of surfaces*

S_1	$(\frac{1}{2}, 0, 0, -\frac{1}{2}, 0, 0)$	distinct points in a fixed H_1
S_2	$(0, 0, \frac{1}{2}, 0, -\frac{1}{2}, 0)$	distinct points in a fixed H_2
V_1	$(0, 0, 0, 1, 0, 0)$	nonreduced, contained, as a set, in a fixed H_1
V_2	$(0, 0, 0, 0, 1, 0)$	nonreduced, contained, as a set, in a fixed H_2
T_1	$(0, 1, 0, 0, -1, 0)$	nonreduced, contained, as a scheme, in any H_1
T_2	$(0, 1, 0, -1, 0, 0)$	nonreduced, contained, as a scheme, in any H_2
FP	$(0, 0, 0, 0, 0, 1)$	support contains a fixed point
$L_{1,2}$	$(0, 1, 0, 0, 0, -1)$	one point in each of a fixed H_1 and a fixed H_2
$L_{1,1}$	$(1, 0, 0, 0, 0, 0)$	one point in each of two fixed H_1's
$L_{2,2}$	$(0, 0, 1, 0, 0, 0)$	one point in each of two fixed H_2's
$M_{1,1}$	$(\frac{1}{2}, 1, \frac{1}{2}, -\frac{1}{2}, -\frac{1}{2}, -1)$	the pullback of $\sigma_{1,1}$ from $\mathrm{Gr}(2, 4)$
M_2	$(\frac{1}{2}, 0, \frac{1}{2}, -\frac{1}{2}, -\frac{1}{2}, 1)$	the pullback of σ_2 from $\mathrm{Gr}(2, 4)$
α_1	$(\frac{1}{2}, 1, 0, -\frac{1}{2}, 0, -1)$	———
α_2	$(0, 1, \frac{1}{2}, 0, -\frac{1}{2}, -1)$	———
α_3	$(\frac{1}{2}, 0, 0, -\frac{1}{2}, 0, 1)$	———
α_4	$(0, 0, \frac{1}{2}, 0, -\frac{1}{2}, 1)$	———

where $f\colon W = \mathrm{Bl}_\Delta(S \times S) \longrightarrow X$ is the double cover defined in Example 4.2.18. In this section, we will show that the exceptional set $\mathsf{Z}_0 \cup \mathsf{Z}_1 \cup \mathsf{Z}_2 \cup \mathsf{Z}_3$ constructed in Section 4.3.2 coincides with a thin set

$$Z' = (E \cup D_1 \cup D_2)(F) \cup f(W(F)).$$

This computation was carried out in [LST22, Example 5.13] using results from [LT17, Section 9.3.1].

First of all, note that we have $\mathbf{B}_+(L) = E$, so $\mathsf{Z}_0 = E(F)$. Since (X, L) is adjoint rigid, we have $\mathsf{Z}_3 = \emptyset$. Thus we only need to compute Z_1 and Z_2.

Let us compute Z_1 first. First, we analyze subvarieties of X. We recall some geometric facts on X. The group $\mathrm{PGL}_2 \times \mathrm{PGL}_2$ acts fiberwise on $\mathbb{P}^1 \times \mathbb{P}^1$, and this action induces the action of $\mathrm{PGL}_2 \times \mathrm{PGL}_2$ on X. The orbits of this action are

$$X \setminus (D_1 \cup D_2 \cup E), \quad D_1 \setminus E \cap D_1, \quad D_2 \setminus E \cap D_2,$$
$$B \setminus ((D_1 \cup D_2) \cap E), \quad E \cap D_1, \quad E \cap D_2.$$

Set $L = -K_X = 2H_1[2] + 2H_2[2]$. Then we have

$$a(X, L) = 1 \quad \text{and} \quad b(F, X, L) = 3.$$

Since $X \setminus (D_1 \cup D_2 \cup E)$ is the unique open dense orbit, any subvariety Y meeting this locus satisfies $a(Y, L|_Y) \leq a(X, L)$ by Lemma 5.1.2. On the other hand, $a(E, L|_E) = \infty$ because $L|_E$ is not big. Each divisor D_i can be realized as a $\mathbb{P}^2$-bundle over $\mathbb{P}^1$, and each fiber isomorphic to $\mathbb{P}^2$ is the class S_i. In particular, intersection computation shows that $L|_{S_i}$ is $\mathcal{O}_{\mathbb{P}^2}(2)$, so we conclude that $a(S_i, L|_{S_i}) = 3/2$. It follows from Lemma 5.1.2 that $a(D_i, L|_{D_i}) \geq 3/2$. We conclude that Z_1 contains $(E \cup D_1 \cup D_2)(F)$.

Thus it suffices to analyze subvarieties Y meeting the open orbit

$$X \setminus (D_1 \cup D_2 \cup E).$$

We begin with curves:

Lemma 6.2.4 ([LT17, Section 9.3.1]) *For any rational curve $C \subset X$ meeting $X \setminus (D_1 \cup D_2 \cup E)$, we have $a(C, L|_C) \leq a(X, L)$. Moreover, $a(C, L|_C) = 1$ holds exactly when C is a curve of the class $F_1(1,2)$ or $F_2(1,2)$.*

Proof After taking a base change, we may assume that $\overline{F} = F$. The first statement has already been discussed, so we prove the second statement. Suppose that there exists a rational curve $C \subset X$ meeting $X \setminus (D_1 \cup D_2 \cup E)$ with $a(C, L|_C) = 1$. Then the numerical class α of C is a nef class and thus can be written as

$$\alpha \equiv aF_1(1,2) + bF_2(1,2) + cR(1,1),$$

with a, b, c non-negative. Since α is an integral class, the intersections of α with $H_1[2], H_2[2], G$ are all integers. This implies that a, b, c are all integers. On the other hand, we have

$$L.\alpha = 2a + 2b + 4c,$$

thus we conclude that

$$a(C, L|_C) = 1 \iff L.\alpha = 2 \iff \alpha = F_1(1,2) \text{ or } F_2(1,2).$$

Therefore our assertion follows. □

Next, we analyze surfaces:

Lemma 6.2.5 ([LT17, Section 9.3.1]) *For any surface $Y \subset X$ meeting the open orbit $X \setminus (D_1 \cup D_2 \cup E)$, we have $a(Y, L|_Y) \leq a(X, L)$. Moreover, $a(Y, L|_Y) = 1$ and (Y, L) is adjoint rigid exactly when Y is a surface of class $L_{1,1}$, $L_{2,2}$, FP, or $L_{1,2}$.*

Proof We may assume that $\overline{F} = F$. The first statement is already verified, hence we will show the second statement. Let $Y \subset X$ be a surface such that $a(Y, L|_Y) = 1$ and (Y, L) is adjoint rigid. Since Y is movable, we write

$$[Y] \equiv aFP + bL_{1,1} + cL_{2,2} + dL_{1,2} + eM_{1,1} + fM_2 + g_1\alpha + g_2\alpha_2 + g_3\alpha_3 + g_4\alpha_4,$$

where $a, b, c, d, e, f, g_1, g_2, g_3, g_4$ are non-negative. This class $[Y]$ is an integral class, so by taking intersection numbers with $S_1, S_2, V_1, V_2, L_{1,2}, FP$, we conclude that

$$c, \quad b, \quad e + f + g_2 + g_4,$$
$$e + f + g_1 + g_3, \quad d + e + g_1 + g_2, \quad a + f + g_3 + g_4$$

are all non-negative integers. Moreover, we have

$$\frac{1}{4}L^2.Y = 2b + 2c + (e + f + g_1 + g_3) + (e + f + g_2 + g_4)$$
$$+ 2(a + f + g_3 + g_4) + 2(d + e + g_1 + g_2).$$

So we conclude that the only classes Y with $L^2.Y < 4 \cdot 3$ are $L_{1,1}, L_{2,2}, FP$, or $L_{1,2}$. Applying Proposition 6.2.1, we can have $a(Y, L|_Y) = 1$ and (Y, L) is adjoint rigid only if Y is a surface of class $L_{1,1}, L_{2,2}, FP$, or $L_{1,2}$.

The surfaces Y of class $L_{1,1}$ or $L_{2,2}$ are fibers of the morphism

$$X \to \mathrm{Sym}^{(2)}(S) \to \mathrm{Sym}^{(2)}(\mathbb{P}^1) \cong \mathbb{P}^2,$$

which is induced by projections of S. Since we are interested only in surfaces meeting $X \setminus (D_1 \cup D_2 \cup E)$, these surfaces are isomorphic to $\mathbb{P}^1 \times \mathbb{P}^1$. Moreover, since $L|_Y = -K_Y$, we conclude that $a(Y, L|_Y) = 1$ and (Y, L) is adjoint rigid. If Y is a surface of class FP or $L_{1,2}$, then the image of Y via

$$X \to \mathrm{Sym}^{(2)}(S) \to \mathrm{Sym}^{(2)}(\mathbb{P}^1) \cong \mathbb{P}^2$$

is a line ℓ_i for each projection. Its product map $X \to \mathbb{P}^2 \times \mathbb{P}^2$ induces a birational morphism from Y to $\ell_1 \times \ell_2$. By intersection calculation, we conclude that $L|_Y$ is the pullback of $\mathcal{O}_{\ell_1 \times \ell_2}(2, 2)$. Thus we deduce that $a(Y, L|_Y) = 1$ and (Y, L) is adjoint rigid. □

Finally, we analyze divisors. To this end, let us consider the morphism

$$\rho_i : X \to \mathrm{Sym}^{(2)}(S) \to \mathrm{Sym}^{(2)}(\mathbb{P}^1) \cong \mathbb{P}^2_i, \ (i = 1, 2)$$

induced by each projection, which is the morphism associated to the complete linear series of $H_i[2]$ on X. The image of the diagonal maps via ρ_i to a smooth conic $C_i \subset \mathbb{P}^2_i$. For any $x \in \mathbb{P}^2_i \setminus C_i$, the fiber $\rho_i^{-1}(x)$ is isomorphic to $\mathbb{P}^1 \times \mathbb{P}^1$. When $x \in C$, the fiber $\rho_i^{-1}(x)$ is a non-reduced scheme whose underlying set is isomorphic to $\mathbb{P}^2$. With this notation, we have:

Lemma 6.2.6 ([LT17, Section 9.3.1]) *For any prime divisor $Y \subset X$ meeting the open orbit $X \setminus (D_1 \cup D_2 \cup E)$, we have $a(Y, L|_Y) \leq a(X, L)$. Moreover, $a(Y, L|_Y) = 1$ and (Y, L) is adjoint rigid exactly when Y is the image of a fiber of one of projections $W \to \mathbb{P}^1$ or the pullback of a line ℓ on $\mathbb{P}^2_i$ tangent to C_i via ρ_i.*

Proof We may assume that $\overline{F} = F$. The first statement has already been verified, so we assume that $Y \subset X$ is a prime divisor meeting the open orbit $X \setminus (D_1 \cup D_2 \cup E)$ such that $a(Y, L|_Y) = 1$ and (Y, L) is adjoint rigid. Then we can write

$$[Y] \equiv aH_1[2] + bH_2[2] + cX_{1,1},$$

where a, b, c are non-negative integers. Then we have

$$L^3.Y = 48a + 48b + 96c.$$

By Corollary 6.2.3, Y is a prime divisor of class $H_1[2]$ or $H_2[2]$, or a prime divisor covered by curves of type $F_1(1, 2)$ or $F_2(1, 2)$.

Suppose that Y is a prime divisor of class $H_1[2]$. The case of $H_2[2]$ is similar. In this case, Y is a member of $H^0(X, \mathcal{O}_X(H_1[2]))$. We consider the Chow morphism $\phi\colon X \to \mathrm{Sym}^{(2)}(S)$. We have

$$H^0(X, \mathcal{O}_X(H_1[2])) = H^0(\mathrm{Sym}^{(2)}(S), \mathcal{O}_{\mathrm{Sym}^{(2)}(S)}(H_1[2])),$$

hence Y is the pullback of a prime divisor Y' on $\mathrm{Sym}^{(2)}(S)$. Since Y is irreducible, the image of the diagonal $\Delta \subset S \times S$ is not contained in Y'.

The group PGL_2 acts on $\mathrm{Sym}^{(2)}(\mathbb{P}^1) \cong \mathbb{P}^2$, and the orbits of this action consist of $\mathbb{P}^2 \setminus C$ and C. The orbits of the action on the dual $(\mathbb{P}^2)^\vee$ consist of the set of lines meeting C transversely and the set of lines tangent to C.

Now Y' is the pullback of a line ℓ on $\mathbb{P}^2$. When ℓ meets with C transversely, Y' is singular along $Y' \cap \Delta$ and Y is a smooth resolution of Y'. By adjunction, one can prove that $a(Y, L|_Y) = 1$ and the canonical fibration of $a(Y, L|_Y)L + K_Y$ is $\rho \circ \phi\colon Y \to \ell$. Thus (Y, L) is not adjoint rigid.

When ℓ is tangent to C, one can show that Y and Y' are not normal. Let $\beta\colon \widetilde{Y}' \to Y'$ be the normalization. Then one can show that $\widetilde{Y}'$ is smooth. Since Y' is movable, we have $a(Y', L|_{Y'}) \leq 1$. Hence $\beta^*L + K_{\widetilde{Y}'}$ is pseudo-effective. Moreover, let F be a general fiber of $\rho|_{Y'}\colon Y' \to \ell$. Then one can show that $(\beta^*L + K_{\widetilde{Y}'})|_F \equiv 0$ so we conclude that $a(Y', L|_{Y'}) = 1$. Note that a local computation shows that $K_{\widetilde{Y}'} \sim \beta^*K_{Y'} - \beta^*H_1[2]|_{Y'} \sim \beta^*K_{\mathrm{Sym}^{(2)}(S)}$. Thus $(\widetilde{Y}', \beta^*L|_{Y'})$ is adjoint rigid.

Next, we assume that Y is covered by curves of type $F_1(1, 2)$. The case of type $F_2(1, 2)$ is similar. Since Y is covered by curves of L-degree 2, we have $a(Y, L|_Y) = 1$. First let us assume that $f^{-1}(Y)$ is reducible. Choose a component Y' of $f^{-1}(Y)$ that is birational to Y. We would like to see when (Y', f^*L) is adjoint rigid. We denote the strict transform of Y' on $S \times S$ by $\widehat{Y}$. Since $f^*L = -\psi^* K_{S\times S}$, we only need to discuss adjoint rigidity of the pair $(\widehat{Y}, -K_{S\times S})$. By our assumption, one can write $\widehat{Y} = \mathbb{P}^1 \times R$ with $R \subset (\mathbb{P}^1)^3$. We let $\beta\colon R' \to R$ be a resolution. Then $\alpha\colon \widehat{Y}' = \mathbb{P}^1 \times R' \to \widehat{Y}$ is a resolution of $\widehat{Y}$. If we denote two projections of $\mathbb{P}^1 \times R'$ by p_1, p_2 then we have

$$K_{\widehat{Y}'} - \alpha^* K_{S\times S} = p_2^*(K_{Z'} + 2H_2 + 2H_3 + 2H_4).$$

Then it follows from Proposition 6.2.1 that $K_{R'} + 2H_2 + 2H_3 + 2H_4$ is not rigid unless $R' = (\mathbb{P}^1)^2 \subset (\mathbb{P}^1)^3$. In this case, Y is the image of a fiber of $W \to \mathbb{P}^1$ and our assertion follows.

Finally, assume that $f^{-1}(Y)$ is irreducible. We consider the projection map

$$X \to \mathrm{Sym}^{(2)}(S) \to \mathrm{Sym}^{(2)}(\mathbb{P}^1) \cong \mathbb{P}^2,$$

which is induced by the first projection. Our assumption implies that the image of Y under this morphism is a curve $C \subset \mathbb{P}^2$. Let C' be the resolution of C. We consider the base change $Y' \to C'$ of $Y \to C$. Let $\beta\colon \widehat{Y} \to Y$ be a resolution of Y'. One can show that the canonical map for $(\widehat{Y}, \beta^* L)$ is given by $\widehat{Y} \to C'$. See [LT17, Section 9.3.1] for more details. Thus our assertion follows. □

Next, we compute Z_1:

Theorem 6.2.7 ([LT17, Section 9.3] and [LST22, Example 5.13]) *Let F be a number field and $S = \mathbb{P}^1 \times \mathbb{P}^1$ be defined over F. Let X be the Hilbert scheme* $\mathrm{Hilb}^{[2]}(S)$ *of length-2 subschemes on S with $L = -K_X$. Then the set Z_1 constructed in Section 4.3.2 is equal to*

$$(E \cup D_1 \cup D_2)(F) \cup f(W(F)).$$

Proof We have already discussed that Z_1 contains $(E \cup D_1 \cup D_2)(F)$. Let $\varphi\colon Y \to X$ be a thin map whose image is not contained in $E \cup D_1 \cup D_2$. Because of group action, we have $a(Y, \varphi^* L) \leq a(X, L) = 1$. Assume that

$$a(Y, \varphi^* L) = a(X, L);$$

otherwise such a φ will not contribute to Z_1.

Assume that the fiber dimension $d(Y, \varphi^* L) = 1$. In this case, it follows from Lemma 5.1.3 that $b(F, Y, \varphi^* L) = 1$. Thus $\varphi\colon Y \to X$ will not contribute to Z_1.

Next, assume that the fiber dimension $d(Y, \varphi^* L) = 2$. Let $R \subset Y$ be a general fiber of the canonical fibration of $(Y, \varphi^* L)$. Let R' be the image of R on X. It follows from Lemma 5.2.2 that $a(R', L) = 1$ and (R', L) is adjoint rigid. Lemma 6.2.5 implies that (R', L) is isomorphic to $(\mathbb{P}^1 \times \mathbb{P}^1, -K_{\mathbb{P}^1\times\mathbb{P}^1})$.

It follows from Corollary 5.2.5 that $\varphi\colon R \to R'$ must be birational so that $b(F, Y, \varphi^*L) \leq 2$. Thus $\varphi\colon Y \to X$ will not contribute to Z_1.

The case of the fiber dimension $d(Y, \varphi^*L) = 3$ is similar. Let $R \subset Y$ be a general fiber of the canonical fibration of (Y, φ^*L). Assume that R' is the image of R. Lemma 6.2.6 implies that W' is the image of a fiber of a projection $W \to \mathbb{P}^1$ or the pullback of a line ℓ on $\mathbb{P}^2_i$ tangent to C_i via ρ_i. In the former case, since ZR admits birational morphism to $(\mathbb{P}^1)^3$, the morphism $\varphi\colon R \to R'$ must be birational. Thus we have $b(F, R, \varphi^*L) = 3$ and $\varphi(R(F))$ is contained in Z_1. Since Y rationally factors through W, our assertion follows. In the latter case, a resolution $\widetilde{R}'$ of R' constructed in Lemma 6.2.6 has Picard rank 2 and the only divisors with higher Fujita invariants are the pullbacks of the non-normal locus of R'. However, the complement of such a locus has a trivial fundamental group, proving that $R \to R'$ is birational. Thus we conclude that $b(F, R, \varphi^*L|_R) = 2$. Thus Y will not contribute to Z_1. □

Finally, we compute Z_2 as well:

Theorem 6.2.8 ([LST22, Example 5.13]) *Let F be a number field and $S = \mathbb{P}^1 \times \mathbb{P}^1$ be defined over F. Let X be the Hilbert scheme* $\mathrm{Hilb}^{[2]}(S)$ *of length-2 subschemes on S with $L = -K_X$. Then the set Z_2 constructed in Section 4.3.2 is equal to*

$$f(W(F)).$$

Proof First, after taking a base change, we will assume that $\overline{F} = F$. Our goal is to show that any a-cover $\varphi\colon Y \to X$ such that (Y, φ^*L) is adjoint rigid is birational to $f\colon W \to X$. Let $U = X \setminus (E \cup D_1 \cup D_2)$. It follows from Corollary 5.2.5 that $\varphi\colon Y \to X$ is étale in codimension 1 over U. We claim that $\pi_1^{\text{ét}}(U) = \mathbb{Z}/2\mathbb{Z}$. To that end, we consider the fibration

$$U \to \mathrm{Sym}^{(2)}(S) \setminus \phi(E \cup D_1 \cup D_2) \to \mathrm{Sym}^{(2)}(\mathbb{P}^1) \setminus C \cong \mathbb{P}^2 \setminus C,$$

where C is the image of Δ that is a smooth conic in $\mathbb{P}^2$. Any fiber of this fibration is an étale double covering of $\mathbb{P}^2 \setminus C$. Since $\pi_1^{\text{ét}}(\mathbb{P}^2 \setminus C) = \mathbb{Z}/2\mathbb{Z}$, our claim follows from Serre's exact sequence. Thus, over $\overline{F}$, $f\colon W \to X$ is the only adjoint rigid a-cover up to birational equivalence.

Now assume that our ground field F is an arbitrary field of characteristic 0. As observed in Example 4.2.18, we have $f(W(F)) \subset \mathsf{Z}_2$. On the other hand, for any non-trivial twist $f^\sigma\colon W^\sigma \to X$, we have $b(F, W^\sigma, (f^\sigma)^*L) = 2$ because the involution of the cover $f : W \to X$ acts on $N^1(W)$ non-trivially. Thus such a $f^\sigma\colon W^\sigma \to X$ will not contribute to Z_2. Our assertion follows. □

We conclude that our exceptional set $\mathsf{Z}_0 \cup \mathsf{Z}_1 \cup \mathsf{Z}_2 \cup \mathsf{Z}_3$ coincides with the exceptional set Z' used in [LR19].

6.3 Quadric Bundles

In this section, we consider the following example of a quadric surface bundle defined over $\mathbb{Q}$:

$$X = \left\{ \sum_{i=0}^{3} x_i y_i^2 = 0 \right\} \subset \mathbb{P}_x^3 \times \mathbb{P}_y^3,$$

with a polarization $L = -K_X$ which is ample by adjunction. This example is analogous to Batyrev–Tschinkel's example described in Example 4.2.17, and a related quadric surface bundle was first considered in [Els11]. In [Els11], Elsenhans conducted a numerical experiment and confirmed numerically that Manin's conjecture (Conjecture 4.2.9) should hold after removing a thin exceptional set.

Later Manin's conjecture for X was proved by Browning–Heath-Brown in [BHB20], and they used the following exceptional set:

$$Z' = \{((x_0 : x_1 : x_2 : x_3), (y_0 : y_1 : y_2 : y_3)) \in X(\mathbb{Q}) \mid x_0x_1x_2x_3 = 0 \text{ or } \square\}.$$

In this section, we compute thin sets $\mathsf{Z}_0, \mathsf{Z}_1, \mathsf{Z}_2, \mathsf{Z}_3$ constructed in Section 4.3.2 and confirm that our conjecture exceptional set coincides with Z' used by Browning and Heath-Brown. This computation was originally carried out in [LT19b, Section 12].

From now on we assume that our ground field is a field F of characteristic 0. We denote projections from X by $\pi_1 : X \to \mathbb{P}_x^3$ and $\pi_2 : X \to \mathbb{P}_y^3$ and denote the pullback of the hyperplane class via π_i by h_i for $i = 1, 2$. Then we have $L = -K_X = 3h_1 + 2h_2$ by adjunction. Also, by the Lefschetz hyperplane theorem, $N^1(X)_{\mathbb{Z}}$ is generated by h_1 and h_2, so we conclude that $a(X, L) = 1$ and $b(F, X, L) = 2$. We denote the union of singular fibers of π_1 by V. This is the pullback of the union of coordinate hyperplanes on $\mathbb{P}_x^3$. To better understand the geometry of X, we first work over $\overline{F}$ and analyze the a, b-invariants of subvarieties of X.

First, using explicit geometry of quadric surface bundles as well as adjunction theory, one can prove the following lemma, whose proof is omitted in this book:

Lemma 6.3.1 ([LT19b, Lemma 12.2 and 12.3]) *Let $\ell \subset \mathbb{P}^3$ be a linear subspace of dimension 1 or 2, and Y be the π_1-preimage of ℓ. Assume that $Y \not\subset V$. Then Y is irreducible and normal, and has only canonical singularities.*

As a corollary, we have the following statement:

Corollary 6.3.2 ([LT19b, Corollary 12.4]) *Let $\ell \subset \mathbb{P}^3$ be a linear subspace of dimension 1 or 2, and Y be the π_1-preimage of ℓ. Assume that $Y \not\subset V$. Then we have $a(Y, L) = 1$ and (Y, L) is not adjoint rigid.*

Proof It follows from Lemma 6.3.1 that Y is normal and has only canonical singularities. Thus by adjunction we have $K_Y + L = (3 - \dim \ell)h_1|_Y$. Hence our claim follows from Proposition 4.1.3. □

Now we analyze adjoint rigid subvarieties whose a-invariants are greater than or equal to $a(X, L) = 1$:

Proposition 6.3.3 ([LT19b, Proposition 12.5]) *Let $Y \subset X$ be an adjoint rigid subvariety with respect to L such that $a(Y, L) \geq a(X, L) = 1$ and $Y \not\subset V$. Then one of the following holds:*

(i) *Y is a smooth fiber of π_1;*
(ii) *Y is a line in a smooth fiber of π_1; or*
(iii) *Y is a smooth fiber of π_2.*

Furthermore any component V' of V satisfies $a(V', L) = 1$. Subvarieties Y in (i) satisfy $a(Y, L) = 1$ and $b(\overline{F}, Y, L) = 2$, and subvarieties Y in (ii) and (iii) satisfy $a(Y, L) = 1$ and $b(\overline{F}, Y, L) = 1$.

Proof We separate our analysis based on dimension of Y:

Divisors. First, let us handle integral divisors on X. Let $Y \subset X$ be an integral divisor. Let $\beta \colon \widetilde{Y} \to Y$ be a smooth resolution and denote $\pi_i \circ \beta$ by f_i for $i = 1, 2$. There are two possibilities: either f_1 is dominant or not.

Suppose that $f_1 : \widetilde{Y} \to \mathbb{P}^3_x$ is dominant. Let Y_1 be a general member of $|\beta^* h_1|$. By Bertini's theorem, this is smooth. It is also connected by [Laz04, Theorem 3.3.3], so it is irreducible. We have a short exact sequence

$$0 \to \mathcal{O}_{\widetilde{Y}}(K_{\widetilde{Y}} + 2\beta^* h_1 + 2\beta^* h_2) \to \mathcal{O}_{\widetilde{Y}}(K_{\widetilde{Y}} + 3\beta^* h_1 + 2\beta^* h_2)$$
$$\to \mathcal{O}_{Y_1}(K_{Y_1} + 2\beta^* h_1|_{Y_1} + 2\beta^* h_2|_{Y_1}) \to 0,$$

which induces the exact sequence

$$0 \to H^0(\widetilde{Y}, \mathcal{O}_{\widetilde{Y}}(K_{\widetilde{Y}} + 2\beta^* h_1 + 2\beta^* h_2)) \to H^0(\widetilde{Y}, \mathcal{O}_{\widetilde{Y}}(K_{\widetilde{Y}} + 3\beta^* h_1 + 2\beta^* h_2))$$
$$\to H^0(Y_1, \mathcal{O}_{Y_1}(K_{Y_1} + 2\beta^* h_1|_{Y_1} + 2\beta^* h_2|_{Y_1}))$$
$$\to H^1(\widetilde{Y}, \mathcal{O}_{\widetilde{Y}}(K_{\widetilde{Y}} + 2\beta^* h_1 + 2\beta^* h_2)).$$

Since $2\beta^* h_1 + 2\beta^* h_2$ is big and nef, it follows from the Kawamata–Viehweg vanishing theorem ([KM98, Theorem 2.64]) that the last cohomology vanishes, so that we have

$$h^0(\widetilde{Y}, \mathcal{O}_{\widetilde{Y}}(K_{\widetilde{Y}} + 3\beta^* h_1 + 2\beta^* h_2)) \geq h^0(Y_1, \mathcal{O}_{Y_1}(K_{Y_1} + 2\beta^* h_1|_{Y_1} + 2\beta^* h_2|_{Y_1})).$$

Now let Y_3 be the intersection of Y_1 with a general member of $|\beta^* h_1|$ and another general member of $|\beta^* h_2|$. As discussed above, Y_3 is a smooth irreducible curve. Moreover, repeating the above argument using the Kawamata–Viehweg vanishing theorem, we prove that

$$h^0(\widetilde{Y}, \mathcal{O}_{\widetilde{Y}}(K_{\widetilde{Y}} + 3\beta^* h_1 + 2\beta^* h_2)) \geq h^0(Y_3, \mathcal{O}_{Y_3}(K_{Y_3} + \beta^* h_1|_{Y_3} + \beta^* h_2|_{Y_3})).$$

Since Y_3 is not contracted by either of the projections f_1 and f_2, we conclude that $\deg(\beta^* h_i|_{Y_3}) \geq 1$ for $i = 1, 2$. By Riemann–Roch, no matter what the genus of Y_3 is, we have

$$h^0(Y_3, \mathcal{O}_{Y_3}(K_{Y_3} + \beta^* h_1|_{Y_3} + \beta^* h_2|_{Y_3})) \geq 1,$$

proving that $a(Y, L) \leq 1$. Moreover, if we have

$$h^0(Y_3, \mathcal{O}_{Y_3}(K_{Y_3} + \beta^* h_1|_{Y_3} + \beta^* h_2|_{Y_3})) = 1$$

then we must have $\deg(\beta^* h_i|_{Y_3}) = 1$ for $i = 1, 2$. However, this implies that

$$Y.h_1^3.h_2 = 1, \quad Y.h_1^2.h_2^2 = 1,$$

but the first equality is not possible because X is a hypersurface of type $(1, 2)$ in $\mathbb{P}^3_x \times \mathbb{P}^3_y$ and $Y \sim ah_1 + bh_2$ with $a \geq 0, b > 0$. This means that

$$h^0(Y_3, \mathcal{O}_{Y_3}(K_{Y_3} + \beta^* h_1|_{Y_3} + \beta^* h_2|_{Y_3})) \geq 2,$$

so we conclude that $a(Y, L) < 1$ or (Y, L) is not adjoint rigid.

Next, let us assume that $f_1 : \widetilde{Y} \to \mathbb{P}^3_x$ is not dominant. In this case, we have $Y \sim ch_1$ for some $c > 0$. Let Y_2 be a complete intersection of two general members of $|\beta^* h_1|$. Since Y is a divisor, Y_2 is birational to the disjoint union of c integral fibers of π_1, so each component is birational to a (possibly singular) quadric surface. Let Y_3 be the intersection of Y_2 with a general member of $|\beta^* h_2|$. Then Y_3 is the disjoint union of c conic sections such that each conic section is embedded into a quadric surface fiber. Using the arguments of the Kawamata–Viehweg vanishing theorem as above, one can prove that

$$h^0(\widetilde{Y}, \mathcal{O}_{\widetilde{Y}}(K_{\widetilde{Y}} + 3\beta^* h_1 + 2\beta^* h_2)) \geq h^0(Y_3, \mathcal{O}_{Y_3}(K_{Y_3} + \beta^* h_2|_{Y_3})) = c \geq 1.$$

Thus we conclude that $a(Y, L) \leq 1$. In particular, if V' is a component of V, then $a(V', L) = 1$ because any quadric cone has a-invariant 1.

Assume that $Y \not\subset V$. If $c > 1$, then we conclude that (Y, L) is not adjoint rigid. Thus we may assume that $c = 1$. This case is handled by Corollary 6.3.2. Altogether we conclude that (Y, L) is not adjoint rigid.

Dimension 3. Next, we assume that Y is a threefold not contained in V. Let $\beta \colon \widetilde{Y} \to Y$ be a smooth resolution and denote $\pi_i \circ \beta$ by f_i for $i = 1, 2$. Again we separate our analysis based on the dimension of images of $\widetilde{Y}$ by f_i for $i = 1, 2$:

(i) both f_1 and f_2 are dominant;
(ii) f_1 is dominant and the image of f_2 is a surface;
(iii) f_1 is dominant and the image of f_2 is a curve;
(iv) the image of f_1 is a surface and f_2 is dominant;
(v) the image of f_1 is a surface and the image of f_2 is also a surface; or
(vi) the image of f_1 is a curve and f_2 is dominant.

In cases (i), (ii), (iv), (v), we may argue as in the case of divisors by cutting out $\widetilde{Y}$ by a general member of $|\beta^* h_1|$ and another general member of $|\beta^* h_2|$. In case (iii), Y is algebraically equivalent to ch_2^2 for some $c > 0$. Then we can argue as the case of divisors by cutting out $\widetilde{Y}$ by two general members of $|\beta^* h_1|$. In the case of (vi), Y is algebraically equivalent to ch_1^2 for some $c > 0$. Then we cut out $\widetilde{Y}$ by a general member of $|\beta^* h_1|$ and another general member of $|\beta^* h_2|$ and denote it by Y_2. Then Y_2 is the disjoint union of c conics. As discussed above, we may conclude that (Y, L) is not adjoint rigid unless $c = 1$. The case of $c = 1$ is handled by Corollary 6.3.2.

Dimension 2. Any fiber P of π_2 is isomorphic to $\mathbb{P}^2$, satisfies $a(P, L) = a(X, L) = 1$, and is adjoint rigid with respect to L. In this case, we have $b(\overline{\mathbb{Q}}, P, L) = 1$. Any smooth fiber Q of π_1 is a quadric surface that satisfies $a(Q, L) = a(X, L) = 1$ and is adjoint rigid. In this case, we have $b(\overline{\mathbb{Q}}, Q, L) = 2$.

Now assume that Y is a surface such that both projections are non-trivial. Then we have

$$Y.(3h_1 + 2h_2)^2 = 9Y.h_1^2 + 12Y.h_1.h_2 + 4h_2^2 > 9.$$

Thus by arguing as explained in the proof of Proposition 6.2.1, we conclude that $a(Y, L) < 1$ or (Y, L) is not adjoint rigid.

Dimension 1. Let C be a rational curve on X. Then we have

$$(3h_1 + 2h_2).C \geq 2,$$

and the equality holds when $h_1.C = 0$ and $h_2.C = 1$. This is exactly when C is a line in a quadric surface fiber of π_1. This concludes our assertion. □

From now on we work over a field F of characteristic 0 and compute $\mathsf{Z}_0, \mathsf{Z}_1, \mathsf{Z}_2$, and Z_3 for (X, L). Since L is ample, we have $\mathsf{Z}_0 = \emptyset$. We also have $\mathsf{Z}_3 = \emptyset$ as

(X, L) is adjoint rigid. Hence we only need to compute Z_1 and Z_2. Here is our theorem:

Theorem 6.3.4 ([LT19b, Section 12.1]) *Let F be a field of characteristic* 0. *Assume that X is defined over F. Then we have $\mathsf{Z}_1 = Z'$ and $\mathsf{Z}_2 = \emptyset$.*

Proof First we claim that $Z' \subset \mathsf{Z}_1$. Indeed, for any smooth quadric surface fiber Q of π_1 at $(x_0 : x_1 : x_2 : x_3) \in \mathbb{P}^3_x(F)$, we have $b(F, Q, L) = 2$ if and only if $x_0x_1x_2x_3$ is a square. For a singular fiber Q, if it is a singular quadric cone then we have $b(F, Q, L) = 2$. (Note that one needs to take the minimal resolution to compute the b-invariant in this case.) When Q is the union of two planes, each plane P satisfies $a(P, L) = 3/2 > 1$. Together these assertions imply that $Z' \subset \mathsf{Z}_1$.

Next, let us prove $\mathsf{Z}_1 \subset Z'$. Let $f : Y \to X$ be a generically finite morphism from a smooth projective variety Y to the image not contained in V such that $a(Y, f^*L) > a(X, L) = 1$, or $a(Y, f^*L) = a(X, L) = 1$ but the fiber dimension $d(Y, f^*L) < \dim X = 5$. In the former case, Lemma 5.2.2 implies that the image $f(Y)$ must satisfy $a(f(Y), L) \geq a(Y, L) > a(X, L) = 1$. Hence it follows from Proposition 6.3.3 that $f(Y) \subset V$. Thus we conclude that $f(Y(F)) \subset Z'$. In the latter case, let $\pi : Y \dashrightarrow W$ be the canonical map. In virtue of Lemma 5.2.12, we may assume that this is a morphism after applying a resolution. Let Y_w be a general fiber and let $S_w = f(Y_w)$. Then (S_w, L) is adjoint rigid with $a(S_w, L) = 1$ by Lemma 5.2.2. Thus it follows from Proposition 6.3.3 that it is a smooth fiber of π_1, a smooth fiber of π_2, or geometrically the union of lines in smooth fibers of π_1. Then Propositions 6.1.2 and 6.1.4 indicate that (S_w, L) does not admit a non-trivial adjoint rigid a-cover proving that $f|_{Y_w} : Y_w \to S_w$ is birational. If we have $b(F, Y, f^*L) \geq b(F, X, L) = 2$, then it follows from Lemma 5.1.3 that $b(F, Y_w, f^*L|_{Y_w}) \geq b(F, Y, f^*L) \geq 2$. This implies that S_w is a split quadric surface fiber of π_1. Thus there is a rational map $\phi : W \dashrightarrow \mathbb{P}^3_x$ such that Y is birational to the main component of the base change of $\pi_1 : X \to \mathbb{P}^3_x$ by $\phi : W \dashrightarrow \mathbb{P}^3_x$. After some birational modification, we may assume that W is smooth and ϕ is a morphism. Moreover, we may assume that $f^{-1}(X \setminus V)$ is isomorphic to the base change. Then the above argument shows that $b(F, S_w, L) \geq 2$ for any rational fiber in $f^{-1}(X \setminus V)$. Thus we conclude that $f(Y(F)) \subset Z'$. Hence our assertion follows.

For Z_2, note that Proposition 6.3.3 states that there is no integral divisor D on X with $a(D, L|_D) > a(X, L) = 1$. Thus if we have an adjoint rigid a-cover $f : Y \to X$, then it follows from Corollary 5.2.5 that $f : Y \to X$ is étale in codimension 1. However, since X has trivial étale fundamental group, such an f must be birational so that it does not satisfy face contracting condition. Thus we conclude that $\mathsf{Z}_2 = \emptyset$. □

Thus we conclude that $\mathsf{Z}_0 \cup \mathsf{Z}_1 \cup \mathsf{Z}_2 \cup \mathsf{Z}_3 = Z'$. A similar computation for Example 4.2.17, that is, Batyrev–Tschinkel's example of a Fano cubic surface bundle was carried out in [Gao24a].

6.4 Surfaces

In this section, we discuss our exceptional sets for smooth projective geometrically uniruled surfaces. Our guiding question is as follows:

Question 6.4.1 Let F be a field of characteristic 0 and S be a smooth projective geometrically uniruled surface defined over F. Let L be a big and nef $\mathbb{Q}$-divisor on S. Let $\mathsf{Z} = \mathsf{Z}_0 \cup \mathsf{Z}_1 \cup \mathsf{Z}_2 \cup \mathsf{Z}_3$ be the exceptional set for (S, L) constructed in Section 4.3.2. Then is the exceptional set Z contained in a proper closed subset?

It turns out that this question has an affirmative answer when S is *split*, that is, $N^1(S) = N^1(\overline{S})$. This follows from the works [LTT18] and [LT17]. We will discuss these results in Section 6.4.1.

However, when S is not split, it is possible that the set Z becomes Zariski dense. Indeed, in [Gao23], Gao constructed an example of a smooth del Pezzo surface of degree 1 and Picard rank 1 where the exceptional set Z is Zariski dense. We will discuss this surprising example in Section 6.4.2.

6.4.1 Split Surfaces

The goal of this section is to prove the following theorem:

Theorem 6.4.2 ([LTT18, Proposition 5.9] and [LT17, Theorem 1.8]) *Let F be a field of characteristic 0 and S be a smooth projective geometrically uniruled surface defined over F. Let L be a big and nef $\mathbb{Q}$-divisor on S. Let $\mathsf{Z} = \mathsf{Z}_0 \cup \mathsf{Z}_1 \cup \mathsf{Z}_2 \cup \mathsf{Z}_3$ be the exceptional set for (S, L) constructed in Section 4.3.2.*

Assume that S is split, that is, $N^1(S) = N^1(\overline{S})$. Then there exists a proper closed subset $V \subset S$ such that $\mathsf{Z} \subset V$. In particular, the exceptional set Z is not Zariski dense.

First we discuss the following result from [LTT18]:

Lemma 6.4.3 ([LTT18, Proposition 5.9]) *Let F be a field of characteristic 0. Let S be a smooth projective geometrically uniruled and integral surface defined over F, and L be a big and nef $\mathbb{Q}$-divisor on S. Suppose that S is split. Then we have the following:*

(i) *When* $\kappa(a(S,L)L + K_S) = 1$*, let* $\pi\colon S \longrightarrow B$ *be the canonical model of* $(S, a(S,L)L)$*, and* C *be a general geometric fiber of* π*. Then we have*

$$(a(S,L), b(F,S,L)) = \left(\frac{2}{L.C}, 1\right).$$

Moreover, there exists a proper closed subset $V \subsetneq S$ *such that for any integral curve* $C' \not\subset V$ *on* S*, we have*

$$(a(C', L|_{C'}), b(F, C', L|_{C'})) \leq (a(S,L), b(F,S,L))$$

in the lexicographic order.

(ii) *When* $\kappa(a(S,L)L + K_S) = 0$*, let* $\phi\colon S \longrightarrow S'$ *be a run of* $(a(S,L)L + K_S)$*-MMP. Then we have*

$$(a(S,L), b(F,S,L)) = (a(S', \phi_* L), \rho(S')).$$

Moreover, there exists a proper closed subset $V \subsetneq S$ *such that, for any integral curve* $C' \not\subset V$ *on* S*, we have*

$$(a(C', L|_{C'}), b(F, C', L|_{C'})) < (a(S,L), b(F,S,L))$$

in the lexicographic order.

Proof In view of Lemma 4.1.5, Remark 4.1.15, and our assumption that S is split, we may assume that F is an algebraically closed field. In this proof, we drop the script F in the notation for b-invariants.

Let us note certain facts about the MMP for smooth projective surfaces. We consider a run of $(a(S,L)L + K_S)$-MMP:

$$S = S_0 \longrightarrow S_1 \longrightarrow \cdots \longrightarrow S_n = S',$$

and we denote a map $S \longrightarrow S'$ by ϕ. Since there is no 2-dimensional flip contraction, each step of MMP is a divisorial contraction. In particular,

$$\phi\colon S \longrightarrow S'$$

is an honest morphism. We denote each divisorial contraction $S_i \longrightarrow S_{i+1}$ by ψ_i and the composition $S \longrightarrow S_i$ by ϕ_i. Let $E_i \subset S_i$ be the exceptional divisor for ψ_i. Then we have $a(S,L)(\phi_i)_* L.E_i + K_{S_i}.E_i < 0$. However, since L is big and nef, its pushforward $(\phi_i)_* L$ is movable therefore nef because any movable divisor on a surface is nef. Together, these imply that $K_{S_i}.E < 0$. Thus each step of a $(a(S,L)L + K_S)$-MMP is a K_{S_i}-negative divisorial contraction. This means that each S_i is smooth, E_i is a (-1)-rational curve, and $\psi_i\colon S_i \longrightarrow S_{i+1}$ is the blowdown of a (-1)-curve E_i.

Moreover, when $a(S,L)L$ is Cartier, one can show that $a(S,L)(\phi_i)_*L$ is also Cartier for each i. Indeed, when $a(S,L)(\phi_i)_*L$ is Cartier, we have

$$a(S,L)(\phi_i)_*L.E_i = 0.$$

Thus we have $a(S,L)\psi_i^*\psi_{i,*}(\phi_i)_*L = a(S,L)(\phi_i)_*L$, proving that $a(S,L)(\phi_{i+1})_*L$ is Cartier. In particular, this shows that we have $\phi^*\phi_*L = L$.
(i): Here we prove the statement (i). In this case, since a map $\phi\colon S \to S'$ to a minimal model is a morphism, the canonical model $\pi\colon S \to B$ is also a morphism and a general fiber C is isomorphic to $\mathbb{P}^1$. We have the Zariski decomposition $a(S,L)L + K_S = P + N$, where P is the pullback of an ample divisor on B and N is a sum of vertical curves, hence we have

$$a(S,L)L.C + K_S.C = 0.$$

Then by adjunction we have $K_S.C = -2$, proving that $a(S,L) = \frac{2}{L.C}$. Moreover it follows from Lemma 5.1.3 that $b(S,L) \leq b(C, L|_C) = 1$. Hence we conclude that $b(S,L) = 1$.

Now it follows from Theorem 5.1.6 that there exists a proper closed subset $V \subsetneq S$ such that for any curve $C' \not\subset V$ on S, we have $a(C', L|_{C'}) \leq a(S,L)$. Then our assertion is clear from the fact that $b(C', L|_{C'}) = 1$.
(ii): It follows from Lemma 4.1.18 that we have

$$(a(S,L), b(F,S,L)) = (a(S',\phi_*L), b(S',\phi_*L)).$$

Then note that we have $a(S,L)\phi_*L + K_{S'} \sim_{\mathbb{Q}} 0$, so we conclude that

$$b(S',\phi_*L) = \rho(S').$$

Hence our first statement of (ii) follows.

For the second statement of (ii), if $\rho(S') \geq 2$, then our assertion follows from Theorem 5.1.6. Thus we may assume that $\rho(S') = 1$ which means that S' is isomorphic to $\mathbb{P}^2$. Let ℓ be a general line on S'. Then we have

$$a(S,L)\phi_*L.\ell + K_{\mathbb{P}^2}.\ell = 0,$$

so that we have

$$a(S,L) = \frac{3}{\phi_*L.\ell}.$$

Let $N \subset S$ be a rational curve on S. Assume that N is not contained in the exceptional locus $\mathrm{Exc}(\phi)$ of ϕ and $L.N > 0$ holds. (Note that rational curves N that violate these assumptions are finite.) Then we have

$$a(N, L|_N) = \frac{2}{L.N}.$$

If N does not meet the exceptional locus of ϕ, then we have

$$L.N = \phi_*L.\phi_*N \geq \phi_*L.\ell.$$

This implies that

$$a(N, L|_N) = \frac{2}{L.N} < \frac{3}{\phi_*L.\ell} = a(S, L).$$

If N meets the exceptional locus of ϕ, then since $a(S, L)L + K_S$ is $\mathbb{Q}$-linearly equivalent to a unique effective $\mathbb{Q}$-divisor D whose support coincides with the exceptional locus $\mathrm{Exc}(\phi)$ of ϕ, we have $(a(S, L)L + K_S).N > 0$, that is,

$$a(S, L)L.N > -K_S.N.$$

If $-K_S.N \geq 2$ holds, then we have

$$a(S, L) > \frac{2}{L.N} = a(N, L|_N).$$

If $L.N > \frac{2}{3}\phi_*L.\ell$, then we also have

$$a(N, L|_N) = \frac{2}{L.N} < \frac{3}{\phi_*L.\ell} = a(S, L).$$

Hence we only need to consider rational curves N satisfying

- $-K_S.N < 2$;
- $L.N \leq \frac{2}{3}\phi_*L.\ell$; and
- $L.N > 0$.

It follows from Lemma 5.1.5 that rational curves N satisfying the second and third condition form a bounded family. On the other hand, rational curves N satisfying the first condition do not deform to cover S. To see this, if such rational curves N deform to cover S, then a general rational curve N is free, so the moduli space has expected dimension $-K_S.N - 1 < 1$. This contradicts the fact that N deforms. See Section 7.2 for more details for freeness of rational curves.

The upshot is that we define a proper closed subset V as

$$V = \mathrm{Exc}(\phi) \cup \left(\bigcup_{L.N=0} N \right) \cup \left(\bigcup_{-K_S.N<2,\ L.N \leq \frac{2}{3}\phi_*L.\ell} N \right).$$

Then for any $C' \not\subset V$ on S, we have $a(C', L|_{C'}) < a(S, L)$. Thus our assertion follows. □

Next, we prove the following theorem, which is proved in [LT17, Corollary 6.6]:

Theorem 6.4.4 ([LT17, Corollary 6.6]) *Let F be a field of characteristic 0 and S be a smooth projective geometrically uniruled and integral surface defined over F. Let L be a big and nef $\mathbb{Q}$-divisor on S. Assume that $N^1(S) = N^1(\overline{S})$. Then we have the following:*

(i) *When $\kappa(a(S, L)L + K_S) = 1$, for any generically finite morphism $f : Y \to S$ from a geometrically integral smooth projective surface Y, we have*

$$(a(Y, f^*L), b(F, Y, f^*L)) \leq (a(S, L), b(F, S, L))$$

in the lexicographic order.

(ii) *When $\kappa(a(S, L)L + K_S) = 0$, for any dominant thin map $f : Y \to S$ from a geometrically integral smooth projective surface Y, we have*

$$(a(Y, f^*L), b(F, Y, f^*L)) < (a(S, L), b(F, S, L))$$

in the lexicographic order.

We first prove the above theorem when S is a weak del Pezzo surface with $L = -K_S$:

Lemma 6.4.5 ([LT17, Proposition 6.3 and Proposition 6.4]) *Let F be a field of characteristic 0 and S be a smooth and geometrically integral weak del Pezzo surface with $L = -K_S$. Assume that $\rho(S) = \rho(\overline{S})$. Then for any dominant thin map $f : Y \to S$ from a geometrically integral smooth projective surface Y, we have*

$$(a(Y, f^*L), b(F, Y, f^*L)) < (a(S, L), b(F, S, L))$$

in the lexicographic order.

Proof Again in view of Lemma 4.1.5, Remark 4.1.15, and our assumption that S is split, we may assume that F is an algebraically closed field.

First of all, we have $a(S, L) = 1$ and $b(S, L) = \rho(S)$. If $\rho(S) = 1$, then S is isomorphic to $\mathbb{P}^2$. In this case, our assertion follows from Proposition 6.1.2.

From now on we suppose that $\rho(S) \geq 2$. It follows from Lemma 5.2.2 that we have

$$a(Y, f^*L) \leq a(S, L).$$

Let us assume that the equality holds. When (Y, f^*L) is not adjoint rigid, it follows from Lemma 6.4.3 that

$$b(Y, f^*L) = 1 < \rho(S) = b(S, L).$$

Thus our assertion follows.

Suppose that (Y, f^*L) is adjoint rigid. We apply a $(f^*L + K_Y)$-MMP and obtain a birational morphism $\phi\colon Y \to Y'$ with Y' being smooth and projective. Then we have

$$\phi_* f^* L + K_{Y'} \sim_{\mathbb{Q}} 0.$$

Since $\phi_* f^* L$ is big and nef, we conclude that Y' is another smooth weak del Pezzo surface. Then we have

$$(-K_{Y'})^2 = \mathrm{Vol}(-K_{Y'}) = \mathrm{Vol}(\phi_* f^* L) = \mathrm{Vol}(f^* L) = (f^* L)^2 = \deg(f)(-K_S)^2.$$

Here the third equality follows from the fact that $\phi^* \phi_* f^* L = f^* L$. Since $\deg f \geq 2$, we conclude that $(-K_{Y'})^2 > (-K_S)^2$. On the other hand, by Lemma 4.1.18, we have $b(Y, f^* L) = \rho(Y')$. Since we have $(-K_{Y'})^2 > (-K_S)^2$, and for any weak del Pezzo surface X, the equality $10 = (-K_X)^2 + \rho(X)$ holds, we conclude that

$$b(Y, f^* L) = \rho(Y') < \rho(S) = b(S, L).$$

Thus our assertion follows. □

The general case can be proved by reducing a problem to the case of a weak del Pezzo surface using the minimal model program:

Proof of Theorem 6.4.4 Again we may assume that F is algebraically closed.
(i) From Lemma 5.2.2, we have

$$a(Y, f^* L) \leq a(S, L).$$

If the equality holds, then since (S, L) is not adjoint rigid, $(Y, f^* L)$ is also not adjoint rigid. Therefore it follows from Lemma 6.4.3 that we have

$$b(Y, f^* L) = 1 = b(S, L).$$

Thus our assertion follows.
(ii) Again we have

$$a(Y, f^* L) \leq a(S, L).$$

Let us assume that the equality holds. Let $\phi\colon S \to S'$ be a run of $(a(S, L)L + K_S)$-MMP. Then S' is a weak del Pezzo surface and it follows from Lemma 6.4.3 that we have

$$(a(S, L), b(S, L)) = (a(S', \phi_* L), b(S', \phi_* L)).$$

We denote $\phi_* L$ by L'. We claim that

$$(a(Y, f^* L), b(Y, f^* L)) = (a(Y, (\phi \circ f)^* L'), b(Y, (\phi \circ f)^* L')). \tag{6.1}$$

Once we prove this claim, it follows from Lemma 6.4.5 that we have

$$(a(Y, f^*L), b(Y, f^*L)) = (a(Y, (\phi \circ f)^*L'), b(Y, (\phi \circ f)^*L'))$$
$$< (a(S', L'), b((S', L')) = (a(S, L), b(S, L)).$$

Thus our assertion follows.

Let us prove (6.1). There exists a unique effective $\mathbb{Q}$-divisor $E \geq 0$ such that $E \sim_{\mathbb{Q}} a(S, L)L + K_S$. Note that the support of E coincides exactly with the exceptional locus $\mathrm{Exc}(\phi)$ of ϕ. It follows from Lemma 3.2.16 that there exists a ϕ-exceptional effective $\mathbb{Q}$-divisor E' on S such that $\phi^*L = L + E'$. Let R be the ramification divisor of f, so that we have $K_Y \sim_{\mathbb{Q}} f^*K_S + R$. Then we have

$$a(Y, f^*L)f^*L + K_Y \sim_{\mathbb{Q}} f^*(a(S, L)L + K_S) + R \sim_{\mathbb{Q}} f^*E + R,$$

and

$$a(Y, f^*L)(\phi \circ f)^*L' + K_Y \sim_{\mathbb{Q}} f^*(a(S, L)L + K_S + E') + R \sim_{\mathbb{Q}} f^*E + f^*E' + R.$$

Then since the support of $f^*E + R$ coincides with the support of $f^*E + f^*E' + R$, $f^*E + R$ being not big implies that $f^*E + f^*E' + R$ is not big. It follows that $a(Y, f^*L) = a(Y, (\phi \circ f)^*L')$. Also the fact that the support of $f^*E + R$ coincides with the support of $f^*E + f^*E' + R$ indicates that $\mathcal{F}(Y, f^*L) = \mathcal{F}(Y, (\phi \circ f)^*L)$. Thus we conclude that $b(Y, f^*L) = b(Y, (\phi \circ f)^*L')$. Hence our claim follows. □

Finally, we prove Theorem 6.4.2:

Proof of Theorem 6.4.2 By definition, Z_0 is contained in a proper closed subset, and Z_3 is also contained in a proper closed subset by Lemma 5.3.1. Thus we only need to show that Z_1 and Z_2 are not Zariski dense.

First let us assume that (S, L) is not adjoint rigid. Then we have $d(S, L) = 1$. Since there is no thin map $f : Y \to S$ such that $d(Y, f^*L) < 1$, we conclude that $\mathsf{Z}_1 = \emptyset$. Suppose that we have a thin map $f : Y \to S$ from a smooth projective variety such that $d(Y, f^*L) = 1$. First suppose that Y is a curve. Let C be the image of Y on S. It follows from Lemma 6.4.3 that there exists a proper closed subset $V \subsetneq S$ such that if $C \not\subset V$, then we have

$$(a(C, L|_C), b(F, C, L|_C)) \leq (a(S, L), b(F, S, L)).$$

Next, it follows from Lemma 5.2.2 that we have

$$(a(Y, f^*L), b(F, Y, f^*L)) \leq (a(C, L|_C), b(F, C, L|_C)) \leq (a(S, L), b(F, S, L)).$$

So we need only consider the case when we have

$$(a(Y, f^*L), b(F, Y, f^*L)) = (a(S, L), b(F, S, L)).$$

This is only possible when Y is birational to C. In this case, $\mathcal{F}(Y, f^*L)$ is 1-dimensional, and the induced map $f_*\colon \mathcal{F}(Y, f^*L) \to \mathcal{F}(S, L)$ is an isomorphism. Thus $f\colon Y \to S$ is not face contracting. When Y is a surface, it follows from Theorem 6.4.4 that we have

$$(a(Y, f^*L), b(F, Y, f^*L)) \leq (a(S, L), b(F, S, L)).$$

So we need only consider the case when the equality holds. In this case, again $\mathcal{F}(Y, f^*L)$ is 1-dimensional, and the induced map $f_*\colon \mathcal{F}(Y, f^*L) \to \mathcal{F}(S, L)$ is again an isomorphism. Thus $f\colon Y \to S$ is not face contracting. Altogether we conclude that $\mathsf{Z}_2 \subset V(F)$. Thus our assertion follows.

When (S, L) is adjoint rigid, it follows from Lemma 6.4.3 and Theorem 6.4.4 that Z_1 is contained in a proper closed subset. Also it follows from Theorem 6.4.4 that Z_2 is empty. Thus our assertion follows. □

6.4.2 Non-split Surfaces

In this section, we discuss Question 6.4.1 when S is a non-split del Pezzo surface. First, we discuss positive results proved in [LT19b, Theorem 9.1]. Then we review a counterexample to Question 6.4.1 that is constructed in [Gao23].

Positive Results

The following theorem shows that for most del Pezzo surfaces, Question 6.4.1 has an affirmative answer:

Theorem 6.4.6 ([LT19b, Theorem 9.1]) *Let F be a number field and S be a smooth and geometrically integral del Pezzo surface defined over F with a polarization $L = -K_S$. When the degree $(-K_S)^2$ of S is equal to 1, we assume that either $\rho(S) \geq 2$ or $\overline{S}$ is general in its moduli.*

Let $\mathsf{Z} = \mathsf{Z}_0 \cup \mathsf{Z}_1 \cup \mathsf{Z}_2 \cup \mathsf{Z}_3$ be the exceptional set for (S, L) constructed in Section 4.3.2. Then Z is not Zariski dense.

To enable us to prove Theorem 6.4.6, we first prove the following lemma classifying low degree rational curves on del Pezzo surfaces:

Lemma 6.4.7 ([LT19b, Lemma 9.2]) *Assume that $F = \overline{F}$. Let S be a smooth del Pezzo surface defined over F with $L = -K_S$. Let $C \subset S$ be a curve. Then the following hold*

(i) *$a(C, L|_C) > a(S, L) = 1$ if and only if $L.C = 1$. In this situation, we have the following possibilities for C:*

  (a) *C is a (-1)-rational curve on S;*

(b) *the degree of* S *is equal to* 1, *and* C *is a rational member of* $|-K_S|$.

(ii) $a(C, L|_C) = a(S, L) = 1$ *if and only if* $L.C = 2$. *In this situation, we have the following possibilities for* C*:*

(a) C *is a smooth fiber of a conic fibration* $\pi\colon S \to \mathbb{P}^1$*;*

(b) *the degree of* S *is equal to* 2 *and* C *is a rational member of* $|-K_S|$*;*

(c) *the degree of* S *is equal to* 1 *and there exists a blow down* $\phi\colon S \to S'$ *to a degree* 2 *del Pezzo surface* S' *such that* C *is a rational member of* $|-\phi^* K_{S'}|$*;*

(d) *the degree of* S *is equal to* 1 *and* C *is a rational member of* $|-2K_S|$.

Proof When C is a rational curve, we have

$$a(C, L|_C) = \frac{2}{L.C}.$$

Thus we have $a(C, L|_C) > a(S, L) = 1$ (resp. $a(C, L|_C) = a(S, L) = 1$) if and only if $L.C = 1$ (resp. $L.C = 2$). Now let us classify such low degree rational curves C on S.

Assume that $L.C = 1$. Then the intersection matrix for L, C is given by

$$\begin{pmatrix} d & 1 \\ 1 & e \end{pmatrix},$$

where $d = L^2$ and $C^2 = e$. By the *Hodge index theorem* [Har77, Chapter 5, Theorem 1.9], we have $de - 1 \leq 0$. On the other hand, the adjunction formula gives us

$$-2 \leq 2p_a(C) - 2 = -1 + e,$$

so we conclude that e must be odd and greater than or equal to -1. Here $p_a(C) = h^1(C, \mathcal{O}_C)$ is the arithmetic genus of C. Therefore the only possibilities are

- $e = -1$ and $p_a(C) = 0$; or
- $d = 1, e = 1$, and $p_a(C) = 1$.

In the former case, C is a (-1)-rational curve. In the latter case, since $de - 1 = 0$, C and $-K_S$ are proportional to each other. This implies that $C \sim -K_S$. Thus our assertion follows.

Next, assume that $L.C = 2$. Then the intersection matrix for L, C is given by

$$\begin{pmatrix} d & 2 \\ 2 & e \end{pmatrix},$$

where $d = L^2$ and $C^2 = e$. By the Hodge index theorem, we have $de - 4 \leq 0$. On the other hand, adjunction formula gives us

$$-2 \leq 2p_a(C) - 2 = -2 + e,$$

so we conclude that e must be even and greater than or equal to 0. Therefore the only possibilities are

- $e = 0$ and $p_a(C) = 0$;
- $d = 2$, $e = 2$, and $p_a(C) = 1$;
- $d = 1$, $e = 2$, and $p_a(C) = 1$, or;
- $d = 1$, $e = 4$, and $p_a(C) = 2$.

In the first case, C is smooth and a fiber of a conic fibration $\pi\colon S \to \mathbb{P}^1$. In the second case, we have $C \sim -K_S$. In the third case, let $E = C + K_S$. Then we have $E^2 = -1$ and $K_S.E = -1$. The Riemann–Roch theorem tells us that

$$\chi(S, O(E)) = 1 + \frac{1}{2}(E - K_S).E = 1.$$

Moreover, the Kawamata–Viehweg vanishing theorem ([KM98, Theorem 2.64]) tells us that

$$h^1(S, O(C + K_S)) = h^2(S, O(C + K_S)) = 0.$$

(Note that since C is irreducible and $C^2 > 0$, C is big and nef.) Thus we conclude that $h^0(S, O(E)) = 1$. This implies that there exists an effective divisor $E' \sim E$. Since $-K_S.E' = 1$ and $-K_S$ is ample, we conclude that E' is integral. This means that E' is a (-1)-rational curve, so there is a birational contraction $\phi\colon S \to S'$ to a degree-2 del Pezzo surface contracting E'. Then we have $C \sim -K_S + E' \sim -\phi^* K_{S'}$. Thus our assertion follows. Finally, in the fourth case, we have $C \sim -2K_S$. Thus our assertion follows. □

Next, we discuss the following theorem from [LT17]:

Theorem 6.4.8 ([LT17, Corollary 6.9]) *Let F be an algebraically closed field of characteristic 0. Let S be a smooth weak del Pezzo surface defined over F, and $\phi\colon S \to S'$ be the anticanonical model of S.*

Suppose that we have an adjoint rigid a-cover $f\colon Y \to S$ from a smooth projective surface. Let $f'\colon Y' \to S'$ be the finite part of the Stein factorization of $\phi \circ f : Y \to S \to S'$. Then $f'\colon Y' \to S'$ is étale in codimension 1.

Here is the definition of the anticanonical model: When X is a smooth weak Fano variety, the basepoint-free theorem ([KM98, Theorem 3.3]) tells us that $-K_X$ is semi-ample. In particular, there exists a positive integer $m \in \mathbb{Z}_{>0}$ such that the complete linear series $|-mK_X|$ defines a birational morphism

$\phi: X \to X'$ to a normal projective variety. Then X' is a Fano variety with only canonical singularities. The Fano variety X' is called the *anticanonical model* of X.

Theorem 6.4.8 shows that an adjoint rigid a-cover of a weak del Pezzo surface corresponds to a finite index subgroup of $\pi_1^{\text{ét}}(S'^{\text{sm}})$ where S'^{sm} is the smooth locus of S'. It follows from [GKP16, Theorem 1.13] that $\pi_1^{\text{ét}}(S'^{\text{sm}})$ is finite. Thus this gives us an alternative proof of Theorem 5.2.7 in the case of weak del Pezzo surfaces.

The proof of Theorem 6.4.8 uses an inductive argument based on the number of steps of $(a(Y, f^*L)f^*L + K_Y)$-MMP. Readers interested in its proof should consult [LT17, Theorem 6.8 and Corollary 6.9]. Assuming this theorem, we can prove the following corollary:

Corollary 6.4.9 *Assume that $\overline{F} = F$. Let S be a smooth del Pezzo surface defined over F. Then there is no adjoint rigid a-cover $f: Y \to S$ of degree ≥ 2.*

Proof The anticanonical model of S is S itself because $-K_S$ is ample. This means that we have $\pi_1^{\text{ét}}(S'^{\text{sm}}) = \pi_1^{\text{ét}}(S)$. Since any smooth projective rationally connected variety is simply connected (see, e.g., [Deb01, Corollary 4.29] in the case of complex varieties), our assertion follows. □

Finally, we prove Theorem 6.4.6:

Proof of Theorem 6.4.6 By definition, the set Z_0 is contained in a proper closed subset of S. The set Z_3 is also not Zariski dense because of Lemma 5.3.1. It follows from Corollary 6.4.9 that Z_2 is the empty set. Thus we only need to show that Z_1 is contained in a proper closed subset.

When $\rho(S) \geq 2$, we have

$$\mathsf{Z}_1 \subset \bigcup_{a(C, -K_S|_C) > a(S, -K_S) = 1} C(F),$$

where $C \subset S$ runs over geometrically integral curves such that

$$a(C, -K_S|_C) > a(S, -K_S) = 1.$$

When the degree $(-K_S)^2$ of S is greater than 1, C is a (-1)-rational curve by Lemma 6.4.7. Since the number of (-1)-rational curves on a smooth del Pezzo surface S is finite, our assertion follows. When the degree of S is equal to 1, again the contribution from (-1)-rational curves is not Zariski dense. Thus by Lemma 5.3.1, we need to consider rational members of $|-K_S|$. However, this complete linear series is 1-dimensional, and a general member is a smooth curve of genus 1, proving that the number of the anticanonical lines in $|-K_S|$ is finite. Thus our assertion follows.

Suppose that $\rho(S) = 1$. Again the contribution from the anticanonical lines is contained in a proper closed subset, so we consider the anticanonical conics. Since the Picard rank is 1, it follows from Lemma 5.3.1 that there is a geometrically integral anticanonical conic C defined over F if and only if either the degree of S is 2 and $C \in |-K_S|$ or the degree of S is 1 and $C \in |-2K_S|$. First, assume that S has degree 2. In this case, $|-K_S|$ is basepoint free, and it defines a degree-2 morphism $\pi\colon S \to \mathbb{P}^2$ which is ramified along a smooth quartic curve D that has genus 3. Then a rational curve $C \in |-K_S|$ is the pullback of a tangent line to D, and such rational curves C are parametrized by D. Mordell–Faltings' theorem tells us that $D(F)$ is finite. Thus our assertion follows.

When S has degree 1, $|-2K_S|$ is also basepoint free. It defines a degree-2 morphism $\pi\colon S \to Q \subset \mathbb{P}^3$, where Q is a quadric cone. The cover π is ramified along a degree-6 curve T and the vertex of Q, and this degree-6 curve is a smooth complete intersection of Q and a cubic surface. Then a rational curve $C \in |-2K_S|$ is the pullback of a bitangent plane to T in $\mathbb{P}^3$, and this bitangent plane is parametrized by a 1-dimensional projective scheme H. One can show that when $\overline{S}$ is general in its moduli, H is irreducible and its geometric genus is ≥ 2. Thus again our claim follows from Faltings' theorem. See [LT19b, Lemma 9.5] for more details. □

A Counterexample

In Theorem 6.4.6, when our del Pezzo surface S has degree 1 and Picard rank 1 over F, we need to assume that $\overline{S}$ is general in its moduli. Indeed, in [Gao23], Gao found the following example which shows that when a degree-1 del Pezzo surface has Picard rank 1, it is possible that Z_1 is Zariski dense:

Example 6.4.10 ([Gao23]) Let F be $\mathbb{Q}(\sqrt{-3})$ and consider the following del Pezzo surface S of degree 1:

$$S : w^2 = 49x^6 + 49y^6 + z^3,$$

in the weighted projective space $\mathbb{P}(1, 1, 2, 3)_{x,y,z,w}$. It is verified in [Gao23, Theorem 1.3, Theorem 1.5, and Proposition 1.6] that this del Pezzo surface S satisfies the following properties:

(i) $\rho(S) = 1$;
(ii) there exists an elliptic curve E of positive Mordell–Weil rank parametrizing rational curves in $|-2K_S|$, that is, we have a family $\pi\colon C \to E$ of geometrically rational curves in $|-2K_S|$ with the evaluation map $s\colon C \to S$, which is dominant; and

(iii) π admits a section defined over F which intersects with a general fiber in a smooth point.

Since E has positive Mordell–Weil rank, $E(F)$ is Zariski dense in E. Thus $\pi: C \to E$ admits infinitely many geometrically integral fibers defined over F. Then since there is a section passing through a smooth point of such a general fiber, a general geometrically integral and rational fiber is birational to $\mathbb{P}^1$ over F so that each fiber admits a Zariski dense set of rational points. Since these rational curves contribute to Z_1, we conclude that Z_1 is Zariski dense. Readers interested in more detail should consult the original paper [Gao23].

6.5 Smooth Prime Fano Threefolds

Let F be an algebraically closed field of characteristic 0 and X be a smooth Fano variety defined over F. We say X is a *prime Fano variety* if X has Picard rank 1. An important invariant of prime Fano manifolds is the following index: let X be a smooth prime Fano variety. The *index* of X is the largest positive integer r such that $-K_X \sim rH$ for some $H \in \mathrm{Pic}(X)$.

The classification of smooth prime Fano threefolds was initiated by Gino Fano and subsequently completed by V. A. Iskovskih in [Isk77], [Isk78], and [Isk79]. Here is the classification:

Theorem 6.5.1 ([Isk77], [Isk78], and [Isk79]) *Let X be a smooth prime Fano threefold defined over F. Then the index r of X satisfies $1 \leq r \leq 4$ and we have $\mathrm{Pic}(X) \cong \mathbb{Z}$. Let H be an ample generator for $\mathrm{Pic}(X)$. In terms of r and H^3 a Fano threefold X is classified into one of the following:*

(i) *When $r = 4$, X is isomorphic to $\mathbb{P}^3$.*
(ii) *When $r = 3$, X is isomorphic to a smooth quadric hypersurface $Q \subset \mathbb{P}^4$.*
(iii) *When $r = 2$, X is isomorphic to one of the following five deformation types:*
  (1) *when $H^3 = 5$, X is a section of the Grassmannian $\mathrm{Gr}(2, 5)$ by a linear subspace of codimension 3;*
  (2) *when $H^3 = 4$, X is a complete intersection of two quadrics in $\mathbb{P}^5$;*
  (3) *when $H^3 = 3$, X is a cubic hypersurface in $\mathbb{P}^4$;*
  (4) *when $H^3 = 2$, X is the double cover of $\mathbb{P}^3$ branched along a smooth quartic surface; or*
  (5) *when $H^3 = 1$, X is a degree 6 hypersurface in the weighted projective space $\mathbb{P}(1, 1, 1, 2, 3)$.*
(iv) *When $r = 1$, X is isomorphic to one of the following 10 deformation types:*

(1) *when* $H^3 = 22$, *X is the zero locus of three sections of the rank-3 bundle* $\wedge^2 \mathcal{U}^\vee$ *where* $\mathcal{U}$ *is the universal subbundle on* $\mathrm{Gr}(3,7)$*;*
(2) *when* $H^3 = 18$, *X is a section of the homogeneous space* $G_2/P \subset \mathbb{P}^{13}$ *by a linear subspace of codimension* 2*;*
(3) *when* $H^3 = 16$, *X is a section of the symplectic Lagrangian Grassmannian* $\mathrm{LGr}(3,6) \subset \mathbb{P}^{13}$ *by a linear subspace of codimension* 3*;*
(4) *when* $H^3 = 14$, *X is a section of* $\mathrm{Gr}(2,6) \subset \mathbb{P}^{14}$ *by a linear subspace of codimension* 5*;*
(5) *when* $H^3 = 12$, *X is a section of a connected component of the orthogonal Lagrangian Grassmannian* $\mathrm{OGr}_+(5,10) \subset \mathbb{P}^{15}$ *by a linear subspace of codimension* 7*;*
(6) *when* $H^3 = 10$, *X is either*
   - *a section of* $\mathrm{Gr}(2,5) \subset \mathbb{P}^9$ *by a linear subspace of codimension* 2 *and a quadric; or*
   - *the double cover of a section of* $\mathrm{Gr}(2,5) \subset \mathbb{P}^9$ *by a linear subspace of codimension* 3 *branched along a divisor in the anticanonical linear system;*
(7) *when* $H^3 = 8$, *X is a complete intersection of three quadrics in* $\mathbb{P}^6$*;*
(8) *when* $H^3 = 6$, *X is a complete intersection of a quadric and a cubic in* $\mathbb{P}^5$*;*
(9) *when* $H^3 = 4$, *X is either*
   - *a quartic hypersurface in* $\mathbb{P}^4$*; or*
   - *the double cover of a smooth quadric hypersurface in* $\mathbb{P}^4$ *branched along an intersection with a quartic;*
(10) *when* $H^3 = 2$, *X is the double cover of* $\mathbb{P}^3$ *branched along a sextic surface.*

The goal of this section is to compute our conjectural exceptional set for smooth prime Fano threefolds. Since $\mathbb{P}^3$ and a smooth quadric threefold are handled by Propositions 6.1.2 and 6.1.4 respectively, we will focus on the cases of $r = 1$ and 2. In Section 6.5.1, we discuss the case of $r = 2$. In Section 6.5.2, we consider the case of $r = 1$.

6.5.1 Del Pezzo Threefolds

Assume that our base field F is an algebraically closed field of characteristic 0. Let X be a smooth prime Fano variety of dimension n with an ample generator H. We say X is a *del Pezzo manifold* if $-K_X \sim (n-1)H$. Note that del Pezzo threefolds are exactly smooth prime Fano threefolds of index 2. In this section, we consider the case of del Pezzo threefolds.

From now on we assume that F is an arbitrary field of characteristic 0 and X is a geometrically integral smooth Fano threefold such that $\overline{X}$ is a del Pezzo threefold over $\overline{F}$. We call such threefolds defined over F del Pezzo threefolds as well.

In this section, we discuss our conjectural exceptional set

$$\mathsf{Z} = \mathsf{Z}_0 \cup \mathsf{Z}_1 \cup \mathsf{Z}_2 \cup \mathsf{Z}_3,$$

for a pair $(X, -K_X)$ constructed in Section 4.3.2. Since $-K_X$ is ample and $(X, -K_X)$ is adjoint rigid, our definition implies that $\mathsf{Z}_0 = \mathsf{Z}_3 = \emptyset$. Thus we will focus our attention on Z_1 and Z_2.

Subvarieties

First, let us describe the exceptional set Z_1 for a del Pezzo threefold X with $L = H$. We analyze curves on X:

Lemma 6.5.2 *Let X be a del Pezzo threefold defined over F. Then for any curve $C \subset X$, $a(C, -K_X|_C) \leq a(X, -K_X) = 1$.*

Proof We may assume that $F = \overline{F}$ so that we may replace L by H, with $-K_X \sim 2H$. First note that, since X has index 2, we have $a(X, H) = 2$. For any rational curve $C \subset X$, we have

$$a(C, H|_C) = \frac{2}{H.C} \leq 2.$$

Thus our assertion follows. □

Next, we consider surfaces $S \subset X$. To this end, we will discuss some results regarding surfaces with higher a-invariants:

Lemma 6.5.3 ([LT21, Lemma 4.5]) *Let S be a smooth geometrically uniruled and integral projective surface defined over F, and L be a big and nef Cartier divisor on S. Then $a(S, L)$ takes the form of $2/n$ or $3/n$ where n is a positive integer.*

Proof We may assume that $F = \overline{F}$. When (S, L) is not adjoint rigid, let $\pi\colon S \to B$ be the canonical map for $(S, a(S, L)L)$. Let C be a general fiber of π. By Lemma 6.4.3 (1), we have $a(S, L) = a(C, L|_C) = \frac{2}{L.C}$. Thus our assertion follows.

Assume that (S, L) is adjoint rigid. We apply a $(K_S + a(S, L)L)$-MMP and obtain a birational morphism $\phi\colon S \to S'$ such that $a(S, L)\phi_*L + K_{S'} \sim_{\mathbb{Q}} 0$. In particular, S' is a smooth weak del Pezzo surface. When the degree of S' is less than 9, S' admits a conic fibration $\pi \colon S' \to \mathbb{P}^1$. Let C be a general fiber of π. Then we have

$$0 = (a(S, L)\phi_* L + K_{S'}).C = a(S, L)\phi_* L.C - 2,$$

which implies $a(S, L) = \frac{2}{\phi_* L.C}$.

Finally, assume that S' has degree 9 so that S' is isomorphic to $\mathbb{P}^2$. Let $\ell \subset \mathbb{P}^2$ be a line. Then we have

$$0 = (a(S, L)\phi_* L + K_{S'}).\ell = a(S, L)\phi_* L.\ell - 3,$$

which implies $a(S, L) = \frac{3}{\phi_* L.\ell}$. Thus our assertion follows. □

Theorem 6.5.4 ([Hör10]) *Assume that $F = \overline{F}$. Let S be a smooth projective surface defined over F. Let L be a big and nef Cartier divisor on S. Assume that $a(S, L) > 1$. Then one of the following holds:*

(i) *We have $a(S, L) = 3$ and there is a birational morphism*

$$\phi \colon S \to \mathbb{P}^2$$

such that $\mathcal{O}(L) = \phi^* \mathcal{O}(1)$.

(ii) *We have$a(S, L) = 2$ and either*

(a) *(S, L) is adjoint rigid, in which case there is a birational morphism*

$$\phi \colon S \to Q$$

to a (possibly singular) quadric surface such that $\mathcal{O}(L) = \phi^* \mathcal{O}(1)$; or

(b) *(S, L) is not adjoint rigid, in which case S is ruled by L-lines;*

(iii) *We have $a(S, L) = \frac{3}{2}$ and there is a birational morphism*

$$\phi \colon S \to \mathbb{P}^2$$

such that $\mathcal{O}(L) = \phi^* \mathcal{O}(2)$.

Proof It follows from Lemma 6.5.3 that $a(S, L)$ is equal to 3, 2 or 3/2. When (S, L) is not adjoint rigid, the proof of Lemma 6.5.3 shows that $a(S, L)$ takes the form of $2/n$. So we conclude that $a(S, L) = 2$. Let $\pi \colon S \to B$ be the canonical map for $(S, a(S, L)L)$. Let C be a general fiber of π. Then we have

$$0 = (a(S, L)L + K_S).C = 2L.C - 2,$$

proving that C is an L-line.

Assume that (S, L) is adjoint rigid. We apply a $(K_S + a(S, L)L)$-MMP and obtain a birational morphism $\phi \colon S \to S'$ such that $a(S, L)\phi_* L + K_{S'} \sim_{\mathbb{Q}} 0$. This means that S' is a smooth weak del Pezzo surface. When $a(S, L) = 3$, $-K_{S'}$ is divisible by 3, so this shows that S' is isomorphic to $\mathbb{P}^2$ and $\mathcal{O}(\phi_* L) = \mathcal{O}(1)$. Let $E \subset S$ be a (-1)-rational curve that is contracted by ϕ as the first step. Then we must have

$$0 > (a(S, L)L + K_S).E = a(S, L)L.E - 1.$$

Since $a(S, L) > 1$, we must have $L.E = 0$. Thus if we denote the first step of the $(K_S + a(S, L)L)$-MMP by $\beta \colon X \to X_1$, then we have $L = \beta^* \beta_* L$. Inductively this implies that $L = \phi^* \phi_* L$. Thus our assertion follows.

When $a(S, L) = 2$, $-K_{S'}$ is divisible by 2, so S' is isomorphic to either a smooth quadric surface Q or the Hirzebruch surface $\mathbb{F}_2$. In the latter case, we replace S' by the anticanonical model so that S' is a (singular) quadric surface. Then, arguing as above, we have $\mathcal{O}(\phi_* L) = \mathcal{O}(1)$ and $L = \phi^* \phi_* L$. Thus our assertion follows.

Finally, when $a(S, L) = 3/2$, S' is isomorphic to $\mathbb{P}^2$, again we have

$$\mathcal{O}(\phi_* L) = \mathcal{O}(2),$$

and $L = \phi^* \phi_* L$. Thus our assertion follows. □

Lemma 6.5.5 ([LTT18, Section 6.3] and [ST22, Lemma 3.1]) *Let X be a smooth del Pezzo threefold defined over F with $L = -K_X$. Let $S \subset X$ be a surface. Then we have $a(S, L|_S) \leq a(X, L)$. Moreover, when $a(S, L|_S) = a(X, L)$, $(S, L|_S)$ is not adjoint rigid.*

Proof We may assume that F is algebraically closed so that we may replace L by H with $-K_X \sim 2H$. Since $a(X, H) = 2$, if $a(S, H|_S) > 2$, then it follows from Theorem 6.5.4 that we have $a(S, H|_S) = 3$. Let $\beta \colon S' \to S$ be the normalization and $\alpha \colon \widetilde{S} \to S'$ be the minimal resolution. We apply a $(K_{\widetilde{S}} + 3\alpha^* \beta^* H)$-MMP, and it follows from Theorem 6.5.4 (i) that we obtain a birational morphism $\phi \colon \widetilde{S} \to \mathbb{P}^2$ such that $\mathcal{O}(\alpha^* \beta^* H) = \phi^* \mathcal{O}(1)$. Since every curve contracted by α is contracted by ϕ, we conclude that a rational map $\phi \circ \alpha^{-1} \colon S' \dashrightarrow \mathbb{P}^2$ is an honest birational morphism. Then since $\beta^* H$ is ample on S', we conclude that S' is isomorphic to $\mathbb{P}^2$. This implies that $(H|_S)^2 = (\beta^* H)^2 = 1$. This is possible only when $H^3 = 1$ and $S \in |H|$. However, it follows from [LTT18, Lemma 6.9] that any surface S in $|H|$ is normal, so we conclude that S itself is isomorphic to $\mathbb{P}^2$. By adjunction we have $-K_S \sim -K_X - H|_S \sim H|_S$. This means that $a(S, H) = 2$, a contradiction. Hence we conclude that $a(S, H|_S) \leq 2$.

Next, assume that $a(S, H|_S) = 2$ and (S, H) is adjoint rigid. Again let

$$\beta \colon S' \to S,$$

be the normalization and $\alpha \colon \widetilde{S} \to S'$ be the minimal resolution. We apply a $(K_{\widetilde{S}} + 2\alpha^* \beta^* H)$-MMP, and it follows from Theorem 6.5.4 (ii) that we obtain a birational morphism $\phi\colon \widetilde{S} \to Q$ such that $\mathcal{O}(\alpha^*\beta^*H) = \phi^*\mathcal{O}(1)$, where Q is a (possibly singular) quadric surface. Using the arguments above, one can show that a rational map from S' to Q is an isomorphism. This implies that $(H|_S)^2 = (\beta^*H)^2 = 2$. This is only possible when $H^3 = 1$ and $S \in |2H|$ or $H^3 = 2$ and $S \in |H|$. When $H^3 = 2$ and $S \in |H|$, it follows from [LTT18, Lemma 6.6] that S is normal, so that S itself is isomorphic to Q. In particular, S comes with only canonical singularities. By adjunction formula, we have $-K_S \sim H|_S$. This means that $(-K_S)^2 = (H|_S)^2 = 2$. However, $(-K_Q)^2 = 8$. Thus this is a contradiction. When $H^3 = 1$ and $S \in |2H|$, it follows from the discussion in the proof of [BLRT22, Theorem 4.1] that any integral member $S \in |2H|$ is normal. Thus S is isomorphic to Q. Then, by the adjunction formula, we have $-K_S \sim -K_X - 2H|_S \sim 0$. However, this contradicts the fact that S is rational. Thus our assertion follows. □

Let X be a geometrically integral del Pezzo threefold defined over F. Let $F_1(X)$ be the Hilbert scheme of $-K_X$-conics on X. We denote its universal family by $\pi\colon \mathcal{C} \to F_1(X)$ with the evaluation map $s\colon \mathcal{C} \to X$. It follows from [Isk79, Remark 1.5, Proposition 1.6, and Remark 1.7] that $F_1(X)$ is geometrically irreducible and 2-dimensional, and the evaluation map $s\colon \mathcal{C} \to X$ is dominant when we have $H^3 \geq 2$. The same thing is true when $H^3 = 1$ assuming X is general in its moduli, by [Tik81]. It follows from Theorem 5.1.1 that $s(\mathcal{C}(F))$ is a thin set. Then we have the following proposition:

Proposition 6.5.6 *Let F be a field of characteristic 0 and X be a geometrically integral smooth del Pezzo threefold defined over F, with $L = -K_X$ as our polarization. Then our exceptional set Z_1 is equal to $s(\mathcal{C}(F))$.*

Proof It follows from Lemmas 6.5.2 and 6.5.5 that, geometrically, only adjoint rigid subvarieties with the same a-invariant as X are $-K_X$-conics. Thus our assertion follows. □

Remark 6.5.7 When X is a complete intersection of two quadrics, $F_1(X)$ is geometrically isomorphic to an abelian surface. In particular, it is possible that $s(\mathcal{C}(F))$ is Zariski dense in X. (See [BL17, Example 3.6] for an explicit example.) On the other hand, when X is a cubic threefold defined over a number field F, $F_1(X)$ is a surface of general type embedded into its Albanese variety. In particular, by Faltings' theorem, $F_1(X)(F)$ is not Zariski dense. Thus $s(\mathcal{C}(F))$ is not Zariski dense on X.

Adjoint Rigid a-covers

In this section, we continue to let F be a field of characteristic 0 and X be a smooth geometrically integral del Pezzo threefold defined over F. We will discuss our exceptional set Z_2 for $(X, -K_X)$. To this end, we have the following proposition:

Proposition 6.5.8 *Let F be a field of characteristic 0 and X be a geometrically integral smooth del Pezzo threefold defined over F, with $L = -K_X$ as our polarization. Then we have $\mathsf{Z}_2 = \emptyset$.*

Before we can prove this, we need the following lemma:

Lemma 6.5.9 *Assume that $\overline{F} = F$. Let $f\colon X \to Y$ be a K_X-negative extremal divisorial contraction between terminal $\mathbb{Q}$-factorial projective threefolds. Let $E \subset X$ be the exceptional divisor of f. Then there exists a family of rational curves C covering E such that*

$$0 < -K_X.C \leq 2.$$

This lemma is stated in [LT17, Proposition 7.2], and a correct proof of this lemma is given by Chen Jiang (unpublished). Here we closely follow Jiang's arguments.

Proof We denote the normalization of E by $\nu\colon E^\nu \to E$. It follows from [Kol13, Proposition 4.5] that there exists an effective $\mathbb{Q}$-divisor Δ on E^ν such that

$$\nu^*((K_X + E)|_E) = K_{E^\nu} + \Delta.$$

Let $\beta\colon S \to E^\nu$ be the minimal resolution. Then there exists an effective $\mathbb{Q}$-divisor Δ_S such that

$$K_S + \Delta_S = \beta^*(K_{E^\nu} + \Delta).$$

Since $-K_X$ and $-E$ are relatively ample over Y, $-(K_S + \Delta)$ is relatively big and nef over $f(E)$. This implies that K_S is not pseudo-effective over $f(E)$. In particular, the outcome of a relative K_S-MMP will be either $\mathbb{P}^2$ or a ruled surface over a smooth curve which maps to $f(E)$. Moreover, in the case of $\mathbb{P}^2$, we may assume that S itself is isomorphic to $\mathbb{P}^2$; otherwise S admits a birational morphism to the blowup of $\mathbb{P}^2$ at a point that admits a conic fibration over $\mathbb{P}^1$.

The upshot is that we need to consider the following two cases:

(i) $S \cong \mathbb{P}^2$; or
(ii) S admits a birational morphism to a smooth ruled surface over a smooth projective curve B that maps to $f(E)$.

In case (ii), let C' be a general fiber of the fibration $S \to B$, and C be the image of C' on E. Then we have

$$0 < -K_X.C \leq -(K_X + E).C = -(K_S + \Delta_S).C' \leq -K_S.C' = 2.$$

Thus our assertion follows.

Next, we consider case (i). In this case, $S = E^\nu \cong \mathbb{P}^2$ and $f(E)$ is a point. As terminal singularities are smooth in codimension 2 by [KM98, Corollary 5.18], singularities on X are isolated. This implies that $K_X + E$ is Cartier outside of finitely many points and therefore we conclude that $\nu^*((K_X + E)|_E)$ is Cartier because E^ν is smooth. Since E^ν is isomorphic to $\mathbb{P}^2$, there exists a family of rational curves ℓ covering E^ν such that $-K_{E^\nu}.\ell = 3$. Let C be the image of ℓ on E.

If $\Delta \neq 0$, then we have

$$\nu^*((K_X + E)|_E).\ell > K_{E^\nu}.\ell = -3.$$

Since $\nu^*((K_X + E)|_E)$ is Cartier, this implies that $-\nu^*((K_X + E)|_E).\ell \leq 2$. Then we have

$$0 < -K_X.C \leq -(K_X + E).C = -\nu^*((K_X + E)|_E).\ell \leq 2.$$

Thus our assertion follows.

When $\Delta = 0$, it follows from the inversion of adjunction ([Kol13, Theorem 4.9]) that (X, E) is purely log terminal. (For the definition of purely log terminal singularities, see [KM98, Definition 2.34].) Then by [KM98, Proposition 5.51], E is normal, so E is isomorphic to $\mathbb{P}^2$. In particular, C avoids $\mathrm{Sing}(X) \cap E$, so $E.C$ is an integer and $E.C \geq -1$. This implies that

$$0 < -K_X.C \leq -(K_X + E).C - 1 = -K_E.C - 1 = 2.$$

Hence the claim. □

With this lemma, we have the following proposition:

Proposition 6.5.10 *Let F be a field of characteristic 0 and X be a smooth geometrically integral del Pezzo threefold defined over F, with anticanonical polarization $L = -K_X$. Then there is no adjoint rigid a-cover $f: Y \to X$ from a smooth projective geometrically integral threefold Y.*

Proof We may assume that F is an algebraically closed field of characteristic 0. Suppose that we have the following commutative diagram

$$\begin{array}{ccc} Y & \xrightarrow{f} & X \\ {\scriptstyle \phi}\,\vdots & & \| \\ Y' & \overset{f'}{\dashrightarrow} & X \end{array}$$

such that

- (Y, f^*L) is an adjoint rigid a-cover of (X, L);
- f' is a generically finite contraction map, that is, if we denote by $\alpha\colon W \to Y'$ a birational morphism such that $\beta = f' \circ \alpha\colon W \dashrightarrow X$ is a morphism, then any prime divisor contracted by α will be contracted by β;
- Y' has terminal singularities and ϕ is a birational contraction as the composition of a sequence of steps in a $(K_Y - f^*K_X)$-MMP.

We claim that the ramification divisor $R = K_Y - f^*K_X$ will be contracted by a suitable choice of ϕ so that if we take a common resolution $\alpha\colon W \to Y'$ with a morphism $\beta\colon W \to X$ and the finite part $\widetilde{f}\colon \widetilde{Y} \to X$ of the Stein factorization of $\beta\colon W \to X$, then $\widetilde{f}$ is étale in codimension 1. In particular, since X is smooth and simply connected, $\widetilde{f}$ does not exist unless it is birational.

Let r be the number of components of R. We prove our claim by induction on $r - \rho(Y/Y')$. When $r - \rho(Y/Y') = 0$, our assertion is clear. Assume that $r - \rho(Y/Y') > 0$. We further continue a $(K_{Y'} - \phi_* f^*K_X)$-MMP on Y'. Since any $(K_{Y'} - \phi_* f^*K_X)$-flip does not change the validity of the assumptions above, we may assume that the next step is a divisorial contraction

$$\psi\colon Y' \to Y_1.$$

Let E be the exceptional divisor of ψ. Lemma 6.5.9 shows that there exists a covering family of rational curves $\Gamma \subset E$ such that

$$-K_{Y'}.\Gamma \leq 2.$$

Since we also have $(K_{Y'} - \phi_* f^*K_X).\Gamma < 0$ and $-\phi_* f^*K_X$ is movable, we conclude that

$$0 \leq -\phi_* f^*K_X.\Gamma < 2.$$

Let $\widetilde{\Gamma}$ be the strict transform of Γ on Y. By Lemma 3.2.16, there exists an effective ϕ-exceptional $\mathbb{Q}$-divisor D on Y such that $-f^*K_X + D = -\phi^*\phi_* f^*K_X$. This implies that

$$0 \leq -f^*K_X.\widetilde{\Gamma} \leq (-f^*K_X + D).\widetilde{\Gamma} = -\phi_* f^*K_X.\Gamma < 2.$$

Since X has index 2, we conclude that $-f^*K_X.\widetilde{\Gamma} = 0$. This means that f' contracts a divisor E, proving that $f_1 = f' \circ \psi^{-1}$ is a generically finite contraction. Thus our assertion follows from the induction hypothesis. □

Proof of Proposition 6.5.8 By the definition of Z_2 in Section 4.3.2, Z_2 is only formed by adjoint rigid a-covers of $(X, -K_X)$. Thus our assertion follows from Proposition 6.5.10. □

6.5.2 Smooth Prime Fano Threefolds of Index 1

We assume that our ground field F is an algebraically closed field of characteristic 0. In this section, we briefly review results for smooth prime Fano threefolds X of index 1 with $-K_X = H$. Recall from the classification in Theorem 6.5.1 that H^3 takes even values between 2 and 22, except 20. We assume that $H^3 \geq 4$ and $-K_X$ is very ample throughout this section.

We begin by summarizing results regarding low degree rational curves on such a threefold. Deformation theory of curves (see Section 7.2) tells us that a $-K_X$-line always deforms, at least in dimension 1. A precise analysis of the normal bundle of C shows that the Hilbert scheme $\Gamma(X)$ parametrizing $-K_X$-lines is 1-dimensional, and a family of $-K_X$-lines always sweeps out a surface. (See, e.g., [IP99, Proposition 4.2.2].) Deformation theory of curves further shows that a $-K_X$-conic always deforms, at least in dimension 2. One can show that the Hilbert scheme $F(X)$ of $-K_X$-conics, which is called the *Fano surface of conics*, is 2-dimensional and there exists a family of $-K_X$-conics for which the evaluation map is dominant. (See, e.g., [KPS18, Proposition 2.3.6] and [IM07, Section 3.1] for these claims.)

Our ground field F is a field of characteristic 0. We say a smooth Fano threefold X defined over F is a *smooth prime Fano threefold of geometric index* 1 if $\overline{X}$ is a smooth prime Fano threefold of index 1. From now on we fix such a threefold defined over F and consider its exceptional set

$$\mathsf{Z} = \mathsf{Z}_0 \cup \mathsf{Z}_1 \cup \mathsf{Z}_2 \cup \mathsf{Z}_3,$$

constructed in Section 4.3.2. First of all, we consider subvarieties of such a threefold. Let us consider curves on X:

Lemma 6.5.11 *Let F be a field of characteristic* 0 *and X be a smooth prime Fano threefold of geometric index* 1 *defined over F. Let $L = -K_X$ be the anticanonical class. Then for any integral curve $C \subset X$, the following statements hold:*

(i) $a(C, -K_X|_C) > a(X, -K_X)$ *if and only if $\overline{C}$ is the union of $-K_{\overline{X}}$-lines;*
(ii) $a(C, -K_X|_C) = a(X, -K_X)$ *if and only if $\overline{C}$ is the union of $-K_{\overline{X}}$-conics.*

We omit a proof of this lemma. Next, we consider surfaces on X:

Lemma 6.5.12 ([LTT18, Section 6.4] and [LT21, Theorem 5.1]) *Let F be a field of characteristic* 0 *and X be a smooth prime Fano threefold of geometric index* 1 *defined over F. Let $L = -K_X$ be the anticanonical class. Assume that $L^3 \geq 4$ and $-K_X$ is very ample. Then, for any surface $S \subset X$, the following statements hold:*

(i) *$a(S, -K_X|_S) > a(X, -K_X)$ if and only if $\overline{S}$ is covered by $-K_{\overline{X}}$-lines.*
(ii) *If $a(S, -K_X|_S) = a(X, -K_X)$ then $(S, -K_X|_S)$ is not adjoint rigid. Moreover, let $\beta\colon \widetilde{S} \to S$ be a smooth resolution and $\pi\colon \widetilde{S} \to B$ be the canonical map for $(\widetilde{S}, -\beta^* K_X|_S)$. Then any general geometric fiber of π is birational to a $-K_{\overline{X}}$-conic.*

Proof The proof of this lemma is similar to the proof of Lemma 6.5.5. For this reason, we omit it here, but refer the reader to relevant papers proving those claims. The first statement follows from the discussion of [LTT18, Section 6.4]. The second statement is proved in [LT21, Theorem 5.1]. □

Let $R(X) \subset X$ be the locus swept out by curves $C \subset X$ such that $\overline{C}$ is the union of $-K_{\overline{X}}$-lines. This is necessarily a 2-dimensional closed subset in X. Let $\pi\colon \mathcal{C} \to F(X)$ be the universal family of $-K_X$-conics with the evaluation map $s\colon \mathcal{C} \to X$. This map is dominant. With this notation, we have the following proposition:

Proposition 6.5.13 *Let F be a field of characteristic 0 and X be a geometrically integral smooth prime Fano threefold of geometric index 1 defined over F. Let $L = -K_X$ be the anticanonical class. Assume that $L^3 \geq 4$ and $-K_X$ is very ample. Let Z_1 be the conjectural exceptional set constructed in Section 4.3.2. Then we have*

$$\mathsf{Z}_1 = R(X)(F) \cup s(\mathcal{C}(F)).$$

Proof This follows from Lemmas 6.5.11 and 6.5.12. □

Next, we consider Z_2 for $(X, -K_X)$. To this end, we have the following proposition:

Proposition 6.5.14 ([LT17, Theorem 1.9]) *Let F be a field of characteristic 0 and X be a geometrically integral smooth prime Fano threefold of geometric index 1 defined over F. Let $L = -K_X$ be the anticanonical class. We assume that $\overline{X}$ is general in its moduli. Then there is no adjoint rigid a-cover $f\colon Y \to X$ from a smooth geometrically integral projective threefold Y.*

The proof of this proposition is similar to the proof of Proposition 6.5.10. For details, we refer interested readers to [LT17, Theorem 1.9]. Here is a generalization of this proposition:

Theorem 6.5.15 ([BLRT22, Theorem 5.5]) *Let F be a field of characteristic 0 and X be a geometrically integral smooth Fano threefold defined over F. Let $L = -K_X$ be the anticanonical class. Then there is no non-trivial adjoint rigid a-cover $f\colon Y \to X$ from a smooth geometrically integral projective threefold Y.*

As a corollary of Theorem 6.5.15, we have the following:

Corollary 6.5.16 *Let F be a field of characteristic 0 and X be a smooth prime Fano threefold of geometric index 1 defined over F. Let $L = -K_X$ be the anticanonical class. Let Z_2 be the conjectural exceptional set constructed in Section 4.3.2. Then we have $\mathsf{Z}_2 = \emptyset$.*

As a summary we have the following:

Corollary 6.5.17 *Let F be a field of characteristic 0 and X be a geometrically integral smooth prime Fano threefold of geometric index 1 defined over F. Let $L = -K_X$ be the anticanonical class. We assume that $L^3 \geq 4$ and $-K_X$ is very ample. Let $\mathsf{Z} = \mathsf{Z}_0 \cup \mathsf{Z}_1 \cup \mathsf{Z}_2 \cup \mathsf{Z}_3$ be the conjectural exceptional set constructed in Section 4.3.2. Then we have*

$$\mathsf{Z} = R(X)(F) \cup s(C(F)).$$

Proof Since $-K_X$ is ample and $(X, -K_X)$ is adjoint rigid, we have $\mathsf{Z}_0 = \mathsf{Z}_3 = \emptyset$. Thus our assertion follows from Propositions 6.5.13 and 6.5.16. □

6.6 Fano Hypersurfaces

In this section, we let F denote an arbitrary field of characteristic 0. Let $X \subset \mathbb{P}^{n+1}$ be a smooth Fano hypersurface of degree d defined over F with the hyperplane class H. In view of Example 4.2.15, we expect that when d is sufficiently small compared to n, our conjectural exceptional set

$$\mathsf{Z} = \mathsf{Z}_0 \cup \mathsf{Z}_1 \cup \mathsf{Z}_2 \cup \mathsf{Z}_3$$

is empty. In fact, we have the following stronger prediction:

Conjecture 6.6.1 Let F denote an arbitrary field of characteristic 0 and $X \subset \mathbb{P}^{n+1}$ be a smooth Fano hypersurface of degree d defined over F with the hyperplane class H. Suppose that either

(i) $d \leq n - 1$ and X is general; or
(ii) $d \leq (n + 1)/2$.

Then there is no proper subvariety $Y \subset X$ such that $a(Y, H|_Y) \geq a(X, H)$.

Assuming this conjecture, we have a result regarding exceptional sets:

Proposition 6.6.2 *Assume that Conjecture 6.6.1 is true. In the setting of Conjecture 6.6.1, our conjectural exceptional set $\mathsf{Z} = \mathsf{Z}_0 \cup \mathsf{Z}_1 \cup \mathsf{Z}_2 \cup \mathsf{Z}_3$ for (X, H) is the empty set.*

Proof Since H is ample and (X, H) is adjoint rigid, we have $\mathsf{Z}_0 \cup \mathsf{Z}_3 = \emptyset$. Also it follows from Conjecture 6.6.1 and Lemma 5.2.2 that $\mathsf{Z}_1 = \emptyset$. Thus we only need to disprove the existence of non-birational adjoint rigid a-covers $f \colon Y \to X$. To this end, it follows from Corollary 5.2.5 and Conjecture 6.6.1 that $f \colon Y \to X$ must be étale in codimension 1. However, since X is smooth and simply connected, this implies that f must be birational. A contradiction. Thus Z_2 also must be empty, proving the claim. □

In general, Conjecture 6.6.1 is widely open, but Hase-Liu proved the statement for the assumption that X is general and $5 \leq d \leq n/4 + 7/4$ using the circle method ([HL25]). The circle method also proves this conjecture for the assumption $n \geq (d-1)2^d$. In this section, we prove this conjecture for cubic and quartic hypersurfaces.

6.6.1 Cubic Hypersurfaces

First we record the following classification result on polarized projective varieties of higher a-invariants. This is a higher dimensional version of Theorem 6.5.4.

Theorem 6.6.3 ([Hör10, Proposition 1.3]) *Let F be an algebraically closed field of characteristic 0 and X be a smooth projective variety of dimension n defined over F equipped with a big and nef divisor L on X. Then we have the following statements:*

(i) *If $a(X, L) > n$, then $a(X, L) = n + 1$ and the pair (X, L) is birationally equivalent to $(\mathbb{P}^n, H)$, where H is the hyperplane class on $\mathbb{P}^n$.*
(ii) *If $a(X, L) = n$ and (X, L) is adjoint rigid, then (X, L) is birationally equivalent to (Q, H), where Q is a (possibly singular) quadric hypersurface and H is the hyperplane class on Q.*
(iii) *If $a(X, L) = n$ and (X, L) is not adjoint rigid, up to birational equivalence the canonical map $\pi \colon X \to C$ realizes X as a $\mathbb{P}^{n-1}$-bundle over C and $L = \mathcal{O}_\pi(1)$.*
(iv) *If $n - 1 < a(X, L) < n$, then (X, L) is birationally equivalent to $X' = \mathbb{P}_{\mathbb{P}^2}(\mathcal{O}(2) \oplus \mathcal{O}^{\oplus n-2})$ with the divisor class $\mathcal{O}_{X'/\mathbb{P}^2}(1)$. In this case, we have $a(X, L) = n - \frac{1}{2}$ and (X, L) is adjoint rigid.*

Here two polarized pairs (X, L) and (X', L') are birationally equivalent to each other if there are birational maps from a smooth projective variety W, $\phi_1 \colon W \to X$ and $\phi_2 \colon W \to X'$, such that $\phi_1^* L$ is linearly equivalent to $\phi_2^* L'$.

Using this result, we can prove Conjecture 6.6.1 for cubic hypersurfaces:

Example 6.6.4 ([LT19a, Example 4.8]) Let F be a field of characteristic 0 and $X \subset \mathbb{P}^{n+1}$ be a smooth cubic hypersurface defined over F. We would like to establish Conjecture 6.6.1 for (X, H), where H is the hyperplane class. This means that when either $n \geq 5$ or $n = 4$ and X is general, there is no proper subvariety $Y \subset X$ such that $a(Y, H|_Y) \geq a(X, H) = n - 1$. Without loss of generality, we may assume that F is algebraically closed.

When $\dim Y \leq n-3$, it follows from Proposition 6.1.1 that $a(Y, H|_Y) \leq n-2$. Thus we need to consider Y such that $\dim Y \geq n - 2$.

When $\dim Y = n - 2$, Proposition 6.1.1 implies that $a(Y, H|_Y) \leq n - 1$. If $a(Y, H|_Y) = n - 1$, then it follows from Theorem 6.6.3 that $H^{n-2}.Y = 1$, so Y is a linear subspace. If $n \geq 5$, then Y is a linear subspace of codimension less than $n/2$. However, a smooth hypersurface cannot contain such a linear space, a contradiction. When $n = 4$, Y is a linear subspace of dimension $n/2 = 2$. However, dimension count shows that a general X cannot contain such a plane. Thus our assertion follows in the case when $\dim Y = n - 2$.

Finally, when $\dim Y = n - 1$, if $a(Y, H|_Y) > n - 1$, then it follows from Theorem 6.6.3 that $H^{n-1}.Y = 1$. However, X cannot contain a linear subspace of codimension 1. Suppose that $a(Y, H|_Y) = n - 1$. If $(Y, H|_Y)$ is not adjoint rigid, then it follows from Lemma 5.1.3 that Y is covered by lower dimensional subvarieties of the same a-invariant. However, we have already excluded this possibility. Thus $(Y, H|_Y)$ must be adjoint rigid. It follows from Theorem 6.6.3 that $H^{n-1}.Y = 2$. On the other hand, the Lefschetz hyperplane theorem implies that $\mathrm{Pic}(X)$ is generated by H and we have $H^n = 3$. Thus $H^{n-1}.Y = 2$ is impossible. Thus Conjecture 6.6.1 holds for (X, H).

6.6.2 Quartic Hypersurfaces

First we will record the following proposition:

Proposition 6.6.5 ([LT19b, Theorem 3.15]) *We assume that our ground field F is an algebraically closed field of characteristic* 0. *Let X be a smooth uniruled projective variety of dimension n defined over F, and L be a big and basepoint-free divisor on X. Let H be a smooth and irreducible member of $|L|$. Then one of the following holds:*

(i) $a(X, L) \leq 1$;
(ii) *X is covered by rational curves C such that $L.C = 1$, and we have $a(X, L) = 2$ and $\kappa(K_X + a(X, L)L) = n - 1$; or*
(iii) $a(H, L|_H) = a(X, L) - 1$ *and* $\kappa(K_H + a(X, L|_H)L|_H) = \kappa(K_X + a(X, L)L)$.

Proof We assume that $a(X, L) > 1$. It follows from [EP12, Corollary D] that for a smooth and irreducible member $H \in |L|$, we have a surjection

$$H^0(X, m(K_X + a(X, L)L)) \to H^0(H, m(K_H + (a(X, L) - 1)L))$$

for sufficiently divisible m. Moreover, this map is also injective as $K_X + a(X, L)L$ is non-big, but H is big. This shows that

$$\kappa(K_H + a(X, L|_H)L|_H) = \kappa(K_X + a(X, L)L).$$

Suppose that $a(H, L|_H) < a(X, L) - 1$ so that $K_H + (a(X, L) - 1)L$ is big. Then the above equality implies that $\kappa(K_X + a(X, L)L) = n - 1$. Thus the canonical map $\pi\colon X \dashrightarrow Z$ for $(X, a(X, L)L)$ is a fibration with a general fiber being a curve. Let $\beta\colon X' \to X$ be a resolution so that $\pi \circ \beta$ is a morphism. Let C' be a general fiber of $\pi \circ \beta$, and C be the image of C' on X. Then we have $a(C', \beta^* L) = a(X, L)$. Also we have $a(C', \beta^* L) = 2/\beta^* L.C'$. Since we assume that $a(X, L) > 1$, we conclude that $\beta^* L.C' = L.C = 1$ and $a(X, L) = a(C', \beta^* L) = 2$. Thus case (ii) follows.

When $a(H, L|_H) = a(X, L) - 1$, we have already shown that

$$\kappa(K_H + a(X, L|_H)L|_H) = \kappa(K_X + a(X, L)L).$$

Hence the proof is complete. □

With this proposition, we are now ready to discuss Conjecture 6.6.1 for smooth quartic hypersurfaces:

Example 6.6.6 ([LT19a, Example 4.9]) Let F be a field of characteristic 0 and $X \subset \mathbb{P}^{n+1}$ be a smooth quartic hypersurface defined over F. Our goal here is to prove Conjecture 6.6.1 for (X, H) where again H is the hyperplane class. This means that when either $n \geq 7$ or $n \geq 5$ and X is general, there is no proper subvariety $Y \subset X$ such that $a(Y, H|_Y) \geq a(X, H) = n - 2$. Again, we may assume that F is algebraically closed.

When $\dim Y \leq n - 4$, it follows from Proposition 6.1.1 that $a(Y, H|_Y) \leq n - 3$. Thus we need to consider Y such that $\dim Y \geq n - 3$.

When $\dim Y = n - 3$, Proposition 6.1.1 implies that $a(Y, H|_Y) \leq n - 2$. If $a(Y, H|_Y) = n - 2$, then it follows from Theorem 6.6.3 that $H^{n-3}.Y = 1$, so Y is a linear subspace. If $n \geq 7$, then Y is a linear subspace of codimension less than $n/2$. However, a smooth hypersurface cannot contain such a linear space, a contradiction. When $n = 5$ or 6, Y is a linear subspace of dimension $\leq n/2$. However, dimension count shows that a general X cannot contain such a linear subspace. Thus our assertion follows in the case when $\dim Y = n - 3$.

When $\dim Y = n - 2$, if $a(Y, H|_Y) > n - 2$, then it follows from Theorem 6.6.3 that $H^{n-1}.Y = 1$. However, X cannot contain a linear subspace of codimension 2. Suppose that $a(Y, H|_Y) = n - 2$. If $(Y, H|_Y)$ is not adjoint rigid, then it follows from Lemma 5.1.3 that Y is covered by lower dimensional subvarieties of the same a-invariant. However, we have already excluded this possibility. Thus $(Y, H|_Y)$ must be adjoint rigid. It follows from Theorem 6.6.3 that $H^{n-2}.Y = 2$ and Y must be contained in a linear subspace $P \subset \mathbb{P}^{n+1}$ of codimension 2. This means that $X \cap P$ cannot be irreducible and it contains Y as an irreducible component. However, this contradicts the fact that X is smooth.

Finally, assume that $\dim Y = n - 1$. If $(Y, H|_Y)$ is not adjoint rigid, then it follows from Lemma 5.1.3 that Y is covered by lower dimensional subvarieties of the same a-invariant. This is impossible. Thus $(Y, H|_Y)$ must be adjoint rigid. It follows from Theorem 6.6.3 that $H^{n-1}.Y \leq 4$. By the Lefschetz hyperplane theorem, the only possibility is $H^{n-1}.Y = 4$ with Y a hyperplane class. Let $\beta\colon Y' \to Y$ be a resolution. Applying Proposition 6.6.5 to X and Y', it follows that, by cutting out by hyperplanes, the existence of such a Y implies that the existence of a smooth quartic threefold $\widetilde{X}$ with a hyperplane section $\widetilde{Y} \subset X$ such that $a(\widetilde{Y}, -K_{\widetilde{X}}|_{\widetilde{Y}}) = a(\widetilde{X}, -K_{\widetilde{X}}) = 1$ and $(\widetilde{Y}, -K_{\widetilde{X}}|_{\widetilde{Y}})$ is adjoint rigid. However, this contradicts Lemma 6.5.12. Thus our assertion follows.

7
Moduli of Rational Curves

So far we have been discussing Manin's conjecture over number fields. One can also develop Manin's conjecture over *global function fields*, that is, the function field $k(B)$ of a smooth geometrically integral projective curve B defined over a finite field $\mathbb{F}_q$. In 1988 in Berlin, Victor Batyrev gave an inspiring lecture ([Bat88]) in which he developed a heuristic argument to prove Manin's conjecture over global function fields based on certain assumptions on the moduli space of rational curves. These assumptions on the moduli space have been refined and formulated as geometric Manin's conjecture in [LT19a], [Tan21], and [LRT25c]. In this chapter, we introduce this conjecture for the space of rational curves, and then discuss its status.

Throughout this chapter we denote the base field by k and it is often either an algebraically closed field of characteristic 0 or a finite field $\mathbb{F}_q$.

7.1 The Space of Rational Curves

Let k be a field and X be a separated scheme of finite type over k. *A rational curve* on X is a non-trivial morphism

$$f\colon \mathbb{P}^1 \to X,$$

defined over k. Note that, by Lüroth's theorem, the image $f(C)$ is also rational, that is, it is birational to $\mathbb{P}^1$. There are two moduli spaces parametrizing such rational curves:

- the morphism scheme $\mathrm{Mor}(\mathbb{P}^1, X)$ parametrizing morphisms from $\mathbb{P}^1$ to X; or
- the moduli space $\overline{M}_{0,0}(X)$ of stable maps of genus 0.

In this section, we introduce both moduli spaces.

7.1.1 Morphism Scheme

Here we assume that k is an arbitrary field. Let Y be a projective scheme and X be a quasi-projective scheme, both defined over k. In this section, we will define the morphism scheme $\mathrm{Mor}(Y, X)$. One of the standard books on the construction of this scheme is [Kol96]; it includes a rigorous treatment using the Hilbert scheme. Another, more accessible book is [Deb01]. We will closely follow the exposition of [Deb01].

Morphism Scheme: General Cases

Here is the definition of the morphism scheme based on [Kol96, Chapter I, 1.9 Definition] and [Deb01, Section 2.2]:

Definition 7.1.1 Let k be a field. Let Y be a projective k-scheme and X be a quasi-projective k-scheme. *The morphism scheme* $\mathrm{Mor}(Y, X)$ parametrizing morphisms from Y to X is a separated k-scheme, which is locally of finite type over k, equipped with a $\mathrm{Mor}(Y, X)$-morphism

$$f^{\mathrm{univ}}: Y \times \mathrm{Mor}(Y, X) \to X \times \mathrm{Mor}(Y, X),\ (y, f') \mapsto (\mathrm{pr}_1 \circ f^{\mathrm{univ}}(y, f'), f'),$$

where $\mathrm{pr}_1 : X \times \mathrm{Mor}(Y, X) \to X$ is the first projection, such that

$$(\mathrm{Mor}(Y, X), f^{\mathrm{univ}})$$

satisfies the following universal property. Suppose that we have a k-scheme T with a T-morphism

$$f : Y \times T \to X \times T.$$

Then there exists a unique k-morphism $g : T \to \mathrm{Mor}(Y, X)$ such that

$$f = (\mathrm{pr}_1 \circ f^{\mathrm{univ}}(y, g(t)), t).$$

It is straightforward to prove from this definition that if $(\mathrm{Mor}(Y, X), f^{\mathrm{univ}})$ exists, then it is unique up to a unique isomorphism.

Let us explain the intuition behind this definition. Let k'/k be any field extension. Then the correspondence

$$\begin{array}{ccc} \mathrm{Mor}(Y, X)(k') & \longrightarrow & \{k'\text{-morphisms } f : Y \otimes_k k' \to X \otimes_k k'\} \\ \rotatebox{90}{$\in$} & & \rotatebox{90}{$\in$} \\ t & \longmapsto & f_t^{\mathrm{univ}} : Y \otimes_k k(t) \to X \otimes_k k(t) \end{array}$$

is a bijection by the universal property in Definition 7.1.1. Thus one may think of $\mathrm{Mor}(Y, X)$ as the moduli space of morphisms $f : Y \to X$.

Theorem 7.1.2 ([Gro95, 4(c)]) *Let k be a field. Let Y be a projective scheme and X be a quasi-projective scheme, both defined over k. Then the morphism scheme* $(\mathrm{Mor}(Y, X), f^{\mathrm{univ}})$ *exists.*

For a modern account of this theorem, readers might consult [Kol96, Chapter I, 1.10 Theorem].

In general a scheme $\mathrm{Mor}(Y, X)$ consists of countably many irreducible components, and one might want to obtain a scheme of finite type over k by fixing some invariants. This is done by *the Hilbert polynomial*: let L be an ample divisor on X. Then the Hilbert polynomial of a morphism $f: Y \to X$ is given by

$$P(m) = \chi(Y, O(mf^*L)).$$

Since the Hilbert polynomial is invariant under flat deformation, the Hilbert polynomial of a morphism $f_t : Y \otimes_k k(t) \to X \otimes_k k(t)$ $(t \in \mathrm{Mor}(Y, X))$ depends only on connected components of $\mathrm{Mor}(Y, X)$. Thus it is safe to consider the open and closed loci $\mathrm{Mor}^P(Y, X)$ of $\mathrm{Mor}(Y, X)$ parametrizing morphisms with the Hilbert polynomial P. With this notation, we have the following theorem:

Theorem 7.1.3 ([Gro95]) *Let k be a field. Let Y be a projective scheme over k, and X be a quasi-projective scheme over k. Suppose that P is a possible Hilbert polynomial. Then the scheme* $\mathrm{Mor}^P(Y, X)$ *is a quasi-projective scheme over k.*

Readers might consult [Kol96, Chapter I, 1.4 Theorem] and [Deb01, Section 2.2] for the proof of this theorem.

Morphism Scheme: Curves

Let k be a field and B be a smooth geometrically integral projective curve defined over k. Let X be a projective scheme defined over k. In this section, we consider the morphism scheme $\mathrm{Mor}(B, X)$.

Let α be an algebraic class (or a numerical class) of an effective 1-cycle on X. Since the algebraic class $f_*[B]$ is invariant under flat deformation, there is the open and closed locus $\mathrm{Mor}(B, X, \alpha)$ of $\mathrm{Mor}(B, X)$ parametrizing morphisms $f_t^{\mathrm{univ}} : B \otimes_k k(t) \to X \otimes_k k(t)$ such that $(f_t^{\mathrm{univ}})_*[B \otimes_k k(t)] = \alpha \otimes_k k(t)$.

Proposition 7.1.4 *Let k be a field and B be a smooth geometrically integral projective curve defined over k. Let X be a projective scheme defined over k with an algebraic class (or a numerical class) α of an effective 1-cycle on X. If* $\mathrm{Mor}(B, X, \alpha)$ *is non-empty, then the scheme* $\mathrm{Mor}(B, X, \alpha)$ *is a quasi-projective scheme over k.*

Proof Let L be an ample divisor on X. Then the Hilbert polynomial of $f_t^{\text{univ}}: B \otimes_k k(t) \to X \otimes_k k(t)$ with $t \in \text{Mor}(B, X, \alpha)$ is given by

$$P(m) = mL.\alpha + 1 - g(B),$$

where $g(B)$ is the genus of B. Thus our assertion follows from Theorem 7.1.3. □

Thus $\text{Mor}(B, X, \alpha)$ consists of finitely many components. [Deb01, Section 2.1] explains how to construct the scheme $\text{Mor}(\mathbb{P}^1, X, d)$ explicitly when $X \subset \mathbb{P}^N$ is a projective scheme over k. We recommend taking a look at [Deb01] to see how one can construct such a scheme in practice.

7.1.2 Moduli Space of Stable Maps

In this section, we assume that k is a field of characteristic 0. Our goal here is to introduce the moduli stack $\overline{\mathcal{M}}_{0,n}(X)$ of stable maps of genus 0 with n marked points and its coarse moduli space $\overline{M}_{0,n}(X)$. This space provides us with a projective compactification of the moduli space of rational curves, which is quite useful for degeneration arguments using Bend and Break. For foundational papers on the construction of this moduli space, see [Kon95], [BF97], [BM96], [Beh97], and [AV02]. For an expository paper, we recommend [FP97] and here we closely follow the exposition of this paper. The first key definition is the notion of stable curves of genus 0:

Definition 7.1.5 (Stable curves of genus 0) First let us assume that our base field k is an algebraically closed field of characteristic 0. *A tree of rational curves* defined over k is a reduced connected projective curve C defined over k such that (i) C is the union of copies of $\mathbb{P}^1_k$ such that every singular point of C is a node, (ii) there is no loop on the graph Γ of C where vertices of Γ correspond to irreducible components of C and each edge corresponds to a node that is the intersection of two components of the end points of the edge. The last condition is equivalent to saying that the arithmetic genus $h^1(C, \mathcal{O}_C)$ is 0.

Let C be a tree of rational curves defined over k. *Marked points on C* are distinct smooth points of C, $p_1, \ldots, p_n$. We call a tree C of rational curves with n marked points $p_1, \ldots, p_n$ *a prestable curve of genus 0 with n marked points*.

Let $(C, p_1, \ldots, p_n)$ be a prestable curve of genus 0 with n marked points. *A special point* is either a node of C or a marked point p_i. An automorphism of $(C, p_1, \ldots, p_n)$ is an automorphism $f: C \to C$ such that $f(p_i) = p_i$ for any i. A prestable curve $(C, p_1, \ldots, p_n)$ is called *a stable curve of genus 0 with n marked points* if any irreducible component of C contains at least three special points. This condition is equivalent to saying that the automorphism group $\mathrm{Aut}(C, p_1, \ldots, p_n)$ is finite.

From now on we assume that k is an arbitrary field of characteristic 0 and let S be a k-scheme. *A family of prestable curves of genus* 0 *with n marked points over* S is a flat projective morphism $\pi: C \to S$ with n disjoint sections $p_1, \ldots, p_n$ such that every geometric fiber of π is a prestable curve of genus 0. When every geometric fiber is a stable curve of genus 0, we say $(\pi: C \to S, p_1, \ldots, p_n)$ is *a family of stable curves of genus* 0 *over* S with n marked points.

Next, we would like to introduce the notion of a stable map of genus 0,

$$f: (C, p_1, \ldots, p_n) \to X,$$

for a projective scheme X defined over k. This provides a natural generalization of the notion of rational curves $f: \mathbb{P}^1 \to X$ on X:

Definition 7.1.6 (Stable maps of genus 0) Let k be a field of characteristic 0 and X be a projective scheme defined over k. *A prestable map of genus* 0 *with n marked points on* X is a pair $((C, p_1, \ldots, p_n), f: C \to X)$ where $(C, p_1, \ldots, p_n)$ is a prestable curve of genus 0 with n marked points over k, and $f: C \to X$ is a k-morphism.

Let $((C, p_1, \ldots, p_n), f: C \to X)$ be a prestable map of genus 0 with n marked points. An automorphism of $((C, p_1, \ldots, p_n), f: C \to X)$ is a k-automorphism $g: C \to C$ such that $g(p_i) = p_i$ for any i and $f \circ g = f$. We say a prestable map $((C, p_1, \ldots, p_n), f: C \to X)$ is *a stable map of genus* 0 *with n marked points on* X if every geometric irreducible component of C contracted by f contains at least three special points. This condition is equivalent to saying that the automorphism group of $((\overline{C}, p_1, \ldots, p_n), f: \overline{C} \to \overline{X})$ is finite. Note that every rational curve $f: \mathbb{P}^1 \to X$ is a stable map of genus 0 on X.

Let S be a k-scheme. *A family of stable maps of genus* 0 *with n marked points over* S is a pair $((\pi: C \to S, p_1, \ldots, p_n), f: C \to X)$ where

$$(\pi: C \to S, p_1, \ldots, p_n)$$

is a family of prestable curves of genus 0 over S, and $f: C \to X$ is a k-morphism such that every geometric fiber is a stable map of genus 0. We say two families of stable maps of genus 0 over S, $((\pi: C \to S, p_1, \ldots, p_n), f: C \to X)$ and $((\pi': C' \to S, p'_1, \ldots, p'_n), f': C' \to X)$, are isomorphic if there exists an S-isomorphism $g: C \cong C'$ such that we have $g \circ p_i = p'_i$ for any $i = 1, \ldots, n$ and $f' \circ g = f$.

With these definitions, we now come to the definition of the moduli functor:

Definition 7.1.7 (Moduli functors of stable maps of genus 0) Let k be a field of characteristic 0. Let X be a projective scheme defined over k with an algebraic (or numerical) class α of an effective 1-cycle on X. We define *the moduli functor of stable maps of genus* 0 *and class* α *with* n *marked points on* X, denoted by $\overline{\mathfrak{M}}_{0,n}(X,\alpha)$, to be a contravariant functor from the category of k-schemes to the category of sets such that for any k-scheme S, we define

$$\overline{\mathfrak{M}}_{0,n}(X,\alpha)(S) = \{((\pi\colon C \to S, p_1, \dots, p_n), f\colon C \to X)\}/\text{up to isomorphisms},$$

where $((\pi\colon C \to S, p_1, \dots, p_n), f\colon C \to X)$ runs over all families of stable maps of genus 0 with n marked points over S such that, for any $t \in S$, its geometric fiber satisfies $f_*[C_t \otimes_{k(t)} \overline{k(t)}] = \alpha \otimes_k \overline{k(t)}$, and we take the quotient of the inner set by the equivalence of being isomorphic. When we have a k-morphism $T \to S$, a map

$$\overline{\mathfrak{M}}_{0,n}(X,\alpha)(S) \to \overline{\mathfrak{M}}_{0,n}(X,\alpha)(T)$$

is given by the base change

$$((\pi\colon C \to S, p_i), f\colon C \to X) \mapsto ((\pi_T : C_T \to T, p_{i,T}), f_T : C_T \to X),$$

where $\pi_T : C_T \to T$ is the base change of $\pi\colon C \to S$ via $T \to S$,

$$p_{i,T} : T = S \times_S T \to C_T = C \times_S T$$

is the base change of $p_i : S \to C$, and f_T is the composition of the projection $C_T \to C$ followed by f.

One of major theorems in modern algebraic geometry is the following:

Theorem 7.1.8 *Let k be a field of characteristic* 0. *Let X be a projective scheme defined over k with an algebraic (or numerical) class α of an effective* 1*-cycle on X. Then the moduli functor $\overline{\mathfrak{M}}_{0,n}(X,\alpha)$ is represented by a proper Deligne–Mumford stack $\overline{\mathcal{M}}_{0,n}(X,\alpha)$ over k.*

For a modern account of this theorem, see [AV02]. We call this stack $\overline{\mathcal{M}}_{0,n}(X,\alpha)$ *the moduli stack of stable maps of genus* 0 *and class* α *with* n *marked points on* X. Unfortunately we are not able to provide a definition of Deligne–Mumford stacks in this book; we refer readers to a beautiful book by Olsson [Ols16]. It turns out that there is a weaker notion of the moduli functor that is represented by a projective scheme. To this end, we introduce the following definition:

Definition 7.1.9 (Coarse moduli spaces) Let k be a field of characteristic 0. Let X be a projective scheme defined over k with an algebraic (or numerical) class α of an effective 1-cycle on X. Let M be a k-scheme. We view any

k-scheme as a contravariant functor from the category of k-schemes to the category of sets by

$$S \mapsto \mathrm{Hom}_k(S, M), \quad \phi\colon T \to S \mapsto \mathrm{Hom}_k(S, M) \to \mathrm{Hom}_k(T, M),$$

where $\mathrm{Hom}_k(S, M)$ is the set of k-morphisms $S \to M$ and the map

$$\mathrm{Hom}_k(S, M) \to \mathrm{Hom}_k(T, M)$$

is given by the composition $g \mapsto g\circ\phi$. This identification is justified by Yoneda's lemma. We say a k-scheme M equipped with a natural transformation

$$\Phi\colon \overline{\mathfrak{M}}_{0,n}(X, \alpha) \to M$$

is *the coarse moduli space of stable maps of genus* 0 *and class* α *with n marked points on* X if the following statements are satisfied:

(i) the map

$$\Phi(\mathrm{Spec}\,\overline{k})\colon \overline{\mathfrak{M}}_{0,n}(X, \alpha)(\overline{k}) \to M(\overline{k})$$

is bijective; and

(ii) for any k-scheme Z with a natural transformation

$$F\colon \overline{\mathfrak{M}}_{0,n}(X, \alpha) \to Z,$$

there exists a unique k-morphism $f\colon M \to Z$ such that $F = f \circ \Phi$.

Our moduli functor $\overline{\mathfrak{M}}_{0,n}(X, \alpha)$ admits a projective coarse moduli space:

Theorem 7.1.10 *Let k be a field of characteristic* 0. *Let X be a projective scheme defined over k with an algebraic (or numerical) class α of an effective* 1*-cycle on X. Then the coarse moduli space of $\overline{\mathfrak{M}}_{0,n}(X, \alpha)$ exists as a projective scheme $\overline{M}_{0,n}(X, \alpha)$ over k.*

For a proof, see [AV02]. In this book, we mainly work with this coarse moduli space $\overline{M}_{0,n}(X, \alpha)$. This coarse moduli space admits a (possibly empty) Zariski open set $M_{0,n}(X, \alpha) \subset \overline{M}_{0,n}(X, \alpha)$ whose geometric points correspond to rational curves $f\colon \mathbb{P}^1 \to X$ on X. Also we have various forgetful morphisms by forgetting a marked point

$$\overline{M}_{0,n}(X, \alpha) \to \overline{M}_{0,n-1}(X, \alpha).$$

In particular, one may view

$$\overline{M}_{0,1}(X, \alpha) \to \overline{M}_{0,0}(X, \alpha)$$

as the universal family over $\overline{M}_{0,0}(X, \alpha)$. Note that forgetting a marked point may destabilize the stable map, and there is the stabilization procedure that involves contracting destabilizing components, that is, contracted components with less than three special points. Finally, we have various evaluation maps

$$\mathrm{ev}_i \colon \overline{M}_{0,n}(X, \alpha) \to X, \quad ((C, p_1, \dots, p_n), f \colon C \to X) \mapsto f(p_i).$$

We frequently use these morphisms without further references throughout this book.

7.1.3 The Space of Sections

Let k be a field and B be a smooth geometrically connected projective curve defined over k. We denote its function field $k(B)$ by K. Suppose that we have a geometrically integral projective variety X defined over K. Algebraic and arithmetic geometers are interested in studying the set of K-rational points:

$$X(K).$$

To this end, we consider the following notion, which enables us to translate arithmetic statements about rational points to geometric statements about curves on varieties:

Definition 7.1.11 Let k be a field and B be a smooth geometrically connected projective curve defined over k. We denote B's function field $k(B)$ by K. Suppose that we have a geometrically integral projective variety X defined over K.

An *integral model of* X is a flat projective morphism $\pi \colon \mathcal{X} \to B$ from a geometrically integral projective variety $\mathcal{X}$ such that its generic fiber is isomorphic to X over K. Since we can embed X into $\mathbb{P}^n_K$, such a model always exists by taking the Zariski closure in $\mathbb{P}^n_B$.

Let $\pi \colon \mathcal{X} \to B$ be an integral model of X. Then by the valuative criterion, there is a natural one-to-one correspondence between $X(K)$ and the set of sections $s \colon B \to \mathcal{X}$

$$\mathrm{Sec}(\mathcal{X}/B).$$

Moreover, one can endow this set with a scheme structure by realizing it as a Zariski open subset of the Hilbert scheme $\mathrm{Hilb}(\mathcal{X})$. We call this scheme the *space of sections*.

Let α be a class of 1-cycles such that $\alpha.\mathcal{X}_b = 1$ where $\mathcal{X}_b$ is a fiber of π. Then we consider the space of sections of class α,

$$\mathrm{Sec}(\mathcal{X}/B, \alpha),$$

parametrizing sections whose classes coincide with α. This also becomes a quasi-projective scheme. Finally, one should note that the space of sections $\mathrm{Sec}(\mathcal{X}/B, \alpha)$ can also be seen as a Zariski open subset in the moduli stack $\overline{\mathcal{M}}_{0,0}(\mathcal{X}, \alpha)$ when B has genus 0.

Sections for Trivial Families

Let k be a field and B be a smooth geometrically connected projective curve defined over k, as before. When a geometrically integral projective variety X is defined over k, the set of rational points $X(K)$ corresponds to the set of K-rational points on the base change $X \otimes_k K$, which corresponds to the space of sections of the trivial family

$$\pi \colon \mathcal{X} = B \times X \to B,$$

where π is the first projection. Note that giving a section $s \colon B \to \mathcal{X}$ is equivalent to giving a morphism $f \colon B \to X$. In this way, there is a natural identification between the following two schemes:

$$\mathrm{Sec}(\mathcal{X}/B), \quad \mathrm{Mor}(B, X).$$

Thus studying $\mathrm{Mor}(B, X)(k)$ is equivalent to studying the set of K-rational points $X(K)$. In view of this, many statements in arithmetic geometry have analogous counterparts in the study of the space of curves on a projective variety X.

7.2 Deformation Theory of Rational Curves

In this section, we discuss deformation theory for morphism schemes and moduli stacks of stable maps of genus 0.

7.2.1 Deformation Theory for Morphism Schemes

Here we discuss deformation theory for the morphism scheme $\mathrm{Mor}(Y, X)$ when Y is a projective scheme and X is a quasi-projective scheme, both defined over a field k. We closely follow the treatment of [Deb01, Section 2.2].

Suppose that we have a k-morphism $f \colon Y \to X$. Our first goal is to describe the Zariski tangent space

$$T_{\mathrm{Mor}(Y,X),[f]}.$$

By definition, this is the set of k-valued points

$$\mathrm{Mor}(\mathrm{Spec}\,(k[\epsilon]/(\epsilon^2)), \mathrm{Mor}(Y, X)(\mathrm{Spec}\, k).$$

By the universal property of the morphism scheme, each element of this set corresponds to a Spec $k[\epsilon]/(\epsilon^2)$-morphism

$$f_\epsilon \colon Y \times \operatorname{Spec}(k[\epsilon]/(\epsilon^2)) \to X \times \operatorname{Spec}(k[\epsilon]/(\epsilon^2)),$$

which should be considered as an infinitesimal deformation of f. Using this description, one can prove the following proposition:

Proposition 7.2.1 *Let k be a field. Suppose that Y is a projective scheme and X is a quasi-projective scheme, both defined over k. Let $f \colon Y \to X$ be a k-morphism such that X is smooth along the image $f(Y)$. Then there is a natural isomorphism of k-vector spaces:*

$$T_{\mathrm{Mor}(Y,X),[f]} \cong H^0(Y, f^*T_X).$$

We call f^*T_X the *restricted tangent bundle.*

Proof The following proof is taken from [Deb01, Proposition 2.4]. Since X is smooth along $f(Y)$, we need to show that

$$T_{\mathrm{Mor}(Y,X),[f]} \cong H^0(Y, \mathcal{H}om(f^*\Omega^1_X, O_Y)).$$

First, let us assume that X and Y are affine so that $X = \operatorname{Spec} A$ and $Y = \operatorname{Spec} B$. Our morphism $f \colon Y \to X$ corresponds to a ring homomorphism $f^* \colon A \to B$. Then our infinitesimal deformation

$$f_\epsilon \colon Y \times \operatorname{Spec}(k[\epsilon]/(\epsilon^2)) \to X \times \operatorname{Spec}(k[\epsilon]/(\epsilon^2))$$

corresponds to a $k[\epsilon]/(\epsilon^2)$-ring homomorphism $f_\epsilon^* \colon A[\epsilon] \to B[\epsilon]$, which is given as

$$f_\epsilon^*(a) = f^*(a) + \epsilon g(a).$$

Using the fact that $f_\epsilon^*(aa') = f_\epsilon^*(a) f_\epsilon^*(a')$, one can check that $g \colon A \to B$ is a k-derivation, which factors through as $A \to \Omega_A \to B$ where Ω_A is the module of differential forms of A over k. Thus such a g is parametrized by

$$\mathrm{Hom}_A(\Omega_A, B) \cong \mathrm{Hom}_B(\Omega_A \otimes_A B, B).$$

Now we come back to the situation when Y is a projective scheme and X is a quasi-projective scheme defined over k. We can take affine open coverings $Y = \cup_i V_i = \cup_i \operatorname{Spec} B_i$ and $X = \cup_i U_i = \cup_i \operatorname{Spec} A_i$ such that $f(V_i) \subset U_i$. Then an infinitesimal deformation

$$f_\epsilon \colon Y \times \operatorname{Spec}(k[\epsilon]/(\epsilon^2)) \to X \times \operatorname{Spec}(k[\epsilon]/(\epsilon^2))$$

gives us the data of $h_i \in \mathrm{Hom}_{B_i}(\Omega_{A_i} \otimes_{A_i} B_i, B_i)$ for each i using the discussion above. Having such a global infinitesimal deformation is equivalent to having gluing conditions

$$h_i|_{V_i \cap V_j} = h_j|_{V_i \cap V_j}.$$

This data satisfying gluing conditions corresponds to a global section of $\mathcal{H}om(f^*\Omega^1_X, \mathcal{O}_Y)$. Thus our assertion follows. □

The following result gives us a lower bound for the dimension of the morphism scheme, and it is quite useful in the study of moduli spaces of rational curves:

Theorem 7.2.2 *Let k be a field. Let X be a quasi-projective variety and Y be a projective variety, both defined over k. Suppose that we have a k-morphism $f: Y \to X$ such that X is smooth along $f(Y)$. Then in a neighborhood of $[f]$, $\mathrm{Mor}(Y, X)$ is defined by $h^1(Y, f^*T_X)$ equations in a nonsingular variety of dimension $h^0(Y, f^*T_X)$. In particular, any irreducible component of $\mathrm{Mor}(Y, X)$ containing $[f]$ has dimension at least*

$$h^0(Y, f^*T_X) - h^1(Y, f^*T_X).$$

This number is called the expected dimension of $\mathrm{Mor}(Y, X)$.

A proof is rather involved, so interested readers are recommended to consult [Deb01, Theorem 2.6]. As a corollary, we have the following conclusion:

Corollary 7.2.3 *Let k be a field. Let X be a quasi-projective variety and C be a geometrically integral smooth projective curve, both defined over k. Suppose that we have a k-morphism $f: C \to X$ such that X is smooth along $f(C)$. Then any irreducible component of $\mathrm{Mor}(C, X)$ containing $[f]$ has dimension at least*

$$h^0(C, f^*T_X) - h^1(C, f^*T_X) = \chi(f^*T_X) = -f^*K_X.C + \dim X(1 - g(C)),$$

where $g(C)$ is the genus of C. Moreover, when $h^1(C, f^*T_X) = 0$, $[f]$ is a smooth point of $\mathrm{Mor}(C, X)$ and there exists a unique component containing $[f]$ whose dimension is equal to the expected dimension.

Proof The first claim follows from Theorem 7.2.2 and the Riemann–Roch theorem. The second claim also follows from Theorem 7.2.2. □

Finally, we review the discussion in [Deb01, Subsection 2.9]. Assume that we fix a closed subscheme $Z \subset Y$ and a morphism $g: Z \to X$. Suppose that we have a morphism $f: Y \to X$ such that its restriction to Z is g. Note that we have the restriction morphism

$$\mathrm{Mor}(Y, X) \to \mathrm{Mor}(Z, X),$$

and we denote its fiber at g as $\mathrm{Mor}(Y, X; g)$. This is the morphism scheme parametrizing $f : Y \to X$ such that its restriction to Z is g. Assume that X is smooth along $f(Y)$. Then the Zariski tangent space of $\mathrm{Mor}(Y, X; g)$ at $[f]$ is given as the kernel of

$$H^0(Y, f^*T_X) \to H^0(Z, g^*T_X),$$

which is isomorphic to $H^0(Y, f^*T_X \otimes I_Z)$ where I_Z is the ideal sheaf of Z. Moreover, using a similar technique to prove Theorem 7.2.2, one can prove that, in a neighborhood of $[f]$, the morphism scheme $\mathrm{Mor}(Y, X; g)$ is defined by $h^1(Y, f^*T_X \otimes I_Z)$ equations in a non-singular variety of dimension

$$h^0(Y, f^*T_X \otimes I_Z).$$

In particular, any irreducible component of $\mathrm{Mor}(Y, X; g)$ containing f has dimension at least

$$h^0(Y, f^*T_X \otimes I_Z) - h^1(Y, f^*T_X \otimes I_Z).$$

7.2.2 Free and Very Free Rational Curves

In this section, we introduce one of the most important properties of rational curves. Throughout this section we assume that our ground field k is an algebraically closed field.

Definition 7.2.4 Let X be a smooth variety of dimension n defined over k. Suppose that we have a rational curve $f \colon \mathbb{P}^1 \to X$ over k. Since the tangent sheaf T_X on X is a locally free sheaf of rank n, its pullback f^*T_X is also a locally free sheaf of the same rank on $\mathbb{P}^1$. By Grothendieck's theorem (see, e.g., [GW20, Theorem 11.53]), one can write f^*T_X as

$$f^*T_X \cong \mathcal{O}(a_1) \oplus \cdots \oplus \mathcal{O}(a_n),$$

where the a_is are integers satisfying $a_1 \leq \cdots \leq a_n$.

We say $f \colon \mathbb{P}^1 \to X$ is *free* if we have $a_1 \geq 0$, and *very free* if we have $a_1 \geq 1$.

When a rational curve $f \colon \mathbb{P}^1 \to X$ is free, we have $H^1(\mathbb{P}^1, f^*T_X) = 0$. In particular, f is a smooth point of $\mathrm{Mor}(\mathbb{P}^1, X)$ and there exists a unique component containing f. Moreover, the dimension of such a component is given by the expected dimension. Below we record further properties of free and very free rational curves:

Lemma 7.2.5 *Let X be a smooth quasi-projective variety over k, and $\pi \colon C \to W$ be a family of rational curves defined over k, that is, π is a flat and projective*

k-morphism with connected fibers of relative dimension 1 such that all geometric fibers are isomorphic to $\mathbb{P}^1$ and it comes with $\mathrm{ev}\colon C \to X$. *(Note that here we do not assume that C is the product of W and $\mathbb{P}^1$. In particular, C may not be induced by $W \to \mathrm{Mor}(\mathbb{P}^1, X)$.) Suppose that* ev *is dominant and separable. Then a general geometric member of π is a free rational curve.*

Proof We may assume that $\bar{k} = k$. We consider $C \times_W C \times_W C$ as the space of rational curves parametrized by W with three ordered points on them. We denote by $\widetilde{W}^\circ \subset C \times_W C \times_W C$ the Zariski open subset parametrizing rational curves with three distinct points. Then note that the morphism $\widetilde{W}^\circ \to W$ is smooth. We consider the base change of $\pi\colon C \to W$ via $\widetilde{W}^\circ \to W$ and denote it by $\widetilde{\pi}\colon \widetilde{C} \to \widetilde{W}^\circ$ with the evaluation map $\widetilde{\mathrm{ev}}\colon \widetilde{C} \to C \to X$. Since $\widetilde{C} \to C$ is smooth, $\widetilde{\mathrm{ev}}$ is dominant and separable.

Now since $\widetilde{\pi}\colon \widetilde{C} \to \widetilde{W}^\circ$ is a family of rational curves with three distinct points on each curve, $\widetilde{C}$ can be identified with $\mathbb{P}^1 \times \widetilde{W}^\circ$. By the universal property of morphism schemes, there is a unique morphism $\phi\colon \widetilde{W}^\circ \to \mathrm{Mor}(\mathbb{P}^1, X)$ such that our family $\widetilde{\pi}\colon \widetilde{C} \to \widetilde{W}^\circ$ is the pullback of the universal family over $\mathrm{Mor}(\mathbb{P}^1, X)$.

Let $f\colon \mathbb{P}^1 \to X$ be a general member of rational curves parametrized by $\widetilde{\pi}$, and let $x \in \mathbb{P}^1$ be a general point. Then since $\widetilde{\mathrm{ev}}$ is dominant and separable, we have a surjection

$$T_{\widetilde{C},x} = T_{\mathbb{P}^1,x} \oplus T_{\widetilde{W}^\circ,\widetilde{\pi}(x)} \to T_{X,f(x)}.$$

Since our family $\widetilde{\pi}$ is induced by $g\colon \widetilde{W}^\circ \to \mathrm{Mor}(\mathbb{P}^1, X)$, we conclude that the homomorphism

$$T_{\mathbb{P}^1,x} \oplus T_{\mathrm{Mor}(\mathbb{P}^1,X),f} = T_{\mathbb{P}^1,x} \oplus H^0(\mathbb{P}^1, f^*T_X) \to T_{X,f(x)}$$

is surjective. Since $H^0(\mathbb{P}^1, f^*T_X)$ contains global sections coming from

$$H^0(\mathbb{P}^1, T_{\mathbb{P}^1}),$$

which generate $T_{\mathbb{P}^1,x}$, it follows that we have a surjection

$$H^0(\mathbb{P}^1, f^*T_X) \to T_{X,f(x)}.$$

Hence f must be free. Thus our assertion follows. □

In particular, one can say that in characteristic 0, if a rational curve passes through a general point of the underlying variety, then its general deformation is free. We frequently use this observation in the next chapter.

Lemma 7.2.6 *Let X be a smooth quasi-projective variety over k. Assume that we have a free rational curve $f\colon \mathbb{P}^1 \to X$. Let $M \subset \mathrm{Mor}(\mathbb{P}^1, X)$ be the unique component containing f. We denote the universal family over M by $\pi\colon C \to M$*

with the evaluation map $\mathrm{ev}\colon C \to X$. *Then* $\mathrm{ev}\colon C \to X$ *is dominant and separable. Moreover, for any point* $x \in \mathbb{P}^1$, *the evaluation map* $\mathrm{ev}\colon C \to X$ *is smooth at the point corresponding to* $(x, [f])$.

Proof The proof is almost backward to Lemma 7.2.5. Since f is free, the restricted tangent bundle f^*T_X is globally generated. In particular, for any closed point $x \in \mathbb{P}^1$, the evaluation map

$$H^0(\mathbb{P}^1, f^*T_X) \to T_{X,f(x)}$$

is surjective. Thus we have a surjection:

$$T_{C,(x,[f])} = T_{\mathbb{P}^1,x} \oplus T_{\mathrm{Mor}(\mathbb{P}^1,X),[f]} \to T_{X,f(x)}.$$

This implies our assertions. □

Corollary 7.2.7 *Let X be a smooth quasi-projective variety over k and let* $Z \subset X$ *be a closed subset whose codimension is greater than or equal to* 2. *Let* $M \subset \mathrm{Mor}(\mathbb{P}^1, X)$ *be a component generically parametrizing free rational curves. Then for a general geometric member* $f\colon \mathbb{P}^1 \to X$ *parametrized by M, the image* $f(\mathbb{P}^1)$ *does not meet Z.*

Proof Let $M^\circ \subset M$ be a Zariski open subset parametrizing free rational curves. Let $\pi\colon C^\circ \to M^\circ$ be the universal family. Then, by Lemma 7.2.6, the evaluation map $\mathrm{ev}\colon C^\circ \to X$ is flat. Thus $\mathrm{ev}^{-1}(Z)$ has codimension ≥ 2 in C°. This implies that $\pi|_{\mathrm{ev}^{-1}(Z)}\colon \mathrm{ev}^{-1}(Z) \to M^\circ$ is not dominant. Thus our assertion follows. □

Lemmas 7.2.5 and 7.2.6 prompt the following definition:

Definition 7.2.8 (Uniruledness) A projective variety X defined over k is *geometrically (separably) uniruled* if there is a family $C \to W$ of rational curves on $\overline{X}$ such that the evaluation map $\mathrm{ev}\colon C \to \overline{X}$ is dominant (and separable).

Corollary 7.2.9 *Let X be a smooth projective variety defined over k. Then X is geometrically separably uniruled if and only if there is a free rational curve* $f\colon \mathbb{P}^1 \to \overline{X}$.

Proof This follows from Lemmas 7.2.5 and 7.2.6. □

Next, we discuss a characterization of very freeness of rational curves.

Lemma 7.2.10 *Let X be a smooth quasi-projective variety over k, and* $\pi\colon C \to W$ *be a family of rational curves defined over k, that is,* π *is a flat and projective k-morphism with connected fibers of relative dimension* 1 *such that all geometric fibers are isomorphic to* $\mathbb{P}^1$*. Let* $\mathrm{ev}\colon C \to X$ *be the evaluation map such that the induced map*

$$\mathrm{ev}_2\colon C \times_W C \to X \times X,\ (x_1, x_2) \mapsto (\mathrm{ev}(x_1), \mathrm{ev}(x_2))$$

is dominant and separable. Then a general geometric member of π *is very free.*

Proof We may assume $\overline{k} = k$. Arguing as in Lemma 7.2.5, one can show that for a general member $f\colon \mathbb{P}^1 \to X$ of π and two general points $x, y \in \mathbb{P}^1$, we have a surjection

$$H^0(\mathbb{P}^1, f^*T_X) \to T_{X,f(x)} \oplus T_{X,f(y)}.$$

This is possible only when $f\colon \mathbb{P}^1 \to X$ is very free. □

In particular, it follows that, in characteristic 0, if a rational curve goes through two general points, then its general deformation is very free.

Lemma 7.2.11 *Let* X *be a smooth quasi-projective variety over* k*. Assume that we have a very free rational curve* $f\colon \mathbb{P}^1 \to X$*. Let* $M \subset \mathrm{Mor}(\mathbb{P}^1, X)$ *be a unique component containing* f*. We denote the universal family over* M *by* $\pi\colon C \to M$ *with the evaluation map* $\mathrm{ev}\colon C \to X$*. Then the induced map* $\mathrm{ev}_2\colon C \times_M C \to X \times X$ *is dominant and separable.*

Proof The proof is almost backward to Lemma 7.2.10. Since f is very free, for any closed points $p_1, p_2 \in \mathbb{P}^1$, the evaluation map

$$H^0(\mathbb{P}^1, f^*T_X) \to T_{X,f(p_1)} \oplus T_{X,f(p_2)},$$

is surjective. Thus we have a surjection:

$$T_{C\times_M C,(p_1,p_2,[f])} = T_{\mathbb{P}^1,p_1} \oplus T_{\mathbb{P}^1,p_2} \oplus T_{\mathrm{Mor}(\mathbb{P}^1,X),[f]} \to T_{X,f(p_1)} \oplus T_{X,f(p_2)}.$$

This implies our assertion. □

Again Lemmas 7.2.10 and 7.2.11 lead us to the following definition and corollary:

Definition 7.2.12 (Rational connectedness) Let X be a projective variety defined over k. We say X is *geometrically (and separably) rationally connected* if there is a family $\pi\colon C \to W$ of rational curves on $\overline{X}$ such that the evaluation map

$$(\mathrm{ev}_1, \mathrm{ev}_2)\colon C \times_W C \to \overline{X} \times \overline{X}$$

is dominant (and separable).

Corollary 7.2.13 *Let X be a smooth projective variety defined over k. Then X is geometrically separably rationally connected if and only if there is a very free rational curve $f : \mathbb{P}^1 \to \overline{X}$.*

Proof This follows from Lemmas 7.2.10 and 7.2.11. □

We record the following lemma regarding the immersion property of free rational curves:

Lemma 7.2.14 ([Kol96, Chapter II, 3.14 Theorem]) *Let X be a smooth quasi-projective variety of dimension n over k. Assume that we have a free rational curve $f : \mathbb{P}^1 \to X$. Let $M \subset \mathrm{Mor}(\mathbb{P}^1, X)$ be a unique component containing f. We denote the universal family over M by $\pi : C \to M$ with the evaluation map $\mathrm{ev} : C \to X$. We write*

$$f^*T_X \cong O(a_1) \oplus \cdots \oplus O(a_n),$$

where the a_is are integers satisfying $a_1 \leq \cdots \leq a_n$. Then

(i) *if $a_{n-1} \geq 1$, a general geometric deformation of f is an immersion $g : \mathbb{P}^1 \to \overline{X}$; and*
(ii) *if $a_{n-2} \geq 1$, a general geometric deformation of f is an embedding $g : \mathbb{P}^1 \to \overline{X}$.*

Proof The following proof is taken from [Kol96, Chapter II, 3.14 Theorem]. We will prove the first claim. The proof of the second claim is similar. Fix $p \in \mathbb{P}^1$ and let $x = f(p)$. We consider the morphism scheme

$$\mathrm{Mor}(\mathbb{P}^1, X; 2p \mapsto x)$$

parametrizing rational curves $g : \mathbb{P}^1 \to X$ that is ramified at p with the image x. Let $U \subset \mathrm{Mor}(\mathbb{P}^1, X; 2p \mapsto x)$ be the Zariski open subset containing $[f]$ such that for any $t \in U$, $h^1(\mathbb{P}^1_{k(t)}, f_t^*T_X(-2p)) \leq h^1(\mathbb{P}^1, f^*T_X(-2p))$ and $h^0(\mathbb{P}^1_{k(t)}, f_t^*T_X) \leq h^0(\mathbb{P}^1, f^*T_X)$ where $f_t : \mathbb{P}^1 \otimes k(t) \to X \otimes k(t)$ is the corresponding morphism.

By our assumption, we have

$$\begin{aligned}\dim U &\leq h^0(\mathbb{P}^1_{k(t)}, f_t^*T_X(-2p)) \leq h^0(\mathbb{P}^1_{k(t)}, f_t^*T_X) - 2n + h^1(\mathbb{P}^1_{k(t)}, f_t^*T_X(-2p)) \\ &\leq h^0(\mathbb{P}^1, f^*T_X) - 2n + h^1(\mathbb{P}^1, f^*T_X(-2p)) \leq h^0(\mathbb{P}^1, f^*T_X) - n - 2.\end{aligned}$$

Here we used the fact that $\#\{i \mid a_i = 0\} \leq n - 2$. Thus since picking p and x contributes $n + 1$ parameters, the dimension of the loci of M parametrizing ramified rational curves is at most

$$h^0(\mathbb{P}^1, f^*T_X) - n - 2 + n + 1 = h^0(\mathbb{P}^1, f^*T_X) - 1,$$

which is strictly less than

$$\dim M = h^0(\mathbb{P}^1, f^*T_X).$$

Thus our assertion follows. □

Next, we review other related notions and results:

Definition 7.2.15 (Rational chain connectedness) Let X be a projective variety defined over k. A *chain of rational curves* on X is a stable map of genus 0 on X such that the graph of the geometric components of the domain contains only two end points.

We say X is *geometrically rationally chain connected* if there is a flat and proper morphism $\pi\colon C \to W$ such that (i) all geometric fibers of π are chains of rational curves and (ii) the evaluation map

$$(\mathrm{ev}_1, \mathrm{ev}_2)\colon C \times_W C \to X \times X$$

is dominant.

Proposition 7.2.16 *Let k be an algebraically closed field of characteristic* 0. *Let X be a smooth projective variety defined over k. Then X is rationally chain connected if and only if it is rationally connected.*

It is straightforward to see that if X is rationally connected, then it is rationally chain connected. The converse is based on a delicate gluing argument. We omit a proof and refer interested readers to [Deb01, Corollary 4.28] or the original source [KMM92b, Section 2.1].

We should emphasize that the above proposition fails if X is not smooth:

Example 7.2.17 Let $S \subset \mathbb{P}^3$ be the cone over a smooth plane cubic. Then S is rationally chain connected since any two points can be connected by a union of two lines of the ruling. However, since S is a cone over an elliptic curve, it is not rationally connected.

Lastly the following lemma holds:

Lemma 7.2.18 *Let k be an algebraically closed field of characteristic* 0. *The following statements hold:*

(i) *Let X be a rationally chain connected projective variety defined over k. Then any two points on X can be connected by a chain of rational curves.*
(ii) *Let $\pi\colon X \to W$ be a flat and projective morphism from an irreducible variety X to a variety W defined over k. Assume that all fibers are irreducible. If the generic fiber is geometrically rationally chain connected, then all geometric fibers are rationally chain connected.*

Proof The following proof is based on [Kol96, Chapter IV 3.5 Corollary]. First, we prove the statement (i). Since X is rationally chain connected, one can find a family $\pi\colon C \to M$ of stable maps of genus 0 with a proper base M such that the evaluation map

$$C \times_M C \to X \times X$$

is dominant. Since $C \times_M C$ is proper, this evaluation map is surjective. This means that, for any two points on X, there exists a stable map $f\colon C \to X$ such that the image $f(C)$ contains two points. While C may not be a chain, C admits a subcurve that is a chain connecting the two points. Thus our assertion follows.

Next, we prove (ii). Here we consider the relative coarse moduli space of stable maps of genus 0, that is,

$$\overline{M}_{0,0}(\mathcal{X}/W) \to W,$$

which parametrizes stable maps $f\colon C \to \mathcal{X}$ that correspond to points on W. This scheme is proper over W and consists of countably many components. In particular, the image of the evaluation map

$$\overline{M}_{0,2}(\mathcal{X}/W) \to \mathcal{X} \times_W \mathcal{X}$$

is the union of countably many closed subsets in $\mathcal{X} \times_W \mathcal{X}$. Let the Z_is be such closed subsets. Then, by the upper semicontinuity of dimension, the set

$$W_i = \{w \in W \mid Z_i \cap (X_w \times X_w) = X_w \times X_w\}$$

is a closed subset of W where X_w is the fiber of π at w. (Note that here the irreducibility of X_w plays a role.) Since the countable union $\cup_i W_i$ contains the generic point of W, one of the W_is is equal to the entire set W. Thus our assertion follows. □

7.2.3 Deformation Theory for Moduli Spaces of Stable Maps

Let us briefly discuss deformation theory for stable maps of genus 0. The following discussion is based on [Tes09, Section 1.1].

Let k be a field of characteristic 0 and X be a smooth projective variety defined over k. Suppose that we have a stable map $f\colon C \to X$ of genus 0 defined over k.

Let $L_f^\bullet$ be the complex $f^*\Omega_X^1 \to \Omega_C^1$ where $f^*\Omega_X^1$ is in degree -1 and Ω_C^1 is in degree 0. We call this complex the *conormal complex of f* and we consider $L_f^\bullet$ as an object in the bounded derived category $\mathrm{D}^b(C)$ of coherent sheaves. (For readers who are not familiar with derived categories, we recommend consulting the first three chapters of [Huy06].) The stability condition guarantees that

$$\mathrm{Hom}(L_f^\bullet, \mathcal{O}_C) = 0.$$

Then the Zariski tangent space of the moduli stack $\overline{\mathcal{M}}_{0,0}(X)$ at $[f]$ is given by the hypercohomology

$$\mathbb{E}\mathrm{xt}^1(L_f^\bullet, \mathcal{O}_C),$$

and the obstruction space is given by the quotient of the hypercohomology

$$\mathbb{E}\mathrm{xt}^2(L_f^\bullet, \mathcal{O}_C).$$

Using this, one can show that we have the following inequality

$$\begin{aligned}\dim_{[f]} \overline{\mathcal{M}}_{0,0}(X) &\geq \dim\, \mathbb{E}\mathrm{xt}^1(L_f^\bullet, \mathcal{O}_C) - \dim\, \mathbb{E}\mathrm{xt}^2(L_f^\bullet, \mathcal{O}_C) \\ &= -f^*K_X.C + \dim\, X - 3.\end{aligned}$$

The right-hand side is called the *expected dimension*. Moreover, when

$$\mathbb{E}\mathrm{xt}^2(L_f^\bullet, \mathcal{O}_C)$$

is zero, the above inequality is an actual equality and the moduli stack $\overline{\mathcal{M}}_{0,0}(X)$ is smooth at $[f]$.

We denote the complexes

$$0 \to \Omega_C^1, \quad f^*\Omega_X^1 \to 0$$

by Ω_C^1 and $f^*\Omega_X^1[1]$ where the first term is in degree -1 and the second term is in degree 0. Then we have a natural distinguished triangle of the complexes

$$\Omega_C^1 \to L_f^\bullet \to f^*\Omega_X^1[1] \to \Omega_C^1[1].$$

Applying the derived Hom functor $R\mathrm{Hom}(\cdot, \mathcal{O}_C)$ to this triangle, we obtain the long exact sequence that can be identified with

$$\begin{aligned}0 &\to \mathrm{Hom}(\Omega_C^1, \mathcal{O}_C) \to H^0(C, f^*T_X) \to \mathbb{E}\mathrm{xt}^1(L_f^\bullet, \mathcal{O}_C) \\ &\to \mathrm{Ext}^1(\Omega_C^1, \mathcal{O}_C) \to H^1(C, f^*T_X) \to \mathbb{E}\mathrm{xt}^2(L_f^\bullet, \mathcal{O}_C) \to 0.\end{aligned}$$

In particular, if $H^1(C, f^*T_X) = 0$, then we have $\mathbb{E}\mathrm{xt}^2(L_f^\bullet, \mathcal{O}_C) = 0$.

When $\mathbb{E}\mathrm{xt}^2(L_f^\bullet, \mathcal{O}_C) = 0$, the moduli stack $\overline{\mathcal{M}}_{0,0}(X)$ is smooth at $[f]$. This implies that the coarse moduli space $\overline{M}_{0,0}(X)$ has only a quotient singularity at $[f]$. Moreover, when f has no nontrivial automorphism, the coarse moduli space $\overline{M}_{0,0}(X)$ is also smooth. Altogether we conclude:

Proposition 7.2.19 *Let k be a field of characteristic 0 and X be a smooth projective variety defined over k. Let $f\colon C \to X$ be a stable map of genus 0 defined over k.*

Suppose that we have $H^1(C, f^*T_X) = 0$. Then the coarse moduli space $\overline{M}_{0,0}(X)$ has only a quotient singularity at $[f]$. In particular, it is normal at $[f]$.

Moreover, when f has no nontrivial automorphism, $[f]$ is a smooth point of $\overline{M}_{0,0}(X)$.

Next, we assume that our stable map $f\colon C \to X$ admits the geometrically irreducible domain, that is, $\overline{C} \cong \mathbb{P}^1$. Note that we have a distinguished triangle

$$\Omega^1_C[-1] \to L^\bullet_f[-1] \to f^*\Omega^1_X \to \Omega^1_C.$$

We apply the derived dual and obtain the triangle

$$T_C \to f^*T_X \to (L^\bullet_f)^\vee[1] \to T_C[1].$$

Then since T_C is locally free, the map $T_C \to f^*T_X$ is injective. In particular, the complex $(L^\bullet_f)^\vee[1]$ is quasi-isomorphic to the *normal sheaf* N_f, which is the cokernel of $T_C \to f^*T_X$. Then we have the following identifications

$$\mathbb{E}\mathrm{xt}^1(L^\bullet_f, \mathcal{O}_C) \cong H^0(C, N_f), \quad \mathbb{E}\mathrm{xt}^2(L^\bullet_f, \mathcal{O}_C) \cong H^1(C, N_f).$$

In particular, we have

$$\dim_{[f]} \overline{\mathcal{M}}_{0,0}(X) \geq h^0(C, N_f) - h^1(C, N_f) = -f^*K_X.C + \dim X - 3.$$

Note that this expected dimension differs from the one in Corollary 7.2.3. A reason for this is that, for $\mathrm{Mor}(\mathbb{P}^1, X)$, composing a map $f\colon \mathbb{P}^1 \to X$ with an element of PGL_2 will produce a different map in general, but in the case of $\overline{\mathcal{M}}_{0,0}(X)$, those are not distinguished.

When $H^1(C, N_f) = 0$, $\overline{\mathcal{M}}_{0,0}(X)$ has expected dimension as its dimension at $[f]$, and moreover $[f]$ is a smooth point of the moduli stack.

Gluing Arguments

In this section, we discuss the gluing arguments described in [Tes05]. We assume that our ground field k is an algebraically closed field of characteristic 0.

Lemma 7.2.20 ([Tes05, Lemma 1.5]) *Let C be a prestable curve of genus 0 and $\mathcal{F}$ be a locally free sheaf on C. Suppose that $C = C_1 \cup C_2$ where the C_is are unions of components of C such that C_1 and C_2 have no common component. Let $\{p_1, \ldots, p_r\} = C_1 \cap C_2$ be the nodes on C lying on both C_1 and C_2. If $h^1(C_1, \mathcal{F}|_{C_1}(-p_1 - \cdots - p_r)) = 0$, then we have $H^1(C, \mathcal{F}) \cong H^1(C_2, \mathcal{F}|_{C_2})$.*

Proof We consider the following component sequence:

$$0 \to \mathcal{F}|_{C_1}(-p_1 - \cdots - p_r) \to \mathcal{F} \to \mathcal{F}|_{C_2} \to 0,$$

where the first map is extension by zero and the second map is restriction. Our assertion follows from the long exact sequence of this sequence. □

Proposition 7.2.21 ([Tes05, Corollary 1.6]) *Let C be a prestable curve of genus 0 and $\mathcal{F}$ be a locally free sheaf on C. Let $R \subset C$ be a connected union of irreducible components of C. Suppose that the restriction of $\mathcal{F}$ to each irreducible component of C not contained in R is globally generated. Then we have $h^1(C, \mathcal{F}) = h^1(R, \mathcal{F}|_R)$.*

Proof Let $C_1 \subset C$ be an irreducible component of C not contained in R such that C_1 is an edge of the tree of C, and let C' be the union of the remaining components. Then C_1 contains only one node p, and we have

$$\mathcal{F}|_{C_1} \cong \bigoplus_i \mathcal{O}(a_i),$$

with $a_i \geq 0$. This implies that $h^1(C_1, \mathcal{F}|_{C_1}(-p)) = 0$. Thus it follows from Lemma 7.2.20 that $h^1(C, \mathcal{F}) = h^1(C', \mathcal{F}|_{C'})$. Now our assertion follows by proof by induction on the number of components of C not contained in R. □

As a corollary, we have:

Corollary 7.2.22 *Let X be a smooth projective variety and $f : C \to X$ be a stable map of genus 0. Suppose that for every component $C' \subset C$, the restricted tangent bundle $f^*T_X|_{C'}$ is globally generated. Then we have $h^1(C, f^*T_X) = 0$.*

Proof This follows from Proposition 7.2.21 applied to $\mathcal{F} = f^*T_X$. □

Smoothing

Let C be the union of two $\mathbb{P}^1$s meeting at a point transversely. We denote two components by C_1 and C_2. Let X be a smooth projective variety and consider a stable map $f : C \to X$ such that $f|_{C_1} : C_1 \to X$ and $f|_{C_2} : C_2 \to X$ are free rational curves. It follows from Corollary 7.2.22 that we have $h^1(C, f^*T_X) = 0$. Thus by Proposition 7.2.19, the moduli stack $\overline{\mathcal{M}}_{0,0}(X)$ is smooth at $[f]$. The goal of this section is to prove the following proposition:

Proposition 7.2.23 *Let C be the union of two $\mathbb{P}^1$s meeting at a point transversely. We denote two components by C_1 and C_2. Let X be a smooth projective variety and consider a stable map $f : C \to X$ such that $f|_{C_1} : C_1 \to X$ and $f|_{C_2} : C_2 \to X$ are free rational curves.*

Then a sufficiently small open neighborhood of $[f]$ in $\overline{\mathcal{M}}_{0,0}(X)$ generically parametrizes free rational curves.

We call such a free rational curve a *smoothing* of $f: C \to X$.

Proof Recall that we have the following exact sequence:

$$\begin{aligned} 0 &\to \mathrm{Hom}(\Omega^1_C, \mathcal{O}_C) \to H^0(C, f^*T_X) \to \mathbb{E}\mathrm{xt}^1(L^\bullet_f, \mathcal{O}_C) \\ &\to \mathrm{Ext}^1(\Omega^1_C, \mathcal{O}_C) \to H^1(C, f^*T_X) \to \mathbb{E}\mathrm{xt}^2(L^\bullet_f, \mathcal{O}_C) \to 0. \end{aligned}$$

By Corollary 7.2.22, we have $H^1(C, f^*T_X) = 0$. Thus Lemma 7.2.24 below shows that $H^0(C, f^*T_X) \to \mathbb{E}\mathrm{xt}^1(L^\bullet_f, \mathcal{O}_C)$ is not surjective. This means that in a sufficiently small open neighborhood of $[f]$ in $\overline{M}_{0,0}(X)$, a general member $f': C' \to X$ is a stable map from $\mathbb{P}^1$.

Now let p be a smooth point of C. It follows from Corollary 7.2.22 that we have $H^1(C, f^*T_X(-p)) = 0$. Then note that by upper semicontinuity of h^1 [Har77, Chapter III Theorem 12.8], we have $H^1(C', f'^*T_X(-1)) = 0$. Thus our assertion follows. □

Lemma 7.2.24 *Let C be the union of two* $\mathbb{P}^1$*s meeting at a point transversely. Then* $\mathrm{Ext}^1(\Omega^1_C, \mathcal{O}_C) \cong k$.

Proof We may realize C as the plane conic defined by $xy = 0$. We denote each closed immersion $C_i \to C$ by ι_i where C_1 is defined by $x = 0$ and C_2 is defined by $y = 0$. Then we have the exact sequence:

$$0 \to k \to \Omega_C \to \iota_{1*}\Omega_{C_1} \oplus \iota_{2*}\Omega_{C_2} \to 0,$$

where k denotes the torsion sheaf supported at the origin $(0:0:1)$. Applying $\mathrm{Ext}^1(\bullet, \mathcal{O}_C)$ and using the fact that $\mathrm{Ext}^1(\Omega_{\mathbb{P}^1}, \mathcal{O}_{\mathbb{P}^1}) = 0$, we conclude that

$$\mathrm{Ext}^1(\Omega_C, \mathcal{O}_C) \cong \mathrm{Ext}^1(k, \mathcal{O}_C).$$

Now, we also have the exact sequence:

$$0 \to \mathfrak{m} \to \mathcal{O}_C \to k \to 0,$$

where $\mathfrak{m}$ is the maximal ideal sheaf of the origin. This implies that we have the exact sequence:

$$0 \to k = \mathrm{Hom}(\mathcal{O}_C, \mathcal{O}_C) \to \mathrm{Hom}(\mathfrak{m}, \mathcal{O}_C) \to \mathrm{Ext}^1(k, \mathcal{O}_C) \to \mathrm{Ext}^1(\mathcal{O}_C, \mathcal{O}_C).$$

Since we have $\mathrm{Ext}^1(\mathcal{O}_C, \mathcal{O}_C) = H^1(C, \mathcal{O}_C) = 0$, it suffices to show that $\mathrm{Hom}(\mathfrak{m}, \mathcal{O}_C)$ is 2-dimensional over k. To this end, we have the exact sequence:

$$0 \to \mathcal{O}_{C_1}(-2) \oplus \mathcal{O}_{C_2}(-2) \to \mathcal{O}_C(-1)^{\oplus 2} \to \mathfrak{m} \to 0,$$

where the first map is given by $(h, g) \mapsto (yh, xg)$ and the second map is given by $(h, g) \mapsto xh - yg$. This induces the exact sequence:

$$0 \to \mathrm{Hom}(\mathfrak{m}, \mathcal{O}_C) \to H^0(C, \mathcal{O}_C(1))^{\oplus 2} \to H^0(C_1, \mathcal{O}_{C_1}(2)) \oplus H^0(C_2, \mathcal{O}_{C_2}(2)).$$

Then the middle vector space is 6-dimensional and its image is 4-dimensional. Thus our assertion follows. □

7.2.4 Deformation Theory for Spaces of Sections

Let k be a field and B be a smooth geometrically connected projective curve of genus $g(B)$ defined over k. Let $\pi\colon \mathcal{X} \to B$ be a flat and projective morphism from a geometrically integral projective variety $\mathcal{X}$ with connected fibers. Let $s\colon B \to \mathcal{X}$ be a section such that the image is contained in the smooth locus of $\mathcal{X}$.

Under these assumptions, the normal sheaf N_s is a locally free sheaf of rank $\dim \mathcal{X} - 1$. Using the theory of the Hilbert scheme as in [Kol96, Chapter I, Section 2], one can conclude that the Zariski tangent space of $\mathrm{Sec}(\mathcal{X}/B)$ at $[s]$ is given by

$$H^0(B, N_s),$$

and its obstruction space is given by

$$H^1(B, N_s).$$

In particular, we have

$$\begin{aligned}\dim_{[s]} \mathrm{Sec}(\mathcal{X}/B) &\geq h^0(B, N_s) - h^1(B, N_s)\\ &= \deg(-s^*K_{\mathcal{X}/B}) + (\dim \mathcal{X} - 1)(1 - g(B)),\end{aligned}$$

where $-K_{\mathcal{X}/B} := -K_{\mathcal{X}} + \pi^*K_B$ is the *relative anticanonical divisor*. Moreover, this is an equality whenever $H^1(B, N_s) = 0$. In this situation one can conclude that $\mathrm{Sec}(\mathcal{X}/B)$ is smooth at $[s]$.

For any section $s\colon B \to \mathcal{X}$, $\deg(-s^*K_{\mathcal{X}/B})$ is called the *anticanonical height of a section s*.

7.3 Batyrev's Heuristics

During a lecture in 1988 in Berlin ([Bat88]), Victor Batyrev developed a heuristic argument to prove Manin's conjecture over finite fields, which led to the original formulation of Manin's conjecture in [FMT89] and [BM90]. This heuristic is described in [Tsc09, Section 4.7], [Bou11, Section 1.2], [Tan21], and [LRT25c, Section 4.1]. In this section, we discuss this heuristic argument by following the exposition of [Tan21] and [LRT25c, Section 4.1].

Let k be a finite field $\mathbb{F}_q$ of size q where $q = p^n$ with p a prime number. Let X be a smooth geometrically integral Fano variety defined over k. Let $L = -K_X$

be the ample anticanonical divisor on X. For any integral numerical curve class $\alpha \in \mathrm{Eff}_1(X) \cap N_1(X)_{\mathbb{Z}}$ and a non-empty Zariski open subset $U \subset X$, let

$$\mathrm{Mor}_U(\mathbb{P}^1, X, \alpha) \subset \mathrm{Mor}(\mathbb{P}^1, X, \alpha)$$

be the Zariski open subset parametrizing rational curves $f\colon \mathbb{P}^1 \to X$ such that $f(\mathbb{P}^1) \cap U \neq \emptyset$. Then we define the counting function in the following manner: let d be any positive integer, then

$$N(U, L, d) = \sum_{\alpha \in \mathrm{Eff}_1(X) \cap N_1(X)_{\mathbb{Z}},\ L.\alpha \leq d} \#\mathrm{Mor}_U(\mathbb{P}^1, X, \alpha)(k).$$

This counts the number of rational curves $f\colon \mathbb{P}^1 \to X$ of L-degree $\leq d$ defined over k such that $f(\mathbb{P}^1) \cap U \neq \emptyset$. Note that since there are only finitely many integral effective classes α such that $L.\alpha \leq d$, the function above is well defined. Regarding this function, Batyrev first assumed the following property for the moduli scheme $\mathrm{Mor}_U(\mathbb{P}^1, X, \alpha)$:

Assumption 7.3.1 There exists a non-empty Zariski open subset $U \subset X$ such that for any integral effective curve class α, every component of $\mathrm{Mor}_U(\mathbb{P}^1, X, \alpha)$ parametrizes a dominant and separable family of rational curves.

Under this assumption, when α is not nef, we have $\mathrm{Mor}_U(\mathbb{P}^1, X, \alpha) = \emptyset$. In particular, this implies that our counting function reduces to

$$N(U, L, d) = \sum_{\alpha \in \mathrm{Nef}_1(X) \cap N_1(X)_{\mathbb{Z}},\ L.\alpha \leq d} \#\mathrm{Mor}_U(\mathbb{P}^1, X, \alpha)(k).$$

The next assumption in Batyrev's heuristic is the following:

Assumption 7.3.2 For any $\alpha \in \mathrm{Nef}_1(X) \cap N_1(X)_{\mathbb{Z}}$, our moduli scheme $\mathrm{Mor}_U(\mathbb{P}^1, X, \alpha)$ is geometrically irreducible.

Assume that $\mathrm{Mor}_U(\mathbb{P}^1, X, \alpha)$ is geometrically irreducible. Since it parametrizes a dominant and separable family of rational curves, a general member is a free rational curve. Thus we conclude that

$$\dim \mathrm{Mor}_U(\mathbb{P}^1, X, \alpha) = -K_X.\alpha + \dim X,$$

where the right-hand side is the expected dimension. Then since the number of rational points on the n-dimensional affine space $\mathbb{A}^n_k$ is q^n, one might expect the following assumption to apply to our moduli scheme:

Assumption 7.3.3 For any $\alpha \in \mathrm{Nef}_1(X) \cap N_1(X)_{\mathbb{Z}}$, we have

$$\#\mathrm{Mor}_U(\mathbb{P}^1, X, \alpha)(k) \approx q^{\dim \mathrm{Mor}_U(\mathbb{P}^1, X, \alpha)},$$

with an error term that is negligible.

Altogether our counting function is approximated by the function

$$N(U, L, d) \approx N_{\text{approx}}(L, d) := \sum_{\alpha \in \text{Nef}_1(X) \cap N_1(X)_{\mathbb{Z}},\ L.\alpha \leq d} q^{-K_X.\alpha + \dim X}.$$

To understand the asymptotic of $N_{\text{approx}}(L, d)$, we introduce the following definitions:

Definitions 7.3.4 Let X be a smooth geometrically integral Fano variety defined over a field k and let $L = -K_X$. We fix the Lebesgue measure on $N_1(X)$ such that the volume of the fundamental domain of the lattice $N_1(X)_{\mathbb{Z}}$ is 1. We define the *α-constant* $\alpha(X, L)$ by

$$\alpha(X, L) := \dim N_1(X) \cdot \text{vol}(\{\alpha \in \text{Nef}_1(X) \mid L.\alpha \leq 1\}).$$

We also define the *index of* (X, L) to be

$$r(X, L) := \min\{-K_X.\alpha \mid \alpha \in N_1(X)_{\mathbb{Z}}, -K_X.\alpha > 0\}.$$

With these definitions, we have the following proposition:

Proposition 7.3.5 *We have*

$$N_{\text{approx}}(L, dr(X, L)) \sim \frac{\alpha(X, L) q^{\dim X}}{1 - q^{-r(X,L)}} q^{dr(X,L)} (dr(X, L))^{\rho(X)-1}$$

as $d \longrightarrow +\infty$.

Proof The following proof is taken from [LRT25c, Proposition 4.3]. We define the function $P(d)$ to be

$$P(d) := \#(N_1(X)_{\mathbb{Z}} \cap \{\alpha \in \text{Nef}_1(X) \mid L.\alpha = d\}).$$

Then by [BR15, Theorem 3.23], $P(d)$ can be expressed as

$$P(d) = c_{\rho(X)-1}(d) d^{\rho(X)-1} + \cdots + c_0(d),$$

where $c_i(d)$ is a periodic function in d. Moreover, it follows from [BR15, Exercise 3.34] that the leading coefficient $c_{\rho(X)-1}(d)$ is a constant function whose value is equal to $\alpha(X, L)$.

Now our function $N_{\text{approx}}(L, dr(X, L))$ can be identified as

$$N_{\text{approx}}(L, dr(X, L)) = \sum_{e=0}^{d} q^{er(X,L)+\dim X} P(er(X, L)).$$

After dividing this function by $q^{dr(X,L)}(dr(X, L))^{\rho(X)-1}$ and letting $d \to \infty$, the above expression converges to

$$\alpha(X, L) q^{\dim X} \left(\sum_{e=0}^{\infty} q^{-er(X,L)} \right).$$

Thus our assertion follows. □

Altogether Batyrev concluded (modulo the heuristics) that

$$N(U, L, dr(X, L)) \sim \frac{\alpha(X, L)q^{\dim X}}{1 - q^{-r(X,L)}} q^{dr(X,L)}(dr(X, L))^{\rho(X)-1},$$

as $d \to \infty$.

7.4 Geometric Manin's Conjecture

In this section, we introduce geometric Manin's conjecture, which was originally proposed in [LT19a] by Brian Lehmann and the author, and has since been further refined and generalized in [Tan21] and [LRT25c]. This conjecture refines Assumptions 7.3.1, 7.3.2, and 7.3.3, and predicts the following properties of the morphism scheme $\mathrm{Mor}(\mathbb{P}^1, X)$ when X is a smooth weak Fano variety defined over an algebraically closed field k of characteristic 0:

(i) Every component of $\mathrm{Mor}(\mathbb{P}^1, X)$ generically parametrizing non-free rational curves does so on a subvariety $Y \subset X$ such that

$$a(Y, -K_X|_Y) > a(X, -K_X) = 1.$$

(ii) For any integral nef algebraic class α of 1-cycles, there is a unique irreducible component of $\mathrm{Mor}(\mathbb{P}^1, X, \alpha)$ which should be counted in Manin's conjecture when α is sufficiently positive. We call this component as the *Manin component.*

(iii) Manin components exhibit homological/motivic stability.

In this section, we provide a precise formulation of these predictions, in which were partially confirmed by many mathematicians; this is a rapidly growing field in mathematics.

7.4.1 Dimension

We assume that our ground field k is an algebraically closed field of characteristic 0. It turns out that, in this situation, Assumption 7.3.1 is true in general. First we record the following proposition:

Proposition 7.4.1 ([LT19a, Proposition 4.2]) *Let X be a smooth uniruled projective variety and let L be a big and nef $\mathbb{Q}$-divisor on X. We fix a numerical class $\alpha \in N_1(X)_{\mathbb{Z}}$ such that $(a(X, L)L + K_X).\alpha = 0$ and let M be an irreducible component of* $\mathrm{Mor}(\mathbb{P}^1, X, \alpha)$. *Let $\pi: C \to M$ be the universal family over M with the evaluation map* $\mathrm{ev}: C \to X$. *Set $Y = \overline{\mathrm{ev}(C)}$. Suppose that either*

(i) *the dimension of M is greater than the expected dimension, that is,*

$$\dim M > -K_X.\alpha + \dim X;$$

or

(ii) $Y \subsetneq X$.

Then $a(Y, L|_Y) > a(X, L)$.

Note that the first case happens only when Y is proper.

Proof If M parametrizes a dominant family, then it follows from Lemma 7.2.5 that a general member of M is a free rational curve. This implies that M has the expected dimension as its dimension. Thus the assumption (i) implies (ii), so we may assume (ii) in the statement.

We may assume that $a(Y, L|_Y) < \infty$. Let $\beta \colon \widetilde{Y} \to Y$ be a smooth resolution. Let $\widetilde{C} \subset \widetilde{Y}$ be the strict transform of a general rational curve $f \colon C \to Y$ parametrized by M. Then since $\widetilde{C}$ is parametrized by a dominant family on $\widetilde{Y}$, a general $\widetilde{C}$ is a free rational curve on $\widetilde{Y}$, by Lemma 7.2.5. Hence we have

$$\dim M = -K_{\widetilde{Y}}.\widetilde{C} + \dim Y.$$

On the other hand, we always have

$$\dim M \geq -K_X.C + \dim X.$$

Thus we conclude that

$$-K_{\widetilde{Y}}.\widetilde{C} + \dim Y \geq -\beta^* K_X.\widetilde{C} + \dim X,$$

which implies

$$(K_{\widetilde{Y}} - \beta^* K_X).\widetilde{C} \leq \dim Y - \dim X < 0.$$

However, since we have $(a(X, L)L + K_X).C = 0$, the above inequality implies that

$$(K_{\widetilde{Y}} + a(X, L)\beta^* L).\widetilde{C} < 0.$$

Thus $K_{\widetilde{Y}} + a(X, L)\beta^* L$ is not pseudo-effective on $\widetilde{Y}$. This implies that

$$a(Y, L|_Y) > a(X, L).$$

Thus our assertion follows. □

Using this proposition, we conclude the following:

Theorem 7.4.2 ([LT19a, Theorem 4.4]) *Let X be a smooth uniruled projective variety with a big and nef $\mathbb{Q}$-divisor on X. By Theorem 5.1.6, the following set*

$$V = \bigcup_{a(Y, L|_Y) > a(X, L)} Y$$

is a proper closed subset of X. Let $U = X \setminus V$ and $\alpha \in \mathrm{Nef}_1(X)_{\mathbb{Z}}$ be a non-zero nef class such that $(K_X + a(X, L)L).\alpha = 0$. If $\mathrm{Mor}_U(\mathbb{P}^1, X, \alpha)$ is non-empty, then we have

$$\dim \mathrm{Mor}_U(\mathbb{P}^1, X, \alpha) = -K_X.\alpha + \dim X,$$

and every irreducible component of $\mathrm{Mor}_U(\mathbb{P}^1, X, \alpha)$ parametrizes a dominant family of rational curves.

Proof This follows from Lemma 7.2.5 and Proposition 7.4.1. □

Remark 7.4.3 The above proof actually shows more. Let

$$\alpha \in \mathrm{Eff}_1(X)_{\mathbb{Z}} \setminus \mathrm{Nef}_1(X)_{\mathbb{Z}}.$$

Then we have $\mathrm{Mor}_U(\mathbb{P}^1, X, \alpha) = \emptyset$. To see this, assume that it is non-empty. Then the above proof shows that every component of $\mathrm{Mor}_U(\mathbb{P}^1, X, \alpha)$ is dominant. This implies that α is nef, a contradiction.

The most compelling statement is the following:

Corollary 7.4.4 *Let X be a smooth weak Fano variety with $L = -K_X$. Let U be the complement of the union of subvarieties $Y \subset X$ such that*

$$a(Y, L|_Y) > a(X, L) = 1.$$

Then for any $\alpha \in \mathrm{Nef}_1(X)_{\mathbb{Z}}$, if $\mathrm{Mor}_U(\mathbb{P}^1, X, \alpha)$ is non-empty, then we have

$$\dim \mathrm{Mor}_U(\mathbb{P}^1, X, \alpha) = -K_X.\alpha + \dim X,$$

and every irreducible component of $\mathrm{Mor}_U(\mathbb{P}^1, X, \alpha)$ parametrizes a dominant family of rational curves.

In particular, this corollary verifies Assumption 7.3.1 in characteristic 0 for any smooth weak Fano variety. Here are examples of applications of this corollary:

Example 7.4.5 ([CS09] and [LT19a, Example 4.8]) Let X be a smooth cubic hypersurface of dimension $n \geq 3$ in $\mathbb{P}^{n+1}$. Then it follows from a modified proof of Example 6.6.4 that there is no subvariety

$$Y \subset X,$$

with $a(Y, -K_X|_Y) > a(X, -K_X) = 1$. (Recall that in Example 6.6.4, we showed that there is no subvariety $Y \subset X$ with $a(Y, -K_X|_Y) \geq a(X, -K_X)$ under the assumption that $n \geq 5$. The above weaker statement can be proved under a weaker assumption that $n \geq 3$ using arguments in Example 6.6.4.) Thus, by Corollary 7.4.4, every irreducible component of $\mathrm{Mor}(\mathbb{P}^1, X)$ is a dominant component and generically parametrizes a free rational curve. In particular, every component has the expected dimension.

Example 7.4.6 ([LT19a, Example 4.9]) Let X be a smooth quartic hypersurface of dimension $n \geq 5$ in $\mathbb{P}^{n+1}$. Then it follows from a modified proof of Example 6.6.6 that there is no subvariety $Y \subset X$ with $a(Y, -K_X|_Y) > a(X, -K_X) = 1$. Thus it follows from Corollary 7.4.4 that every irreducible component of $\mathrm{Mor}(\mathbb{P}^1, X)$ is a dominant component.

7.4.2 Manin Components

It turns out that Assumption 7.3.2 fails even in characteristic 0. This is due to the existence of Zariski dense exceptional sets in Manin's conjecture studied in previous chapters. Here is an example of this phenomenon:

Example 7.4.7 Here we consider the example studied in Section 6.2. Let $S = \mathbb{P}^1 \times \mathbb{P}^1$ and $X = \mathrm{Hilb}^{[2]}(S)$ with $L = -K_X$. Then we have

$$a(X, L) = 1, \quad b(X, L) = \dim N_1(X) = 3.$$

We denote the blowup of $S \times S$ along the diagonal by W. This admits a degree-2 finite and dominant morphism

$$g\colon W \to X.$$

As explained in Section 6.2, we have

$$a(W, g^*L) = 1, \quad b(W, g^*L) = 4.$$

Let us consider the following pushforward map:

$$g_*\colon \mathcal{F}(W, g^*L) \cap N_1(W)_{\mathbb{Z}} \to \mathcal{F}(X, L) \cap N_1(X)_{\mathbb{Z}} = \mathrm{Nef}_1(X) \cap N_1(X)_{\mathbb{Z}}.$$

The image of this map consists of integral nef classes α such that $E.\alpha = 0$ where E is the divisor parameterizing non-reduced schemes. In particular, the image of this map spans a rank-2 lattice in $N_1(X)_{\mathbb{Z}}$. Let α be an integral nef class in this image. Then the cardinality of $(g_*)^{-1}(\alpha)$ grows quadratically as the anticanonical degree of α grows. For each $\alpha' \in (g_*)^{-1}(\alpha)$, $\mathrm{Mor}(\mathbb{P}^1, W, \alpha')$ is irreducible and parametrizing a dominant family of rational curves. Moreover, a natural morphism

$$\mathrm{Mor}(\mathbb{P}^1, W, \alpha') \to \mathrm{Mor}(\mathbb{P}^1, X, \alpha)$$

is a generically finite and dominant morphism to a component M of $\mathrm{Mor}(\mathbb{P}^1, X, \alpha)$ because we have

$$\dim \mathrm{Mor}(\mathbb{P}^1, W, \alpha') = -K_W.\alpha' + \dim W = -K_X.\alpha + \dim X = \dim M.$$

This follows from the fact that α' has a vanishing intersection with the ramification divisor of g. Finally, a natural morphism

$$\bigsqcup_{\alpha' \in (g_*)^{-1}(\alpha)} \mathrm{Mor}(\mathbb{P}^1, W, \alpha') \to \mathrm{Mor}(\mathbb{P}^1, X, \alpha)$$

is a degree-2 morphism. Hence there are many dominant components in $\mathrm{Mor}(\mathbb{P}^1, X, \alpha)$ as the anticanonical degree grows.

Based on such examples, we propose the following definitions from [LT19a], [BLRT22], and [LRT25c]:

Definition 7.4.8 Let X be a smooth weak Fano variety defined over k with $L = -K_X$. Let $g\colon Y \to X$ be a generically finite morphism to the image from a smooth projective variety Y. We say that g is an *exceptional map* if g satisfies one of the following conditions:

(i) $a(Y, g^*L) > a(X, L)$;
(ii) g is an a-cover such that $\kappa(-g^*K_X + K_Y) > 0$; or
(iii) g is an a-cover such that $\kappa(-g^*K_X + K_Y) = 0$ and g is face-contracting.

With these definitions we introduce the following components of $\mathrm{Mor}(\mathbb{P}^1, X)$:

Definition 7.4.9 Let X be a smooth weak Fano variety defined over k with $L = -K_X$. Let $M \subset \mathrm{Mor}(\mathbb{P}^1, X)$ be an irreducible component. We say M is an *exceptional component* if there is an exceptional map $g\colon Y \to X$ and a dominant component $N \subset \mathrm{Mor}(\mathbb{P}^1, Y)$ such that f induces a dominant morphism $g_*\colon N \to M$.

We say M is a *Manin component* if M is not an exceptional component.

With this definition, here is a refined version of Assumption 7.3.2:

Conjecture 7.4.10 (Geometric Manin's conjecture II) Let X be a smooth weak Fano variety defined over k with $L = -K_X$. Then there exists an integral nef class $\tau \in \mathrm{Nef}_1(X)$ such that every algebraic equivalence class of rational curves whose numerical class is contained in $\tau + \mathrm{Nef}_1(X)$ is represented by a unique Manin component.

Remark 7.4.11 Note that it is an algebraic equivalence, but not a numerical equivalence, which plays a role in Conjecture 7.4.10. Let us discuss a conjectural relationship of two equivalences over $\mathbb{C}$. The Griffith group of a complex smooth projective variety is the quotient of the group of homologically trivial integral 1-cycles by the group of algebraically trivial integral 1-cycles.

Suppose that X is a smooth Fano variety. Then we have the following outstanding questions:

(i) Is the Griffith group of 1-cycles for X trivial?
(ii) Does the integral Hodge conjecture hold for 1-cycles on X?

(Both questions have positive answers in dimension ≤ 3 by [BS83] and [Voi06].) Suppose that the answers to these questions are affirmative. Then the number of algebraic equivalence classes representing a single numerical class is given by

$$|H_2(X, \mathbb{Z})_{\text{tors}}|,$$

where $H_2(X, \mathbb{Z})_{\text{tors}}$ is the torsion subgroup of the second singular homology $H_2(X, \mathbb{Z})$. Since $H^{2,0}(X) = 0$, Poincaré duality implies that

$$|H_2(X, \mathbb{Z})_{\text{tors}}| = |H^3(X, \mathbb{Z})_{\text{tors}}| = |\text{Br}(X)|,$$

where $\text{Br}(X)$ is the *Brauer group* of X. (See [CTS21] for the definition of the Brauer group.) In particular, we expect that the number of algebraic equivalence classes representing a single numerical class is given by the size of the Brauer group of X.

We now list classes of Fano varieties where Conjecture 7.4.10 has been confirmed:

- Fano hypersurfaces ([HRS04], [BK13], [RY19], and [BV17]);
- homogeneous spaces ([Tho98] and [KP01]);
- toric varieties ([Bou12] and [Bou16]);
- del Pezzo surfaces ([Tes05], [Tes09], and [BLRT23]);
- moduli spaces of vector bundles ([Cas04] and [MTiB25]);
- Fano threefolds ([Sta00], [Cas04], [LT19a], [LT21], [BLRT22], [ST22], [BJ22], [JLT25], [Oka25b], and [Oka25a]);
- del Pezzo fibrations ([LT24] and [LT22]);
- del Pezzo manifolds ([CS09] and [Oka24]); and
- Fano manifolds of coindex 3 ([JO24]).

In the next chapter, we will discuss proofs of Conjecture 7.4.10 for del Pezzo threefolds and del Pezzo fibrations.

Finally, we conclude this section by introducing a proposition useful for proving Conjecture 7.4.10 using inductive arguments:

Proposition 7.4.12 ([LT19a, Proposition 5.15]) *Let X be a smooth weak Fano variety. Suppose that M is a component of* $\mathrm{Mor}(\mathbb{P}^1, X)$ *parametrizing a dominant family of rational curves $\pi\colon C \to M$ with the evaluation map* $\mathrm{ev}\colon C \to X$. *Let $\widetilde{C}$ be a resolution of a projective compactification of C with a morphism* $\mathrm{ev}'\colon \widetilde{C} \to X$ *extending* ev. *Let*

$$\widetilde{C} \to Y \to X$$

be the Stein factorization of ev' *where we denote the morphism $Y \to X$ by g. Then we have $a(Y, -g^*K_X) = a(X, -K_X) = 1$.*

Proof Let $\beta\colon \widetilde{Y} \to Y$ be a smooth resolution of Y and consider a family of strict transforms $\widetilde{f}\colon C \to \widetilde{Y}$ of general rational curves $f\colon C \to Y$ parametrized by M. Then the dimension of this family is given by both quantities

$$-K_{\widetilde{Y}}.\widetilde{f}_*C + \dim \widetilde{Y}, \quad -g^*K_X.f_*C + \dim X,$$

which means that they must coincide. This implies that we have

$$(K_{\widetilde{Y}} - \beta^*g^*K_X).\widetilde{f}_*C = 0.$$

Thus we conclude that $K_{\widetilde{Y}} - \beta^*g^*K_X$ is pseudo-effective, but not big. Hence our assertion follows. □

7.4.3 Homological Stability

Grothendieck–Lefschetz Trace Formula

As before, let $\mathbb{F}_q$ be a finite field of size q with $q = p^n$ and p a prime number. Let M be a quasi-projective scheme defined over $\mathbb{F}_q$. Assumption 7.3.3 suggests that when M is geometrically irreducible, the number of $\mathbb{F}_q$-points on M is approximated as follows:

$$\#M(\mathbb{F}_q) \approx q^{\dim M},$$

with a negligible error term. One might ask how this assumption can be justified. One prominent approach to this problem is to appeal to the Grothendieck–Lefschetz trace formula, which we describe now: we fix an algebraic closure $\overline{\mathbb{F}}_q \supset \mathbb{F}_q$ and denote its base change

$$M \otimes_{\mathbb{F}_q} \overline{\mathbb{F}}_q,$$

by $\overline{M}$. Let ℓ be a prime number that is different from p. Then one can associate the ℓ-adic cohomology with compact supports:

$$H_c^i(\overline{M}, \mathbb{Q}_\ell) \quad (0 \le i \le 2\dim M).$$

This is a finite dimensional $\mathbb{Q}_\ell$-vector space, where $\mathbb{Q}_\ell$ is the completion of $\mathbb{Q}$ with respect to the ℓ-adic norm. Moreover, it comes with a continuous action of the Galois group $\mathrm{Gal}(\overline{\mathbb{F}}_q/\mathbb{F}_q)$, which is a profinite group topologically isomorphic to

$$\widehat{\mathbb{Z}} = \varprojlim \mathbb{Z}/n\mathbb{Z}.$$

It is well known that this profinite group is generated by the geometric Frobenius Frob_q. Here is the *Grothendieck–Lefschetz trace formula*:

Theorem 7.4.13 (Grothendieck–Lefschetz trace formula [SGA72a, SGA72b, SGA73]) *Let M be a quasi-projective scheme defined over $\mathbb{F}_q$. Then we have the following equality:*

$$\#M(\mathbb{F}_q) = \sum_{i=0}^{2\dim M} (-1)^i \mathrm{tr}(\mathrm{Frob}_q \mid H_c^i(\overline{M}, \mathbb{Q}_\ell)),$$

where tr *is the trace of a linear map* $\mathrm{Frob}_q : H_c^i(\overline{M}, \mathbb{Q}_\ell) \to H_c^i(\overline{M}, \mathbb{Q}_\ell)$.

For a reference, one might consult [FK88, 3.15 Corollary] or [Fu11, Theorem 10.4.5].

Now one might ask how one can use this formula to estimate the number of $\mathbb{F}_q$-points on a quasi-projective scheme M over $\mathbb{F}_q$. For simplicity, let us assume that M is geometrically irreducible and we denote its dimension by d. Then the highest cohomology is given by

$$H_c^{2d}(\overline{M}, \mathbb{Q}_\ell),$$

which is isomorphic to $\mathbb{Q}_\ell$ as a $\mathbb{Q}_\ell$-vector space. Indeed, this follows from a fact that the top cohomology is generated by the fundamental class of M, which is well known in the setting of singular cohomology for complex varieties. The action of Frob_q on this cohomology is scalar multiplication by q^d. The actions on other cohomologies are explained by the following generalization of the *Weil conjecture*:

Theorem 7.4.14 ([Del80]) *Let M be a quasi-projective scheme defined over $\mathbb{F}_q$. Then the characteristic polynomial of* Frob_q *on $H_c^i(\overline{M}, \mathbb{Q}_\ell)$ is a polynomial with integer coefficients and independent of ℓ.*

Moreover, any root α of this polynomial satisfies

$$|\alpha| \le q^{i/2},$$

where $|\cdot|$ *is induced by any embedding* $\overline{\mathbb{Q}} \hookrightarrow \mathbb{C}$.

For a reference, we refer readers to [FK88, 5.2 Theorem] or the original sources: [Dwo60] for being a polynomial and [Del80] for the bounds for eigenvalues.

Combining these results, we conclude that when M is geometrically irreducible, we have

$$|\#M(\mathbb{F}_q) - q^d| \leq \sum_{i=0}^{2d-1} q^{i/2} b_i,$$

where b_i is the *ith Betti number*, which is the dimension of $H^i_c(\overline{M}, \mathbb{Q}_\ell)$.

However, this is still not enough to establish Assumption 7.3.3 when

$$M = \mathrm{Mor}(\mathbb{P}^1, X, \alpha),$$

where X is a smooth Fano variety over $\mathbb{F}_q$. The first step towards Assumption 7.3.3 is to understand how b_i behaves when $-K_X.\alpha$ goes to ∞. Jordan Ellenberg and Akshay Venkatesh suggested that this should be achieved by homological stability of the moduli space $\mathrm{Mor}(\mathbb{P}^1, X, \alpha)$ (see, e.g., [VE10] and [EVW16]). Another important ingredient is a subexponential bound on the Betti numbers (in addition to homological stability).

Homological Stability

We work over $\mathbb{C}$. Let M be a quasi-projective scheme over $\mathbb{C}$. We denote an analytic space $M(\mathbb{C})$ with the Euclidean topology by M^{an}.

We fix $B = \mathbb{P}^1$ with a fixed basepoint. Let X be a smooth projective variety with a basepoint. We denote the morphism scheme of pointed maps $f: B \to X$ mapping the basepoint of B to the basepoint of X by

$$\mathrm{Mor}_*(B, X, \alpha),$$

where α is an effective integral homology class of 1-cycles on X.

We also denote the *space of pointed continuous maps* $f: (\mathbb{P}^1)^{\mathrm{an}} \to X^{\mathrm{an}}$,

$$\mathrm{Top}_*((\mathbb{P}^1)^{\mathrm{an}}, X^{\mathrm{an}}, \alpha),$$

which parametrizes $f: (\mathbb{P}^1)^{\mathrm{an}} \to X^{\mathrm{an}}$ such that $f_*[(\mathbb{P}^1)^{\mathrm{an}}] = \alpha$ and f maps the basepoint of B to the basepoint of X. This space is equipped with the compact-open topology. (See [AGP02, Section 1.2] for more details.) Then we have an embedding of topological spaces:

$$\mathrm{Mor}_*(\mathbb{P}^1, X, \alpha)^{\mathrm{an}} \hookrightarrow \mathrm{Top}_*((\mathbb{P}^1)^{\mathrm{an}}, X^{\mathrm{an}}, \alpha).$$

Inspired by the fact that continuous functions are approximated by polynomial functions, Cohen–Jones–Segal conjectured the following:

Conjecture 7.4.15 (Cohen–Jones–Segal's conjecture [CJS00]) Let X be a smooth Fano variety. Then there exists a rational homogeneous continuous piecewise linear function $\ell(\alpha)$ on homology classes of effective 1-cycles and a constant $c \geq 0$ such that $\ell(\alpha)$ is positive in the interior of the nef cone of curves, and for any integral class $\alpha \in \mathrm{Nef}(X)^\circ$ and $i \leq \ell(\alpha) - c$, the above inclusion induces an isomorphism

$$H_i(\mathrm{Mor}_*(\mathbb{P}^1, X, \alpha)^{\mathrm{an}}, \mathbb{Q}) \cong H_i(\mathrm{Top}_*((\mathbb{P}^1)^{\mathrm{an}}, X^{\mathrm{an}}, \alpha), \mathbb{Q}).$$

Note that the disjoint union

$$\bigsqcup_{\alpha:\ \text{integral homology classes}} \mathrm{Top}_*((\mathbb{P}^1)^{\mathrm{an}}, X^{\mathrm{an}}, \alpha)$$

can be considered as the *double loop space* $\Omega^2 X^{\mathrm{an}} = \Omega\Omega X^{\mathrm{an}}$, where

$$\Omega X^{\mathrm{an}} = \mathrm{Top}_*(S^1, X^{\mathrm{an}})$$

and S^1 is the unit circle. Up to homotopy, there is a natural action of $\pi_2(X^{\mathrm{an}})$ on this double loop space, and this shows that the homotopy type of

$$\mathrm{Top}_*((\mathbb{P}^1)^{\mathrm{an}}, X^{\mathrm{an}}, \alpha)$$

is independent of α. So Cohen–Jones–Segal's conjecture implies homological stability of the space $\mathrm{Mor}_*(\mathbb{P}^1, X, \alpha)^{\mathrm{an}}$. See [AGP02] for more information on the loop space.

Conjecture 7.4.15 was extensively studied in the 1990s for homogeneous spaces admitting group actions of solvable groups with dense orbits, (see, e.g., [Seg79], [Gue95]). More recently Browning–Sawin studied Conjecture 7.4.15 for low degree affine hypersurfaces ([BS20]) and Das–Tosteson proved this for del Pezzo surfaces of degree ≥ 5 ([DT24]). The later work has been extended to quartic del Pezzo surfaces by Das–Lehmann–Tosteson and the author in [DLTT25].

7.5 Weak Manin's Conjecture as the Prime Tends to Infinity

In this section, we explain one arithmetic application of results from geometric Manin's conjecture that was developed in [LRT25c]. Let F be a number field and X be a smooth weak Fano variety defined over F. Let S be a finite set of places of F including all archimedean places; we denote the ring of S-integers in F by $\mathfrak{o}_{F,S}$.

We fix a projective flat integral model $\widetilde{\pi}\colon \mathcal{X} \to \operatorname{Spec} \mathfrak{o}_{F,S}$ of X, that is, $\widetilde{\pi}$ is flat and projective such that the generic fiber $\mathcal{X}_\eta$ is isomorphic to X. After enlarging S, we may assume that $\widetilde{\pi}$ is smooth and the relative anticanonical divisor $-K_{\mathcal{X}/\operatorname{Spec}\mathfrak{o}_{F,S}}$ is relatively ample.

Let $V \subset X$ be the Zariski closure of the union of images of non-dominant families of rational curves in $\operatorname{Mor}(\mathbb{P}^1, X)$. If we take the base change $V \otimes \overline{F}$ of V to $\overline{F}$, then it follows from Theorem 5.1.6 and Corollary 7.4.4 that $V \otimes \overline{F}$ is contained in a proper closed subset, showing that V is itself a proper closed subset. We denote the flat closure of V in $\mathcal{X}$ by $\mathcal{V}$.

Let v be a non-archimedean place of F that is not contained in S, and denote the reduction of $\mathcal{X}, \mathcal{V}$ at v by X_v, V_v which are defined over a finite field k_v. Let

$$\operatorname{Mor}(\mathbb{P}^1, X_v, V_v)_{\leq d}$$

be the open subset of $\operatorname{Mor}(\mathbb{P}^1, X_v)$ parametrizing rational curves $f\colon \mathbb{P}^1 \to X_v \otimes \overline{k}_v$ of anticanonical degree $\leq d$ such that $f(\mathbb{P}^1) \not\subset V_v \otimes \overline{k}_v$.

We consider the following counting function:

$$N(X_v \setminus V_v, -K_{X_v}, d) = \#\operatorname{Mor}(\mathbb{P}^1, X_v, V_v)_{\leq d}(k_v),$$

which counts the number of k_v-rational curves on X_v of anticanonical degree $\leq d$ not contained in V_v. Here is a conjecture regarding this counting function:

Conjecture 7.5.1 (Weak Manin's conjecture over $k_v(t)$) For any $\epsilon > 0$, we have

$$N(X_v \setminus V_v, -K_{X_v}, d) = o(q_v^{d(1+\epsilon)}),$$

as $d \to \infty$ where $q_v = \#k_v$.

In [LRT25c], Lehmann, Riedl, and the author proved the following approximation of the above conjecture:

Theorem 7.5.2 ([LRT25c, Theorem 1.11]) *Let $F, S, \mathcal{X}$ be as above and fix $\epsilon > 0$. Assume that $d\epsilon > \dim X$. Then we have*

$$\frac{N(X_v \setminus V_v, -K_{X_v}, d)}{q_v^{d(1+\epsilon)}} \to 0$$

as $v \to \infty$.

This result fits into the recent trend in analytic number theory over global function fields where the problems considered become easier by letting $v \to \infty$, (see, e.g, [ETW17] and [LWZB24]). Note that these papers achieved something more sophisticated: estimating not only the dimension of moduli spaces, but also the dimension of cohomology. See the relevant papers for more details.

Proof Let us consider the open subscheme of the relative morphism scheme of $\mathcal{X} \to \operatorname{Spec}\mathfrak{o}_{F,S}$ parametrizing rational curves of anticanonical degree $\leq d$

not contained in $\mathcal{V}$. By generic flatness, one can find a finite set $S_d \supset S$ of places such that this open subscheme is flat over $\operatorname{Spec} \mathfrak{o}_{F,S_d}$ after the base change. Let $\psi\colon \mathcal{H}_d \to \operatorname{Spec} \mathfrak{o}_{F,S_d}$ be the closure of this scheme in the relative Hilbert scheme of $\mathbb{P}^1 \times \mathcal{X} \to \operatorname{Spec} \mathfrak{o}_{F,S}$. It follows from the construction that ψ is projective and flat. Let H_d be the generic fiber of ψ. By the construction of V, every irreducible component of H_d generically parametrizes a dominant family of rational curves, so by Lemma 7.2.5, every component has expected dimension equal to $d + \dim X$. We define

$$C_d = \sum_{i=0}^{2(d+\dim X)} h^i_{\text{sing}}(H_d^{\text{an}}, \mathbb{Q}),$$

where $h^i_{\text{sing}}(H_d^{\text{an}}, \mathbb{Q})$ is the dimension of the singular cohomology $H^i_{\text{sing}}(H_d^{\text{an}}, \mathbb{Q})$. Using the pro-étale topology introduced in [BS15], one can show that the ℓ-adic sheaf $R^i\psi_*\underline{\mathbb{Q}}_\ell$ is constructible in the pro-étale topology. By proper base change, the stalk of this sheaf at v is identified with $H^i_{\text{ét}}(H_{d,\overline{v}}, \mathbb{Q}_\ell)$ where $H_{d,\overline{v}}$ is the base change of the reduction $H_{d,v} \to \operatorname{Spec} k_v$ to an algebraic closure. On the other hand, the stalk at the generic point has dimension equal to $h^i_{\text{sing}}(H_d^{\text{an}}, \mathbb{Q})$ by the comparison theorem. These imply that there exists a finite set $S'_d \supset S_d$ such that for any i and $v \notin S'_d$ we have

$$h^i_{\text{ét}}(H_{d,\overline{v}}, \mathbb{Q}_\ell) = h^i_{\text{sing}}(H_d^{\text{an}}, \mathbb{Q}),$$

where $h^i_{\text{ét}}(H_{d,\overline{v}}, \mathbb{Q}_\ell)$ is the dimension of étale cohomology $H^i_{\text{ét}}(H_{d,\overline{v}}, \mathbb{Q}_\ell)$. Using the Grothendieck–Lefschetz trace formula (Theorem 7.4.13) and Weil's conjecture (Theorem 7.4.14) we conclude that

$$N(X_v \setminus V_v, -K_{X_v}, d) \leq \#H_{d,v}(k_v) \leq C_d q_v^{d+\dim X}.$$

Thus, assuming $d\epsilon > \dim X$, we conclude that

$$\frac{N(X_v \setminus V_v, -K_{X_v}, d)}{q_v^{d(1+\epsilon)}} \to 0$$

as $v \to \infty$.

□

8

Examples of Geometric Manin's Conjecture II

In this chapter, we discuss proofs of geometric Manin's conjecture II (Conjecture 7.4.10) for certain classes of algebraic varieties. The main idea is to use Mori's famous Bend and Break techniques to prove the irreducibility of the moduli space of rational curves of a given class based on induction on the degree of that class. This is pioneered in [HRS04], and further developed in subsequent papers. We first discuss Bend and Break techniques and their proofs.

8.1 Bend and Break

In this section, we discuss Mori's *Bend and Break* techniques, which were invented by Shigefumi Mori in his proof of the cone theorem [Mor82]. We closely follow the exposition of [KM98]. In this section, our ground field k is an algebraically closed field, but we allow arbitrary characteristic.

First, we discuss a key lemma to prove the first Bend and Break result:

Lemma 8.1.1 (Rigidity lemma [KM98, Lemma 1.6]) *Let X, Y, Z be irreducible varieties and $h\colon Y \to Z$ be a proper surjective morphism with every fiber connected and of the same dimension n. Let $g\colon Y \to X$ be another morphism such that $g(h^{-1}(z_0))$ consists of a single point for some $z_0 \in Z$. Then for any $z \in Z$, $g(h^{-1}(z))$ consists of a single point on X.*

Proof We closely follow the exposition of the proof of [KM98, Lemma 1.6]. We set $W = \mathrm{im}(h \times g) \subset Z \times X$ and obtain the following diagram:

$$h\colon Y \to W \to Z,$$

where we denote $Y \to W$ and $W \to Z$ by ξ and ζ. Note that $Y \to W$ is proper and hence surjective. Thus we have $\zeta^{-1}(z) = \xi(h^{-1}(z))$, and in particular, $\zeta^{-1}(z_0)$

consists of a single point. By upper semicontinuity, there exists a Zariski open subset $U \subset Z$ such that $\dim \zeta^{-1}(z) = 0$ for any $z \in U$. Then ξ has fiber dimension n over $\zeta^{-1}(U)$, therefore any fiber of ξ has dimension $\geq n$. For any $w \in W$, we have $\xi^{-1}(w) \subset h^{-1}(\zeta(w))$, $\dim \xi^{-1}(w) \geq n$, and $\dim h^{-1}(\zeta(w)) = n$. Thus $\xi^{-1}(w)$ is the union of irreducible components of $h^{-1}(\zeta(w))$. Hence we conclude that $\xi(h^{-1}(\zeta(w))) = \zeta^{-1}(\zeta(w))$ is a finite set. Finally, since $h^{-1}(\zeta(w))$ is connected, $\zeta^{-1}(\zeta(w))$ consists of a single point, proving the claim. □

As a corollary, we have the first Bend and Break result:

Corollary 8.1.2 (Bend and Break I [KM98, Corollary 1.7]) *Let C be a smooth projective curve and X be a proper variety. Let $f_0: C \to X$ be a non-trivial morphism. Assume that there is a smooth connected curve D with a point $0 \in D$ and a morphism $G: C \times D \to X$ such that we have*

(i) $G|_{C\times\{0\}} = f_0$;
(ii) $G(\{c\} \times D) = \{f_0(c)\}$; *and*
(iii) *for a general $t \in D$, $G|_{C\times\{t\}}$ is different from f_0.*

Let $\overline{D}$ be a smooth projective compactification of D. Then the rational map $G: C \times \overline{D} \dashrightarrow X$ has a point of indeterminacy on $\{c\} \times \overline{D}$. In particular, after successively blowing up the indeterminacy to obtain a birational morphism $\beta: S \to C \times \overline{D}$ with the induced morphism $\widetilde{G}: S \to X$, there is some fiber F of $S \to \overline{D}$ such that F consists of C and the union of rational curves Z whose image $\widetilde{G}(Z)$ in X is 1-dimensional and contains $f_0(c)$.

Informally, this result may be phrased as follows: if one deforms a curve with a fixed point on X, then it degenerates to the union of a curve $f_1 : C \to X$ (possibly trivial) and rational curves Z on X.

Proof Assume that G is well defined along $\{c\} \times \overline{D}$. Then there exists a Zariski open neighborhood U of c such that G is well defined on $U \times \overline{D}$. Let $h: U \times \overline{D} \to U$; we apply Lemma 8.1.1 to h with $g = G|_{U\times\overline{D}}: U \times \overline{D} \to X$. Since $G(h^{-1}(c))$ consists of only one point $f_0(c)$, we conclude that

$$G(z, t) = g_0(z),$$

for every $z \in U$ and $t \in D$. This contradicts our assumption. □

When $C = \mathbb{P}^1$, this result may not increase the number of rational curves in a degeneration as it is possible that C in F is contracted by $\widetilde{G}$. For this reason, we will need the following improvement of Bend and Break for rational curves:

Lemma 8.1.3 (Bend and Break II [KM98, Lemma 1.9]) *Let X be a projective variety and $f_0: \mathbb{P}^1 \to X$ be a non-constant morphism. Suppose that there* *exists a smooth connected curve D with a point $0_D \in D$ and a morphism $G\colon \mathbb{P}^1 \times D \to X$ such that*

(i) $G|_{\mathbb{P}^1 \times \{0_D\}} = f_0$;
(ii) $G(\{0\} \times D) = \{f_0(0)\}$, $G(\{\infty\} \times D) = \{f_0(\infty)\}$; *and*
(iii) $G(\mathbb{P}^1 \times D)$ *is a surface.*

Let $\overline{D}$ be the projective compactification of D, and S be a smooth projective surface with a morphism $\pi\colon S \to \overline{D}$ such that $\pi\colon S \to \overline{D}$ is a $\mathbb{P}^1$-bundle and contains $\mathbb{P}^1 \times D \to D$ as a Zariski open subset. (For example, one may take $S = \mathbb{P}^1 \times \overline{D}$.) Let $\beta : \widetilde{S} \to S$ be a successive blowup of the indeterminacy so that the induced rational map $\widetilde{G} = G \circ \beta\colon \widetilde{S} \dashrightarrow X$ becomes a morphism. Then there exists a point $c \in \overline{D}$ such that $\widetilde{G}_*(\beta^*\pi^*(c))$ is a sum of rational curves which is not integral.

Informally, one may phrase this lemma in the following way: if one deforms a rational curve while fixing two points, then it degenerates to a union of two or more rational curves.

Proof First we show that a rational map $G\colon S \dashrightarrow X$ is not a morphism. Assume otherwise. Let H be an ample divisor on X and C_0, C_∞ be the closures of $\{0\} \times D$ and $\{\infty\} \times D$ in S. Then we have $(G^*H)^2 > 0, G^*H.C_0 = G^*H.C_\infty = 0$. Thus by the Hodge index theorem [Har77, Chapter 5, Theorem 1.9], we have $C_0^2 < 0, C_\infty^2 < 0$. We also have $C_0.C_\infty \geq 0$, showing that G^*H, C_0, and C_∞ are linearly independent in $N^1(S)$. However, since $\pi\colon S \to \overline{D}$ is a $\mathbb{P}^1$-bundle, we have $\dim N^1(S) = 2$, a contradiction.

We prove our assertion by induction on the number of blowups to obtain $\beta\colon \widetilde{S} \to S$. Suppose that our assertion fails, that is, $\widetilde{G}_*(\beta^*\pi^*(c)))$ is integral for any $c \in \overline{D}$. Let $r\colon S' \to S$ be the first blowup of $\beta\colon \widetilde{S} \to S$ that happens in a fiber $\pi^{-1}(c)$. We write $\widetilde{r}\colon \widetilde{S} \to S'$ for the remaining blowup. Let F_1 be the exceptional divisor of r and write $r^*\pi^*(c) = F_1 + F_2$ where F_1, F_2 are (-1)-curves. Note that two curves F_1, F_2 meet at a single point $x_0 \in F_1 \cap F_2$.

We claim that $G'\colon S' \dashrightarrow X$ is a morphism along F_2. If G is not defined at two distinct points $x, x' \in \pi^{-1}(c)$, then we have

$$\widetilde{G}_*(\beta^*\pi^*(c)) \geq \widetilde{G}_*(\beta^{-1}(x)) + \widetilde{G}_*(\beta^{-1}(x')),$$

which contradicts our assumption. If G' is not well defined at x_0, then every irreducible component of $\widetilde{r}^{-1}(x_0)$ has multiplicity at least 2 in $\beta^*\pi^*(c)$. This contradicts our assumption too.

Our assumption implies that F_2 is contracted by G'. Thus let $r'\colon S' \to S''$ be the blowdown of F_2. Then the number of blowups to resolve the indeterminacy of $S'' \dashrightarrow X$ is strictly smaller than the number of successive blowups $\beta\colon \widetilde{S} \to S$. Then our assumption contradicts the inductive hypothesis. □

Next, we introduce a slight improvement of the previous lemma, Bend and Break II, by Lehmann and the author:

Lemma 8.1.4 (Bend and Break II+α [LT24, Lemma 4.1]) *Let $\pi\colon S \to D$ be a morphism with a general fiber isomorphic to $\mathbb{P}^1$ from a smooth projective surface S to a smooth projective curve D. Let X be a projective variety and $G\colon S \to X$ be a morphism such that* $\dim G(S) = 2$. *Assume that there exist two sections C_1 and C_2 of π such that these sections are contracted to two distinct points x_1, x_2 by G. Then there exists a singular fiber of π with two distinct components F_1, F_2 such that*

(i) *neither F_1 nor F_2 is contracted by G; and*
(ii) *$G(F_1)$ contains x_1 and $G(F_2)$ contains x_2.*

To prove this lemma, we will need the following negativity result:

Lemma 8.1.5 ([Mat02, Lemma 1-2-10]) *Let S be a smooth projective surface and C a smooth projective curve. Suppose that we have a morphism $\pi\colon S \to C$ with connected fibers. Then for any divisor D on S supported in fibers of π, we have $D^2 \leq 0$. Moreover, the following equivalence holds:*

$$D^2 = 0 \iff D \text{ is a (rational) linear combination of full fibers of } \pi.$$

Proof Let H be an ample divisor on C. Then we have $(\pi^*H)^2 = 0$ and $\pi^*H.D = 0$. Let A be an ample divisor on S. By the Hodge index theorem [Har77, Chapter 5, Theorem 1.9], the determinant of the intersection matrix of A, π^*H, D must be non-negative, showing that $D^2 \leq 0$.

Suppose that $D^2 = 0$. For any $c \in C$, let D_c be the part of D supported on $\pi^{-1}(c)$. Then since D is supported in fibers of π, we must have $D_c^2 = 0$. So without loss of generality, we may assume that D is supported in $\pi^{-1}(c)$. Now assume, for a contradiction, that our assertion fails so that D is not proportional to a full fiber $F = \pi^*(c)$. Let t be a rational number such that $D + tF$ is effective and we have

$$\mathrm{Supp}(D + tF) \subsetneq \mathrm{Supp}(F).$$

Our assumption implies that F is connected, so that there exists a component $E \subset \pi^{-1}(c)$ such that $(D + tF).E > 0$. Since we have $(D + tF)^2 = D^2 = 0$, for any $0 < \epsilon \ll 1$, it follows that

$$0 \geq ((D + tF) + \epsilon E)^2 = 2\epsilon(D + tF).E + \epsilon^2 E^2 > 0.$$

This is a contradiction. □

Proof of Lemma 8.1.4 Assume that our assertion fails. Let D_1 be the connected component of the G-fiber over x_1 that contains C_1. Let D_2 be the connected component of the G-fiber over x_2 which contains C_2. Note that D_1 and D_2 do not intersect. Since every fiber of π is a tree of rational curves, we can assign coefficients to the components of D_1 so that the resulting divisor $\widetilde{D}_1$ satisfies the following properties:

- C_1 occurs in $\widetilde{D}_1$ with coefficient 1;
- any other π-horizontal component of D_1 has coefficient 0 in $\widetilde{D}_1$; and
- for any π-vertical component E of D_1, we have $\widetilde{D}_1.E = 0$.

In the same way, we construct $\widetilde{D}_2$ with respect to C_2. By our assumption, for any fiber F of π, there exists a unique component $F_0 \subset F$ such that

- F_0 is not contracted by G;
- F_0 is the unique component of F that meets D_1, but is not contained in D_1; and
- F_0 is the unique component of F that meets D_2, but is not contained in D_2.

If F_0 has coefficient m_0 in F, then it follows from our construction that

$$m_0 F_0.\widetilde{D}_1 = F.\widetilde{D}_1 = 1, \quad m_0 F_0.\widetilde{D}_2 = F.\widetilde{D}_2 = 1$$

This implies that $\widetilde{D}_1 - \widetilde{D}_2$ has vanishing intersection with every π-vertical curve. Since a general fiber is a rational curve, this implies that $\widetilde{D}_1 - \widetilde{D}_2$ is numerically equivalent to a linear combination of π-vertical curves. Thus, by Lemma 8.1.5, $\widetilde{D}_1 - \widetilde{D}_2$ is numerically proportional to a fiber of π. Hence we conclude that

$$0 = (\widetilde{D}_1 - \widetilde{D}_2)^2 = \widetilde{D}_1^2 + \widetilde{D}_2^2 - 2\widetilde{D}_1.\widetilde{D}_2.$$

However, since $\widetilde{D}_i$ is contracted by G, the Hodge index theorem shows that $\widetilde{D}_i^2 < 0$. Also since D_1 and D_2 do not intersect, we have $\widetilde{D}_1.\widetilde{D}_2 = 0$. This is a contradiction. □

Finally, we prove that these degenerations can be realized in the moduli space of stable maps:

Corollary 8.1.6 (Bend and Break for stable maps [LT24, Corollary 4.2]) *Assume that our ground field k has characteristic 0. Let X be a projective variety and let $f : \mathbb{P}^1 \to X$ be a rational curve on X with two distinct points $p_1, p_2 \in \mathbb{P}^1$ such that $f(p_1) \neq f(p_2)$. Suppose that there is an irreducible curve $B^\circ \subset \mathrm{Mor}(\mathbb{P}^1, X, f|_{p_1,p_2})$ containing f such that the images of maps parametrized by B° sweep out a surface. Then f deforms as a stable map to a morphism $f' : C \to X$ where C has (at least) two non-contracted components C_1, C_2 such that $f'(C_i) \ni f(p_i)$ for $i = 1, 2$.*

Proof Let $G\colon \mathbb{P}^1 \times B^\circ \to X$ be the pullback of the universal family on B°. By the construction, this morphism contracts two sections to two distinct points $f(p_1) \neq f(p_2)$. Let B be the normalization of a proper compactification of B°. Then we obtain a rational map $G\colon \mathbb{P}^1 \times B \dashrightarrow X$. Let $\beta\colon S \to \mathbb{P}^1 \times B$ be a birational morphism that resolves the indeterminacy of G. Now we apply Lemma 8.1.4 to $\widetilde{G} = G \circ \beta\colon S \to X$ and we find a reducible fiber of $S \to B$ which is a tree of rational curves with two distinct components T_1, T_2 such that the images $\widetilde{G}(T_1), \widetilde{G}(T_2)$ are two distinct rational curves containing $f(p_1)$ and $f(p_2)$ respectively. While the family $\widetilde{G}\colon S/B \to X$ may not be a family of stable maps, one can stabilize this surface using the procedure described in [BM96, Proposition 3.10]. Our assertion follows from this discussion. □

8.2 Del Pezzo Threefolds

Let k be an algebraically closed field of characteristic 0. Let X be a smooth del Pezzo threefold as defined in Section 6.5.1 so that we have an ample generator H for $\mathrm{Pic}(X)$ and we have $-K_X \sim 2H$. Recall from the classification in Theorem 6.5.1 that we have $1 \leq H^3 \leq 5$ and such threefolds are classified into five deformation types according to the degree H^3. Let $\overline{M}_{0,0}(X, d)$ be the coarse moduli space of stable maps of genus 0 and H-degree d. Our goal of this section is to prove the following theorem:

Theorem 8.2.1 ([CS09, Cas04, LT19a, ST22]) *Let X be a smooth del Pezzo threefold with $\mathrm{Pic}(X) = \mathbb{Z}H$ and $-K_X \sim 2H$. We assume that X is general in its moduli when $H^3 = 1$ or 2. Then*

(i) $\overline{M}_{0,0}(X, 1)$ *consists of one irreducible component R_1 of dimension 2 which parametrizes H-lines; and*
(ii) *for $d \geq 2$, $\overline{M}_{0,0}(X, d)$ consists of two irreducible components R_d and N_d of dimension $2d$ such that the former space generically parametrizes birational stable maps from $\mathbb{P}^1$ and the latter space generically parametrizes degree-d multiple covers of H-lines.*

Let us briefly explain the history of this theorem. Originally, this statement was obtained by Starr for cubic threefolds, that is, when $H^3 = 3$, which is described in [CS09], then extended to complete intersections of two quadrics, that is, when $H^3 = 4$, by Castravet in [Cas04]. Later this was generalized by Lehmann and the author for the cases of $H^3 \geq 2$ ([LT19a]), and finally, the case of $H^3 = 1$ settled as in [ST22]. In this book, we mainly prove the case of $H^3 = 3$. We follow the proof in [LT19a], which in turn was modeled on [CS09].

8.2.1 First Observations

Let $\mathrm{Mor}(\mathbb{P}^1, X, d)$ be the morphism scheme of rational curves of H-degree d on X. The first thing to observe is the following result:

Lemma 8.2.2 *Let X be a smooth del Pezzo threefold and d be a positive integer. Then every component of* $\mathrm{Mor}(\mathbb{P}^1, X, d)$ *has expected dimension $2d+3$, and parametrizes a dominant family of rational curves.*

Proof Note that we have $a(X, H) = 2$. It follows from Lemmas 6.5.2 and 6.5.5 that for any subvariety $Y \subset X$, we have $a(Y, H|_Y) \leq a(X, H)$. Thus our assertion follows from Corollary 7.4.4. □

Next, we establish certain auxiliary results following [LT19a, Section 7], which is motivated by [CS09]:

Lemma 8.2.3 ([LT19a, Lemma 7.5]) *Let Y be a projective variety of dimension n and fix a smooth resolution $\beta \colon \widetilde{Y} \to Y$. We further fix a point $y \in Y$ and assume that we have a dominant family of rational curves passing through y on Y, which is parametrized by a variety $W \to \overline{M}_{0,0}(Y)$ that is finite to its image. Let C' be the strict transform on $\widetilde{Y}$ of a general curve parametrized by W. If the fiber of β at y has dimension ℓ, then we have*

$$\dim(W) \leq -K_{\widetilde{Y}}.C' - 2 + \ell.$$

Proof Let $W' \subset W$ be a Zariski open subset parametrizing the strict transforms C' on $\widetilde{Y}$. Since this family is dominant and $\widetilde{Y}$ is smooth, it follows from Lemma 7.2.5 that a general strict transform C' is free on $\widetilde{Y}$. Let y' be a general point of $\beta^{-1}(y)$ which C' passes through. Then the dimension of the sublocus of W' parametrizing C' passing through y' is given by the expected dimension $-K_{\widetilde{Y}}.C' - 2$. Since the dimension of each component of $\beta^{-1}(y)$ is at most ℓ, our assertion follows. □

Using the above lemma, we prove the following proposition:

Proposition 8.2.4 ([LT19a, Theorem 7.6] and [ST22, Theorem 4.1]) *Let X be a smooth del Pezzo threefold and d be a positive integer. Let M be a component of $\overline{M}_{0,0}(X, d)$. Let $x \in X$ be a point on X, and M_x denote the sublocus of M parametrizing stable maps through x.*

Then there exists a finite set $\mathsf{K} \subset X$ such that:

(i) *when $x \notin \mathsf{K}$, M_x has expected dimension $2d - 2$; and*
(ii) *when $x \in \mathsf{K}$, M_x has dimension at most $2d - 1$.*

Moreover, for $x \notin \mathsf{K}$, a general curve parametrized by M_x is irreducible.

Proof It follows from Lemma 7.2.5 that there exists a proper closed subset $V \subset X$ such that any non-free rational curve of H-degree $\leq d$ is contained in V.

The proof is by induction on d. When $d = 1$, let $\pi: \mathcal{C} \to M$ be the universal family. Then the evaluation map $\mathrm{ev}: \mathcal{C} \to X$ is dominant and generically finite. Let K be the locus of points x such that $\mathrm{ev}^{-1}(x)$ has positive dimension. If K has positive dimension, then the preimage $\mathrm{ev}^{-1}(\mathsf{K})$ has dimension ≥ 2 and must map dominantly to M because M has dimension 2. This contradicts the fact that a general curve parametrized by M is free. Thus K consists of finitely many points, and for any $x \in \mathsf{K}$, M_x has dimension 1, proving the claim.

Next, we prove the induction step. Let M' be a component of M_x. First we suppose that a general curve parametrized by M' is irreducible. If such a general curve is free, then M' has expected dimension $2d - 2$. So we may assume that all such curves are non-free. Let $Y \subset V$ be the closure of the locus swept out by curves parametrized by M'. Suppose that $\dim Y = 1$. Then a general curve C parametrized by M' maps to Y dominantly and we denote its degree by r. Then the dimension of M' is at most $2r - 2$. On the other hand, the expected dimension of M' is $-K_X.C - 2 = 2rH.Y - 2$, which is a lower bound for the dimension of W'. This forces us to have $H.Y = 1$, and M' has the expected dimension.

Now suppose that $\dim Y = 2$. Let $\beta: \widetilde{Y} \to Y$ be a smooth resolution. Since the dimension of $\beta^{-1}(x)$ is at most 1, it follows from Lemma 8.2.3 that we have

$$\dim M' \leq -K_{\widetilde{Y}}.C' - 1,$$

where C' is the strict transform of a curve C parametrized by M'. If $\dim M' \geq -K_X.C$, then we have $(K_{\widetilde{Y}} - \beta^* K_X).C' \leq -1 < 0$. This implies that

$$a(Y, -K_X|_Y) > a(X, -K_X) = 1,$$

but this contradicts Lemma 6.5.5. Thus we conclude that we always have $\dim M' \leq -K_X.C - 1 = 2d - 1$. If we have $\dim M' = -K_X.C - 1$, then we have $-K_X.C = -K_{\widetilde{Y}}.C'$. This implies that $a(Y, -K_X|_Y) = a(X, -K_X) = 1$. Lemma 6.5.5 claims that, in this case, $K_{\widetilde{Y}} - \beta^* K_X|_Y$ has Iitaka dimension 1, so we have the associated canonical fibration $p: \widetilde{Y} \to B$ whose general fiber is isomorphic to $\mathbb{P}^1$. Since C' has vanishing intersection with $K_{\widetilde{Y}} - \beta^* K_X|_Y$, we conclude that C' is a general fiber of p. This implies that

$$2d = -K_X.C = -K_{\widetilde{Y}}.C' = 2.$$

This means that $d = 1$ and $x \in \mathsf{K}$.

Next, we suppose that a general curve C parametrized by M' is reducible. We will use the dimension counting arguments of [CS09, Proposition 2.5] in this situation. First let us assume that $x \notin \mathsf{K}$ and let $f\colon C \to X$ be a general member parametrized by M'.

(i) Suppose that a node of C maps to $x \notin \mathsf{K}$. Let D be the maximal connected subcurve of C which contains the node mapping to x and is contracted by f. Let $C_1, \ldots, C_u$ be the closures of connected components of $C \setminus D$. Let $d_i = H.f_*C_i$. Note that we have $d = \sum_{i=1}^{u} d_i$. By the induction hypothesis, the dimension of W' is bounded by

$$\sum_{i=1}^{u}(2d_i - 2) + u - 2,$$

where $u - 2$ accounts for the moduli of the marked point and points where the C_is are attached to D. This is equal to $2d - u - 2$, which is smaller than the expected dimension $2d - 2$. Hence impossible.

(ii) Suppose that a node of C maps to $x_i \in \mathsf{K}$. As before, let D be the maximal connected subcurve of C which contains the node mapping to x_i and is contracted by f. Let $C_1, \ldots, C_u$ be the closures of connected components of $C \setminus D$. Let $d_i = H.f_*C_i$. Again we have $d = \sum_{i=1}^{u} d_i$. We assume that x is contained in the image of C_1. By the induction hypothesis, the dimension of W' is bounded by

$$2d_1 - 2 + \sum_{i=2}^{u}(2d_i - 1) + \max\{0, u - 3\},$$

where $\max\{0, u-3\}$ accounts for the moduli of points where the C_is are attached to D. This is equal to $2d - u - 1 + \max\{0, u - 3\}$, which is smaller than the expected dimension $2d - 2$. Again this is impossible.

(iii) Suppose that a node of C maps to $y \notin \{x\} \cup \mathsf{K}$. Again let D be the maximal connected subcurve of C which contains the node mapping to y and is contracted by f. Let $C_1, \ldots, C_u$ be the closures of connected components of $C \setminus D$. Let $d_i = H.f_*C_i$. We assume that x is contained in the image of C_1. By the induction hypothesis, the dimension of W' is bounded by

$$\sum_{i=1}^{u}(2d_i - 2) + \max\{0, u - 3\} + 1,$$

where $\max\{0, u-3\} + 1$ accounts for the moduli of $y \in f(C_1)$ and points where the C_is are attached to D. This is equal to $2d - 2u + \max\{0, u - 3\} + 1$, which is smaller than the expected dimension $2d - 2$. A contradiction.

In particular, this analysis shows that when $x \notin \mathsf{K}$, a general curve parametrized by M' must be irreducible. The case when $x \in \mathsf{K}$ is similar. We refer interested readers to [CS09, Proposition 2.5] for more details. □

An immediate consequence of Proposition 8.2.4 is the following corollary:

Corollary 8.2.5 ([LT19a, Corollary 7.7] and [ST22, Corollary 4.2]) *Let X be a smooth del Pezzo threefold and d be a positive integer. Assume that $\overline{M}_{0,0}(X,d)$ is non-empty and let M be a component of $\overline{M}_{0,0}(X,d)$. Then M generically parametrizes a free rational curve so that the dimension of M is equal to the expected dimension $2d$.*

Proof It follows from Proposition 8.2.4 that a general curve parametrized by W is irreducible. Thus our assertion follows from Lemma 8.2.2. □

Finally, we prove the existence of H-degree d rational curves for every $d \geq 1$:

Proposition 8.2.6 *Let X be a smooth del Pezzo threefold with an ample generator H for* $\mathrm{Pic}(X)$ *and d be a positive integer. Then there exists a birational free rational curve $f\colon \mathbb{P}^1 \to X$ of H-degree d.*

Proof We prove this statement by induction on d. First we prove our claim when $d = 1$. Let $S \in |H|$ be a general member of the complete linear system. When $H^3 \geq 2$, the linear system $|H|$ is basepoint free, so by Bertini's theorem, S is smooth. When $H^3 = 1$, the fact that X is a double cover of the Veronese cone ramified along the intersection with a cubic hypersurface implies that S is smooth. By adjunction, we have $-K_S \sim -K_X|_S - H|_S \sim H|_S$ which is ample. Thus S is a del Pezzo surface of degree H^3. Then H-lines on S correspond to (-1)-rational curves on S (in the case of $H^3 = 1$, there are additional rational curves in $|-K_S|$), and the existence of such curves is well known. For a cubic surface this is the famous theorem of 27 lines. See [Har77, Chapter V Theorem 4.9]. It follows from Corollary 8.2.5 that a general H-line on X is free.

Now suppose that our assertion holds for d. Then we choose a general point $x \in X$. One can find a free H-line $f_1\colon C_1 \to X$ and a birational free rational curve $f_2\colon C_2 \to X$ of H-degree d, both passing through x. We glue these two curves C_1, C_2 at x and obtain a birational stable map $f\colon C \to X$. It follows from Proposition 7.2.19 and Corollary 7.2.22 that this is a smooth point of $\overline{M}_{0,0}(X)$ and by Proposition 7.2.23 one can smooth this stable map; the resulting curve is a birational free rational curve of H-degree $d+1$. Thus our assertion follows. □

8.2.2 Spaces of Low Degree Rational Curves

Our proof of Theorem 8.2.1 is based on induction on the H-degree, and in this section, we consider the base cases, that is, the spaces of rational curves of H-degree 1 and 2. We state our theorems for all degrees $H^3 = 1, \dots, 5$; however for simplicity, we restrict our proofs for $H^3 = 3$ whenever this assumption simplifies the proofs.

The Space of H-lines

Our main result here is the following theorem:

Theorem 8.2.7 *Let X be a del Pezzo threefold with an ample generator H. When $H^3 = 1$ or 2, we assume that X is general in its moduli. Then the moduli space $\overline{M}_{0,0}(X, 1)$ is an irreducible smooth projective surface.*

Let us explain the history of this theorem. The moduli space $\overline{M}_{0,0}(X, 1)$ is isomorphic to the Hilbert scheme of lines on X, and this scheme has been extensively studied under the name of the *Fano scheme of lines*. For $H^3 = 5$, this theorem is [Isk79, Chapter III Proposition 1.6]. For $H^3 = 4$, this was established in [New68, Rei72]. For $H^3 = 3$, the Fano schemes were studied in [Tju71, AK77]. The statements for $H^3 = 1, 2$ can be found in [Isk79, Chapter III Remark 1.7], and a detailed study can be found in [Tik80, Tik81]. We briefly explain a proof for $H^3 = 3$ following [AK77].

Proof It follows from Corollary 8.2.5 that $\overline{M}_{0,0}(X, 1)$ is an equidimensional projective surface. Let us prove that $\overline{M}_{0,0}(X, 1)$ is smooth when $H^3 \geq 3$. Under this assumption the divisor class H is very ample, so in particular, any H-line ℓ is a smooth rational curve. Thus its normal sheaf $N_{\ell/X}$ is a locally free sheaf of rank 2 and its degree is equal to 0. Let $S \in |H|$ be a general member containing ℓ. As explained before, S is a smooth del Pezzo surface and ℓ is a (-1)-rational curve on S. We have the exact sequence of locally free sheaves:

$$0 \to N_{\ell/S} \to N_{\ell/X} \to N_{S/X}|_{\ell} \to 0,$$

which is equal to

$$0 \to \mathcal{O}(-1) \to N_{\ell/X} \to \mathcal{O}(1) \to 0.$$

This implies that the only possibilities of $N_{\ell/X}$ are either

$$\mathcal{O}(-1) \oplus \mathcal{O}(1) \text{ or } \mathcal{O} \oplus \mathcal{O}.$$

We conclude that $H^1(\ell, N_{\ell/X}) = 0$ in either case. This shows that the moduli space $\overline{M}_{0,0}(X, 1)$ is smooth.

Finally we briefly explain how to prove $\overline{M}_{0,0}(X, 1)$ is connected when $H^3 = 3$. The irreducibility of $\overline{M}_{0,0}(X, 1)$ follows from this connectedness and the smoothness of $\overline{M}_{0,0}(X, 1)$. The moduli space $\overline{M}_{0,0}(X, 1)$ is a closed subscheme of the Grassmannian

$$\mathrm{Gr}(2, 5).$$

Then one may realize $\overline{M}_{0,0}(X, 1)$ as a zero locus of a section of a certain vector bundle on Gr(2, 5). Applying Koszul complex techniques to this vector bundle, one can prove that

$$h^0(\overline{M}_{0,0}(X, 1), \mathcal{O}) = 1,$$

showing the connectedness of $\overline{M}_{0,0}(X, 1)$. Interested readers should consult [AK77] for more details. □

From now on, we assume that X is general in its moduli when $H^3 = 1$ or 2. Let us discuss a few geometric facts on the moduli space $\overline{M}_{0,0}(X, 1)$ and its universal family. Under the assumption we made, the moduli space $R_1 := \overline{M}_{0,0}(X, 1)$ is an irreducible smooth projective surface. Let $R_1^{(1)} \to R_1$ be the universal family of lines. Then its dimension is 3, and it follows from Corollary 8.2.5 that the evaluation map $\mathrm{ev}_1 : R_1^{(1)} \to X$ is dominant and generically finite. In particular, Lemma 5.2.2 implies that

$$a(R_1^{(1)}, \mathrm{ev}_1^* H) \leq a(X, H) = 2.$$

On the other hand, any H-line $\ell \subset R_1^{(1)}$ satisfies $a(\ell, H|_\ell) = 2/H.\ell = 2$, so it follows from Lemma 5.1.2 that we have $a(R_1^{(1)}, \mathrm{ev}_1^* H) = 2$. We conclude that $\mathrm{ev}_1 : R_1^{(1)} \to X$ is an a-cover of X. Similarly for any generically finite dominant morphism $Z \to R_1$ from a projective surface Z, one can prove that its base change $C = R_1^{(1)} \times_{R_1} Z \to R_1^{(1)} \to X$ is an a-cover of X. The following lemma shows that the converse is also true:

Lemma 8.2.8 *We assume that X is general in its moduli when $H^3 = 1$ or 2. Let $f : Y \to X$ be an a-cover of X with respect to H from a smooth projective threefold. Then $f : Y \to X$ is birational to a base change of $R_1^{(1)} \to R_1$ via a dominant and generically finite morphism $Z \to R_1$ from a projective surface Z.*

Proof It follows from Proposition 6.5.10 that the Iitaka dimension of

$$2f^*H + K_Y$$

is positive. Suppose that its dimension is equal to 1. Let $Y \dashrightarrow Z$ be the canonical fibration associated to $2f^*H + K_Y$ and S be a general fiber of this fibration. Let $\beta : \widetilde{S} \to S$ be a smooth resolution. Then $a(S, f^*H|_S) = 2$ and $(\widetilde{S}, \beta^* f^*H)$ is adjoint rigid. By Lemma 5.2.2, the image $(\Gamma = f(S), H|_\Gamma)$ is also adjoint rigid with $a(\Gamma, H|_\Gamma) = 2$. However, this contradicts Lemma 6.5.5.

Thus the Iitaka dimension of $2f^*H + K_Y$ is equal to 2. Let $Y \dashrightarrow Z$ be the canonical fibration again. Let $C \subset Y$ be a general fiber of this fibration. Then we must have $a(C, f^*H|_C) = a(Y, f^*H) = 2$. On the other hand, we also have $a(C, f^*H|_C) = 2/f^*H.C$ showing that $f|_C : C \to X$ is a birational stable map to an H-line. By the universal property, there exists a generically finite and dominant rational map $Z \dashrightarrow R_1$. After resolving indeterminacy we obtain a morphism $Z' \to R_1$. Then $f\colon Y \to X$ is birational to the fiber product $R_1^{(1)} \times_{R_1} Z' \to R_1^{(1)} \to X$. Thus our assertion follows. □

These geometric facts enable us to state useful properties about the moduli space $\overline{M}_{0,0}(X, d)$. Let us explain these in detail. First let us consider a degree-d cover from $\mathbb{P}^1$ to an H-line $\ell \subset X$:

$$f\colon \mathbb{P}^1 \to \ell \subset X.$$

The locus of $\overline{M}_{0,0}(X, d)$ parametrizing such curves has dimension

$$2 + 2d - 2 = 2d,$$

where 2 accounts for the dimension of the space of H-lines, and $2d-2$ accounts for the moduli of degree-d stable maps to ℓ. In particular, this coincides with the dimension of $\overline{M}_{0,0}(X, d)$, so such a locus forms a component of the moduli space. Let us denote the component generically parametrizing degree-d covers of H-lines in $\overline{M}_{0,0}(X, d)$ by N_d. We prove that this is the only component generically parametrizing non-birational stable maps:

Lemma 8.2.9 *Assume that X is general when $H^3 = 1$ or 2. Let $M \subset \overline{M}_{0,0}(X, d)$ be a component generically parametrizing non-birational stable maps. Then $M = N_d$.*

Proof Let $f\colon C \to X$ be a general member of M. It follows from Corollary 8.2.5 that C is irreducible. Let $C' = f(C)$; we denote $H.C'$ by e and the degree of $C \to C'$ by $r > 1$. Then we have $d = er$. It follows from Corollary 8.2.5 that the dimension of M is given by

$$2e + 2r - 2,$$

which is supposed to be equal to $2d$. This is only possible when $e = 1$. Thus our assertion follows. □

The next proposition is an application of analysis of a-covers, and this is a key statement for our inductive proof of Theorem 8.2.1:

Proposition 8.2.10 *Assume that X is general when $H^3 = 1$ or 2. Let $d \geq 2$ be an integer and $M \subset \overline{M}_{0,0}(X, d)$ be a component generically parametrizing birational stable maps; we denote its universal family by $M^{(1)}$ with the evaluation map $\mathrm{ev}\colon M^{(1)} \to X$. Let $\beta\colon \widetilde{M} \to M^{(1)}$ be a smooth resolution. Then a general fiber of $\mathrm{ev} \circ \beta$ is irreducible.*

Proof Suppose that a general fiber of $\mathrm{ev} \circ \beta$ is not irreducible. Since this fiber is smooth, our assumption implies that it is not connected. Let $g: Y \to X$ be a finite part of the Stein factorization of $\mathrm{ev} \circ \beta: \widetilde{M} \to X$. Then as explained in Proposition 7.4.12, our construction indicates that $g: Y \to X$ is an a-cover. It follows from Lemma 8.2.8 that, after applying a birational modification $\beta: \widetilde{Y} \to Y$, the morphism $g \circ \beta: \widetilde{Y} \to X$ factors through the evaluation map $\mathrm{ev}_1: R_1^{(1)} \to X$. Let $f: C \to \widetilde{Y}$ be the strict transform of a general member of M parametrizing curves on Y. Then we have

$$\dim M = -f^* K_{\widetilde{Y}}.C = -f^* \beta^* g^* K_X.C.$$

This implies that $(2\beta^* g^* H + K_{\widetilde{Y}}).f_* C = 0$. It follows from Lemma 8.2.8 that the image of C on X is an H-line, but this contradicts the fact that a general member of M is a birational stable map. Thus our assertion follows. □

The Space of H-conics

Next, we consider the space of H-conics. Here is the main theorem of this section:

Theorem 8.2.11 *Let X be a del Pezzo threefold with an ample generator H for* Pic(X). *When $H^3 = 1$ or 2, we assume that X is general in its moduli. Then the moduli space $\overline{M}_{0,0}(X, 2)$ consists of two components R_2 and N_2; the former generically parametrizes birational stable maps from $\mathbb{P}^1$ and the latter generically parametrizes double covers of H-lines.*

The proof of this theorem is explained in [LT19a, Proposition 7.4] when $H^3 \geq 2$ and the proof for the case of $H^3 = 1$ is given in [ST22, Theorem 3.9].

Proof We prove this statement when $H^3 = 3$, that is, X is a smooth cubic threefold in $\mathbb{P}^4$. By Lemma 8.2.9, we only need to consider birational stable maps, that is, an H-conic $C \subset X$ such that $H.C = 2$. Such a curve C is contained in a unique plane $P \subset \mathbb{P}^4$. Then the intersection $X \cap P$ is the union of a conic C and a line ℓ. So the locus parametrizing H-conics admits a map to R_1. Conversely, for each H-line $\ell \subset X$, planes $P \subset \mathbb{P}^4$ containing ℓ are parametrized by $\mathbb{P}^2$ showing that the space of H-conics is birational to a $\mathbb{P}^2$-bundle over R_1. Thus this locus is irreducible, proving the claim. □

8.2.3 Movable Bend and Break

Our proof of Theorem 8.2.1 is based on an induction on the H-degree, and a key to this part of the argument is Bend and Break techniques developed in Section 8.1. However, we need a more controlled Bend and Break that we call the *movable Bend and Break lemma*. First let us state a movable Bend and Break lemma for arbitrary smooth Fano threefolds:

Theorem 8.2.12 ([BLRT22, Theorem 1.4]) *Let k be an algebraically closed field of characteristic 0. Let X be a smooth Fano threefold defined over k. Let $M \subset \overline{M}_{0,0}(X)$ be a component generically parametrizing free rational curves. Assume that the anticanonical degree of rational curves parametrized by M is ≥ 6. Then M contains a stable map $f : C \to X$ where C has exactly two components and the restriction of f to each component realizes this curve as a free rational curve on X.*

However, instead of proving this general statement, we focus on giving a proof for the following statement:

Theorem 8.2.13 *Let X be a del Pezzo threefold and $d \geq 2$ be an integer. We assume that X is general in its moduli when $H^3 = 1$ or 2. Let*

$$M \subset \overline{M}_{0,0}(X, d)$$

be a component generically parametrizing birational stable maps. Then M contains a birational stable map $f : C \to X$ where C has exactly two components and the restriction of f to each component realizes this curve as a free rational curve on X.

To this end, first we prove the following lemma:

Lemma 8.2.14 ([LT21, Lemma 8.1] and [ST22, Proposition 3.2]) *Let X be a del Pezzo threefold. We assume that X is general in its moduli when $H^3 = 1$ or 2. Let $d \geq 2$ be an integer and $M \subset \overline{M}_{0,0}(X, d)$ be a component generically parametrizing birational stable maps. Then a general member*

$$[f : C \to X] \in M$$

is a very free rational curve.

Such a statement was first proved in [LT21, Lemma 8.1] for prime Fano threefolds of index 1, then [ST22, Proposition 3.2] used this argument to prove the above statement for del Pezzo threefolds of degree 1.

Proof We prove this statement when $H^3 = 3$. It follows from Lemma 7.2.10 that it suffices to show that the evaluation map

$$\mathrm{ev}_2 : M^{(2)} \to X \times X$$

is dominant where $M^{(2)} \subset \overline{M}_{0,2}(X, d)$ is the component above M.

Suppose not. Let $x_1 \in X$ be a general point and consider the fiber of the evaluation map $\mathrm{ev}_1 : M^{(1)} \to X$ at $x_1 \in X$ where $M^{(1)} \subset \overline{M}_{0,1}(X, d)$ is the component above M. Since M generically parametrizes free rational curves, M has expected dimension $2d$ and $\mathrm{ev}_1 : M^{(1)} \to X$ is dominant. We conclude that the fiber $\mathrm{ev}_1^{-1}(x_1)$ is equidimensional and we have

$$\dim \mathrm{ev}_1^{-1}(x_1) = 2d - 2 \geq 2.$$

Now ev_2 is not dominant, and M generically parametrizes birational stable maps. These imply that stable maps parametrized by a component T of $\mathrm{ev}_1^{-1}(x_1)$ sweep out a surface $S \subset X$. Let $\beta \colon \widetilde{S} \to S$ be a smooth resolution such that $\beta^{-1}(x_1)$ is divisorial. Let $C \subset S$ be a rational curve parametrized by T, and $\widetilde{C}$ be the strict transform of C on $\widetilde{S}$. Since any deformation of $\widetilde{C}$ will meet $\beta^{-1}(x_1)$, the expected dimension of the moduli of $\widetilde{C}$ on $\widetilde{S}$ coincides with the dimension of T. This means that we have

$$-K_{\widetilde{S}}.\widetilde{C} - 1 = 2d - 2,$$

hence we have

$$-K_{\widetilde{S}}.\widetilde{C} = 2d - 1.$$

On the other hand, it follows from Lemma 6.5.5 that we have

$$a(S, H|_S) \leq a(X, H) = 2.$$

Now the adjoint divisor

$$a(S, H|_S)\beta^* H + K_{\widetilde{S}}$$

is pseudo-effective, hence we conclude that

$$(a(S, H|_S)\beta^* H + K_{\widetilde{S}}).\widetilde{C} \geq 0.$$

This inequality is simplified to

$$a(S, H|_S) \geq \frac{2d - 1}{d}.$$

It follows from Lemma 6.5.3 that $a(S, H|_S)$ takes the form of $2/n$ or $3/n$ where n is a positive integer. This discussion implies that when $d \geq 3$, we have $a(S, H|_S) = 2$, and when $d = 2$, we have $a(S, H|_S) = 2$ or $3/2$.

Assume that $a(S, H|_S) = 2$. Then by Lemma 6.5.5, $(S, H|_S)$ is not adjoint rigid. The canonical fibration associated to $2\beta^* H + K_{\widetilde{S}}$ defines a fibration $\pi \colon \widetilde{S} \to B$. Moreover, by Theorem 6.5.4, general fibers of π are the strict transforms of H-lines. Now since we have

$$(2\beta^* H + K_{\widetilde{S}}).\widetilde{C} = 1,$$

the strict transform $\widetilde{C}$ maps dominantly to B. This implies that B is isomorphic to $\mathbb{P}^1$. However, it is known that the Fano variety R_1 of lines does not contain any rational curves. Over $\mathbb{C}$, the Abel–Jacobi map embeds R_1 into the intermediate Jacobian, which is a 5-dimensional principally polarized abelian variety. Since no abelian variety contains a rational curve, our assertion follows. See [CG72] for more details. Over an arbitrary algebraically closed field of characteristic 0, note that any scheme of finite type over k is defined over a finitely generated field over $\mathbb{Q}$. Using this observation, one can reduce the problem to the case over $\mathbb{C}$.

Finally, let us assume that $a(S, H|_S) = 3/2$. This case happens only when $d = 2$. By Theorem 6.5.4, there exists a birational morphism $\phi\colon \widetilde{S} \to \mathbb{P}^2$ such that $O(\beta^* H) = \phi^* O(2)$. This means that $S.H^2 = (\beta^* H)^2 = 4$, but this is impossible because $H^3 = 3$. □

As a corollary we have the following:

Corollary 8.2.15 *Let X be a del Pezzo threefold. We assume that X is general in its moduli when $H^3 = 1$ or 2. Let $d \geq 1$ be an integer and $M \subset \overline{M}_{0,0}(X, d)$ be a component generically parametrizing birational stable maps. Then a general member $[f\colon C \to X] \in M$ is a closed embedding from $\mathbb{P}^1$ to X.*

Proof When $d = 1$, the assertion follows from Theorem 8.2.7 and the fact that a general line is smooth. (This is obvious in the cases of $H^3 \geq 3$. When $H^3 = 1$ or 2, X is realized as a double cover of a projective threefold Y ramified along a smooth divisor D, and a general H-line on X is one component of the pullback of a smooth rational curve on Y, which is everywhere tangent to D.)

When $d \geq 2$, Lemma 8.2.14 implies that a general member $[f\colon C \to X] \in M$ is very free. Then by Lemma 7.2.14, a general member f is an embedding. □

Let $d \geq 1$ be a positive integer and $M \subset \overline{M}_{0,0}(X, d)$ be a component generically parametrizing birational stable maps. The previous corollary indicates that for a general member $[f\colon C \to X] \in M$, f is a closed embedding, so the normal sheaf N_f is a locally free sheaf of rank 2. Since $C = \mathbb{P}^1$, we can write

$$N_f \cong O(a) \oplus O(b),$$

where a, b are integers. Moreover, since f is free, we must have $a \geq 0$ and $b \geq 0$. The following theorem was proved by Shen and significantly simplifies the proof of Theorem 8.2.13:

Theorem 8.2.16 ([She12, Theorem 1.4]) *Let X be a del Pezzo threefold. We assume that X is general in its moduli when $H^3 = 1$ or 2. Let d be a positive integer and $M \subset \overline{M}_{0,0}(X, d)$ be a component generically parametrizing* *birational stable maps. Then for a general* $[f\colon C \to X] \in M$, *we have*

$$N_f \cong \mathcal{O}(d-1) \oplus \mathcal{O}(d-1).$$

Proof As explained above, the normal sheaf N_f is a locally free sheaf of rank 2 and degree $2d-2$. Thus our assertion easily follows when $d = 1$.

Suppose that $d \geq 2$. Lemma 8.2.14 shows that f is very free so we can apply [She12, Theorem 1.4] stating that $|a-b| \leq 1$, where we write

$$N_f \cong \mathcal{O}(a) \oplus \mathcal{O}(b).$$

Since we also have $a+b = 2d-2$, our assertion follows. Note that [She12, Theorem 1.4] is stated over $\mathbb{C}$, but one might reduce the case over arbitrary algebraically closed fields of characteristic 0 to the case over $\mathbb{C}$ by using the standard arguments explained in the proof of Lemma 8.2.14 combined with constancy of Chern classes in a flat family of locally free sheaves of constant rank. □

As a corollary, we have

Corollary 8.2.17 *Let X be a del Pezzo threefold. We assume that X is general in its moduli when $H^3 = 1$ or 2. Let d be a positive integer and $M \subset \overline{M}_{0,0}(X,d)$ be a component generically parametrizing birational stable maps. Let $M^{(n)} \subset \overline{M}_{0,n}(X,d)$ be the component above M; we denote its evaluation map by $\mathrm{ev}_n \colon M^{(n)} \to X^n$. Then ev_n is dominant if and only if $n \leq d$.*

Proof Since M has the expected dimension, the dimension of $M^{(n)}$ is equal to $2d+n$. The dimension of X^n is $3n$, so it is easy to see that $n \leq d$ is a necessary condition for ev_n to be dominant.

Assume that $n \leq d$. Let $f\colon C \to X$ be a general member of M; we choose distinct general points $p_1, \ldots, p_n$ on C. Let $Z \subset M$ be the subscheme parametrizing $f'\colon C' \to X$ such that $f'(C')$ contains $f(p_i)$ for $i = 1, \ldots, n$. Then $H^0(C, N_f(-p_1 - \cdots - p_n))$ is the Zariski tangent space of Z at $[f]$ and its obstruction space is given by $H^1(C, N_f(-p_1 - \cdots - p_n))$. By Theorem 8.2.16 and our assumption, the obstruction space vanishes, so that Z has expected dimension $2(d-n)$. This implies that the fiber of ev_n at $(f(p_1), \ldots, f(p_n))$ has dimension $2(d-n)$ at $[f, (p_1, \ldots, p_n)]$. This means that the image has to have dimension $3n$, proving that ev_n is dominant. □

Now we prove our main theorem of this section:

Proof of Theorem 8.2.13 When $d = 2$, our assertion follows from Theorem 8.2.11 because R_2 must contain a gluing of two distinct free H-lines. Thus let us assume that $d \geq 3$.

Let L be a very ample divisor on X and let D be a general complete intersection of two divisors in $|L|$. By Bertini's theorem, this is a smooth irreducible curve. We also fix $d-1$ general points $x_1, \ldots, x_{d-1} \in X$. It follows from Corollary 8.2.17 that we can find birational stable maps $f\colon C \to X$ parametrized by M such that $x_i \in f(C)$ for $i = 1, \ldots, d-1$ and $f(C)$ intersects with D. Moreover, such birational stable maps form 1-dimensional loci on M. Since $d-1 \geq 2$, Corollary 8.1.6 implies that there exists a stable map $f\colon C' \to X$ in these loci such that C' contains at least two non-contracted components C_1, C_2 such that $x_i \in f(C_i)$ for $i = 1, 2$. Let $C_1, \ldots, C_s$ be non-contracted components of C' whose images contain our general points. and let $C_{s+1}, \ldots, C_{s+t}$ be non-contracted components of C' which do not contain general points. Let us denote the H-degree of C_i by d_i and the number of general points contained in $f(C_i)$ by n_i. It follows from Corollary 8.2.17 that we have

$$d-1 \leq \sum_{i=1}^{s} n_i \leq \sum_{i=1}^{s} d_i \leq \sum_{i=1}^{s+t} d_i = d.$$

This implies that $t = 0$ or 1. Suppose first that $t = 0$. Then we may assume that $n_i = d_i$ for $i = 1, \ldots, s-1$ and $n_s = d_s - 1$. For $i = 1, \ldots, s-1$, C_i contains the maximum number of general points, and by Corollary 8.2.17, there are only finitely many possibilities for C_i. It follows from Corollary 7.2.7 that those curves are disjoint if $s-1 \geq 2$. Also those curves do not intersect D. On the other hand, C_s contains $d_s - 1$ general points and it must intersect with D. Such curves form 1-dimensional loci. Among them there are only finitely many C_s which intersect with exactly one of the C_is for $i = 1, \ldots, s-1$. Since $f(C')$ has to be connected, we conclude that $s = 2$. Since C_1, C_2 contain general points, those curves are free.

The case of $t = 1$ is similar. We must have $d_{s+1} = 1$. Curves C_i for $i = 1, \ldots, s$ must contain the maximum number of general points, so they are disjoint and C_{s+1} must meet D. Such curves C_{s+1} form 1-dimensional loci, and among them there are only finitely many curves such that they meet exactly one of the C_is for $i = 1, \ldots, s$. This implies that $s = 1$. But this contradicts the fact that $s \geq 2$. Thus our assertion follows. □

8.2.4 Geometric Manin's Conjecture II

Finally, we prove Theorem 8.2.1:

Proof of Theorem 8.2.1 We prove this statement by induction on d. The cases $d = 1, 2$ follow from Theorems 8.2.7 and 8.2.11.

We assume that the statement holds up to $d \geq 2$. Let d_1, d_2 be positive integers such that $d + 1 = d_1 + d_2$. We consider the fiber product $R^{(1)}_{d_1} \times_X R^{(1)}_{d_2}$. We claim that there is a unique component $R_{d_1,d_2} \subset R^{(1)}_{d_1} \times_X R^{(1)}_{d_2}$ that maps dominantly to both $R^{(1)}_{d_1}$ and $R^{(1)}_{d_2}$. Let $U \subset X$ be a non-empty Zariski open subset such that any rational curve of H-degree $\leq d + 1$ meeting U is free. Such a U exists by Lemmas 7.2.5 and 8.2.2. Let $R_{d_i,\circ} \subset R_{d_i}$ be the Zariski open subset parametrizing free birational stable maps from $\mathbb{P}^1$; we denote the associated universal curve by $R^{(1)}_{d_i,\circ} \subset R^{(1)}_{d_i}$. Then it follows from Lemma 7.2.6 that the evaluation map $\mathrm{ev}_{d_i} : R^{(1)}_{d_i,\circ} \to X$ is flat. Thus it follows from Proposition 8.2.10 that

$$R^{(1)}_{d_1,\circ} \times_X R^{(1)}_{d_2,\circ}$$

is irreducible. Note that we have either $d_1 \geq 2$ or $d_2 \geq 2$ by $d_1 + d_2 = d + 1 \geq 3$ so Proposition 8.2.10 applies. Moreover, our proof shows that we have

$$R^{(1)}_{d_1,\circ} \times_X R^{(1)}_{d_2,\circ} \subset R_{d_1,d_2}.$$

Let $M \subset \overline{M}_{0,0}(X, d + 1)$ be a component that is different from N_d. It follows from Corollary 8.2.5 and Lemma 8.2.9 that M generically parametrizes free birational stable maps from $\mathbb{P}^1$. By Theorem 8.2.13, M contains a birational stable map $f : Z \to X$ consisting of two free rational curves of H-degrees d_1 and d_2. Such a stable map is a smooth point of $\overline{M}_{0,0}(X, d+1)$ by Corollary 7.2.22 and Proposition 7.2.19. This implies that M contains the entire image of $R_{d_1,d_2} \to \overline{M}_{0,0}(X, d + 1)$. Now we claim that M contains the entire image of $R_{1,d}$. Once we prove this claim, the uniqueness of M follows from the irreducibility of $R_{1,d}$, Corollary 7.2.22, and Proposition 7.2.19. We may assume that $d_1 \geq 2$. Since R_{d_1,d_2} contains $R^{(1)}_{d_1,\circ} \times_X R^{(1)}_{d_2,\circ}$, M contains a birational stable map consisting of two free rational curves of H-degree d_1 and d_2 such that the image x of the node is a general point contained in U. Since R_{d_1} is unique, it contains the image of a component of $R^{(1)}_{1,\circ} \times_X R^{(1)}_{d_1-1,\circ}$ generically parametrizing the unions of two distinct birational free rational curves. Thus we conclude that M contains a birational stable map which is a chain of three free rational curves $f_i : C_i \to X (i = 1, 2, 3)$ of H-degrees $1, d_1 - 1, d_2$ respectively where the degree of the middle curve is $d_1 - 1$ and the image of the intersection of C_2, C_3 is x. Since such a stable map is also a smooth point of $\overline{M}_{0,0}(X, d + 1)$, we conclude that M contains $R_{1,d}$. Thus our claim follows. □

As a corollary, we have the following:

Corollary 8.2.18 *Let X be a smooth del Pezzo threefold with* $\mathrm{Pic}(X) = \mathbb{Z}H$ *and $-K_X \sim 2H$. We assume that X is general in its moduli when $H^3 = 1$ or 2. Then Conjecture 7.4.10 holds for X.*

Proof When $d \geq 2$, there are two components R'_d, N'_d of $\mathrm{Mor}(\mathbb{P}^1, X, d)$ corresponding to R_d, N_d respectively. Since N'_d parametrizes rational curves on the a-cover $R_1^{(1)} \to X$, N'_d is an exceptional component. The component R'_d is a Manin component by Lemmas 8.2.8 and 8.2.9. □

Remark 8.2.19 Our main result shows that any algebraic class of 1-cycles is represented by a multiple of an H-line, implying that the Brauer group of X is trivial.

8.3 Del Pezzo Fibrations

Let k be an algebraically closed field of characteristic 0. Let B be a smooth projective curve defined over k; we denote its function field by $K = k(B)$. We are interested in a geometrically integral smooth Fano variety X defined over K and its set of K-rational points $X(K)$. To this end, it is natural to consider the following integral models of X over B:

Definition 8.3.1 (Fano fibrations) Let X be a geometrically integral smooth Fano variety defined over $K = k(B)$ where B is a smooth projective curve defined over k. Let $\pi\colon \mathcal{X} \to B$ be an integral model of X over B as defined in Definition 7.1.11. We say $\pi\colon \mathcal{X} \to B$ is a *Fano fibration* if $\mathcal{X}$ is a normal $\mathbb{Q}$-Gorenstein projective variety such that $\pi_* \mathcal{O}_{\mathcal{X}} \cong \mathcal{O}_B$. We say π is a *smooth model* if $\mathcal{X}$ is also smooth. Such a model always exists by a resolution of singularities. When X has dimension 2, we say $\pi\colon \mathcal{X} \to B$ is a *del Pezzo fibration*. Note that Fano fibrations as defined in [LRT25c] are called smooth models in this book.

Let $\pi\colon \mathcal{X} \to B$ be a Fano fibration. Then, by the valuative criterion, the set $X(K)$ of K-rational points is identified with the space of sections

$$\mathrm{Sec}(\mathcal{X}/B).$$

In this section, we study the geometry of the space of sections when $\pi\colon \mathcal{X} \to B$ is a del Pezzo fibration. Our main reference for this section is [LT24]. Before going into details, we recall the following important theorem, which guarantees the existence of sections:

Theorem 8.3.2 ([GHS03, Theorem 1.1]) *Let $\pi\colon \mathcal{X} \to B$ be an algebraic fiber space from a smooth projective variety defined over k. Assume that the generic fiber $\mathcal{X}_\eta$ is geometrically rationally connected. Then $X(K)$ is non-empty.*

For weak approximation of jets by sections, one might consult [HT06].

8.3.1 Preliminaries

Here we record some preliminaries for this section.

Rational Curves on Gorenstein Terminal Threefolds

Here we recall some deformation theory of rational curves on singular threefolds. First, we introduce some definitions:

Definition 8.3.3 Let X be a variety defined over k that may not be smooth. We say a rational curve $f\colon \mathbb{P}^1 \to X$ is *free* if the image $f(\mathbb{P}^1)$ is contained in the smooth locus X_{sm} of X and $f\colon \mathbb{P}^1 \to X_{\mathrm{sm}}$ is free.

We say f is *very free* if again the image $f(\mathbb{P}^1)$ is contained in the smooth locus X_{sm} of X and $f\colon \mathbb{P}^1 \to X_{\mathrm{sm}}$ is very free.

Next, we record the following lemma, omitting its proof:

Lemma 8.3.4 ([Kol96, Chapter II 1.3 Theorem]) *Let X be a locally complete intersection variety. Let $f\colon \mathbb{P}^1 \to X$ be a rational curve such that the image $f(\mathbb{P}^1)$ meets the smooth locus X_{sm} of X. Then for any component $M \subset \mathrm{Mor}(\mathbb{P}^1, X)$ containing $[f]$ we have*

$$\dim M \geq \deg(-f^*K_X) + \dim X.$$

As before, the right-hand side is called the *expected dimension* of M.

Corollary 8.3.5 ([LT24, Lemma 2.2]) *Let X be a Gorenstein terminal threefold. Then for any component $M \subset \mathrm{Mor}(\mathbb{P}^1, X)$, we have*

$$\dim M \geq \deg(-f^*K_X) + 3,$$

where $[f\colon \mathbb{P}^1 \to X] \in M$ is a general member of M.

Proof It follows from [KM98, Corollary 5.38] that a Gorenstein terminal threefold X has only isolated cDV singularities as defined in [KM98, Definition 5.32]. In particular, any rational curve $f\colon \mathbb{P}^1 \to X$ meets the smooth locus of X. Thus our assertion follows from Lemma 8.3.4 once we verify that X has only locally complete intersection singularities. By [Kaw24, Proposition 2.3.3], any cDV singularity is analytically isomorphic to a hypersurface singularity. For a local ring of varieties, being locally complete intersection can be detected on the completion (see, e.g., the discussion in [Eis95, Section 18.5, p. 462]), so we conclude our assertion. □

Our main lemma in this section is the following:

Lemma 8.3.6 ([LT24, Lemma 2.3]) *Let X be a Gorenstein terminal projective threefold. Let $M \subset \mathrm{Mor}(\mathbb{P}^1, X)$ be a component which parametrizes a dominant family of rational curves. Then a general member $f \in M$ is free.*

Proof Suppose that a general member $f\colon \mathbb{P}^1 \to X$ avoids the singular locus of X. Then our assertion follows from Lemma 7.2.5. So it suffices to show that this assumption holds.

Assume that every member of M meets the singular locus of X. Let

$$\beta\colon X' \to X$$

be a smooth resolution. Since X has only terminal singularities, there exists an effective exceptional divisor $E \geq 0$ such that

$$K_{X'} \sim \beta^* K_X + E.$$

Moreover, the support of E coincides with the exceptional locus of β. Let $C \in M$ be a general member and $\widetilde{C}$ be the strict transform of C. Then our assumption implies that we have $E.\widetilde{C} > 0$, so that we conclude

$$-K_{X'}.\widetilde{C} < -K_X.C. \tag{8.1}$$

On the other hand, since M parametrizes a dominant family, it follows from Lemma 7.2.5 that a general $\widetilde{C}$ is free on X'. Hence we conclude

$$\dim M = -K_{X'}.\widetilde{C} + 3.$$

However, Lemma 8.3.5 indicates that we have

$$\dim M \geq -K_X.C + 3.$$

This contradicts (8.1). Thus our assertion follows. □

Finally, note the difference between the above lemma and Lemma 7.2.5. While Lemma 7.2.5 considers an arbitrary dominant family of rational curves, Lemma 8.3.6 concerns only a dominant family of rational curves that is parametrized by a component of the morphism scheme. The following example shows that this restriction is necessary:

Example 8.3.7 Let $Q \subset \mathbb{P}^4$ be a quadric cone with the vertex as a unique singularity. This is a Gorenstein terminal projective threefold. Any hyperplane section cuts out a quadric surface, and lines on such surfaces form a 3-dimensional dominant family of rational curves on Q. Any line on a hyperplane section avoiding the vertex does not pass through the unique singular point, and such a general line is a free rational curve. However, lines passing through the vertex form a 2-dimensional dominant family of rational curves, and any member of this family passes through a singular point. Therefore these lines are not free.

Further Remarks on Fano Fibrations

Here we record a few properties of Fano fibrations. In this section, B is a smooth projective curve defined over k.

Lemma 8.3.8 *Let $\pi\colon \mathcal{X} \to B$ be a Fano fibration. Then we have*

(i) *any smooth fiber of π is rationally connected;*
(ii) *for a smooth fiber $\mathcal{X}_b$, the rank of the Néron–Severi space $N^1(\mathcal{X}_b)$ stays constant. In particular, let $U \subset B$ be a non-empty Zariski open subset such that $\pi^{-1}(U) \to U$ is smooth. Then $N^1(\mathcal{X}_b)$ forms a local system on U in the étale topology.*

Proof First we prove (i). From Lemma 7.2.18, we know that every smooth fiber is rationally chain connected. Thus our assertion follows from Proposition 7.2.16.

For (ii), our Fano fibration can be defined over a finitely generated field over $\mathbb{Q}$. Then such a field can be embedded into $\mathbb{C}$. Hence our claim follows from the corresponding claim over $\mathbb{C}$. Since every smooth fiber $\mathcal{X}_b$ is rationally connected, we have $H^1(\mathcal{X}_b, O) = 0$. Thus Hodge theory implies that we have

$$N^1(\mathcal{X}_b) \cong H^2(\mathcal{X}_b, \mathbb{R}).$$

Any proper smooth fibration is topologically isotrivial (see, e.g., [Voi02, Theorem 9.3]). Thus our assertion follows. □

Finally, we record the following proposition:

Proposition 8.3.9 ([Wiś09, Theorem 1] and [dFH11, Theorem 6.8 and preceding discussion]) *Let $\pi\colon \mathcal{X} \to B$ be a Fano fibration. Let L be a $\mathbb{Q}$-Cartier $\mathbb{Q}$-divisor on $\mathcal{X}$. Then the following statements are equivalent:*

(i) *$L|_{\mathcal{X}_b}$ is ample for some smooth Fano fiber $\mathcal{X}_b$ of π;*
(ii) *$L|_{\mathcal{X}_b}$ is ample for all smooth Fano fibers $\mathcal{X}_b$ of π; and*
(iii) *$L|_{\mathcal{X}_\eta}$ is ample.*

Moreover, the same statements are true for nefness, bigness, and pseudo-effectiveness as well.

Proof The statements for ampleness and nefness follow from [Wiś09, Theorem 1] and the claims for bigness and pseudo-effectiveness follow from [dFH11, Theorem 6.8 and preceding discussion]. Both results are about the constancy of the relevant cones in a family of Fano manifolds over $\mathbb{C}$. Let us explain our statement for ampleness.

First we prove the statement over $\mathbb{C}$. Being ample is an open condition, so $L|_{\mathcal{X}_\eta}$ is ample if and only if $L|_{\mathcal{X}_b}$ is ample for any general smooth Fano fibers

$\mathcal{X}_b$ of π. Then [Wiś09, Theorem 1] establishes the constancy of the nef cone of divisors in a family of smooth Fano manifolds, so $L|_{\mathcal{X}_b}$ being ample for one smooth Fano fiber implies the same for all smooth Fano fibers. Thus our assertion follows.

Over an arbitrary algebraically closed field of characteristic 0, again one can find a finitely generated field over $\mathbb{Q}$ such that $\pi\colon \mathcal{X} \to B$ is defined over such a field. Then one can reduce the problem to the case over $\mathbb{C}$. This concludes the proof. □

Height Functions

Let B be a smooth projective curve defined over k. Let $\pi\colon \mathcal{X} \to B$ be a dominant projective morphism from a projective scheme $\mathcal{X}$ over k. Given a $\mathbb{Q}$-Cartier $\mathbb{Q}$-divisor L on $\mathcal{X}$ such that its restriction to the generic fiber is ample, we establish the following Northcott property in the setting of function fields:

Proposition 8.3.10 *Let $\pi\colon \mathcal{X} \to B$ be a dominant projective morphism from a projective scheme $\mathcal{X}$ over k. Let L be a $\mathbb{Q}$-Cartier $\mathbb{Q}$-divisor on $\mathcal{X}$. Assume the restriction L to the generic fiber $L|_{\mathcal{X}_\eta}$ is ample. For any constant t, the set of classes of sections of π satisfying $L.C \leq t$ lies in a bounded subset of $N_1(\mathcal{X})$.*

This proposition is referred to as the *Northcott property* for sections of fibrations.

Proof The following proof is based on the proof of [LT24, Lemma 2.6]. We prove this statement by induction on the relative dimension of $\pi\colon \mathcal{X} \to B$. The base case is when the relative dimension is 0. In this case, there are only finitely many sections on π, so our assertion is obvious.

Suppose that the dimension of the generic fiber $\mathcal{X}_\eta$ is $n > 0$. Let the $\mathcal{X}_i$s be the reduced irreducible components of $\mathcal{X}$ that map dominantly to B. Then any section is contained in some $\mathcal{X}_i$, so the class of such a section is contained in the image of

$$\oplus_i N_1(\mathcal{X}_i) \to N_1(\mathcal{X}).$$

Thus it suffices to prove our statement for each $\mathcal{X}_i$s so we may assume that $\mathcal{X}$ is an irreducible projective variety of dimension $n+1$. Let us fix an ample divisor A on $\mathcal{X}$. By our assumption, there exists a non-empty Zariski open subset $U \subset B$ such that $L|_{\pi^{-1}(U)}$ is relatively ample over $\pi^{-1}(U)$. Let F be a general fiber of π. After adding a multiple of F, we may assume that $(L+mF)|_{\pi^{-1}(U)}$ is ample on $\pi^{-1}(U)$. Then for a sufficiently large $k > 0$, $(L+mF-\frac{1}{k}A)|_{\pi^{-1}(U)}$ is still ample on $\pi^{-1}(U)$. This implies that after increasing m if necessary, we can conclude that $L+mF-\frac{1}{k}A$ is $\mathbb{Q}$-linearly equivalent to an effective $\mathbb{Q}$-divisor $E \geq 0$ on $\mathcal{X}$.

The upshot is that for a sufficiently large m, we can write

$$L + mF = H + E,$$

where H is an ample $\mathbb{Q}$-divisor and E is an effective $\mathbb{Q}$-divisor.

Now for any section $C \subset \mathcal{X}$ such that $L.C \leq t$, there are two possibilities:

- $E.C \geq 0$; or
- $E.C < 0$.

In the first case, we can conclude that $H.C \leq t + m$. Such a C is contained in a bounded subset in $N_1(\mathcal{X})$. In the second case, C is contained in E, so our claim follows by the induction hypothesis. Thus our assertion follows. □

By the above proposition, there is the minimal value of $L.C$ as C varies over all sections of $\pi\colon \mathcal{X} \to B$. This enables us to adopt the following definition:

Definition 8.3.11 Let $\pi\colon \mathcal{X} \to B$ be a Fano fibration. Let L be a $\mathbb{Q}$-Cartier $\mathbb{Q}$-divisor on $\mathcal{X}$. Assume the restriction of L to the generic fiber $L|_{\mathcal{X}_\eta}$ is ample. We define $\operatorname{neg}(\mathcal{X}, L)$ as the minimal value of $L.C$ as C runs over all sections of π. We call this value the *minimal height of sections* of π.

Rational Curves on del Pezzo Surfaces

Finally, we explain a few facts about rational curves on smooth del Pezzo surfaces. First we review the geometry of the moduli spaces of rational curves on del Pezzo surfaces. The following theorem is mainly due to Testa:

Theorem 8.3.12 ([Tes09, Theorem 5.1], [Tes05, Theorem 7.1.2], and [BLRT23, Theorems 1.5 and 1.6]) *Let S be a smooth del Pezzo surface of degree d defined over k. Let α be a nef class on S such that $-K_S.\alpha \geq 3$. Then the following hold:*

(i) *If α is not a multiple of a $-K_S$-conic, then there exists a unique component M of $\overline{M}_{0,0}(S,\alpha)$ generically parametrizing stable maps with irreducible domains. A general map parametrized by M is a birational stable map onto a free rational curve.*

(ii) *If α is a multiple of a smooth rational conic, then there exists a unique component M of $\overline{M}_{0,0}(S,\alpha)$ generically parametrizing stable maps with irreducible domains. A general map parametrized by M is a finite cover to a smooth conic.*

(iii) *If $d = 2$ and α is a multiple of $-K_S$ or $d = 1$ and there is a contraction of a (-1)-curve $\beta\colon S \to S'$ such that α is a multiple of $-\beta^*K_{S'}$, then there exist exactly two components of $\overline{M}_{0,0}(S,\alpha)$ generically parametrizing stable maps with irreducible domains. One component generically parametrizes* *birational stable maps to free rational curves, and the other component generically parametrizes multiple covers of $-K_S$-conics.*

(iv) *If $d = 1$ and α is a multiple of $-2K_S$, then there exist at least two components of $\overline{M}_{0,0}(S, \alpha)$ generically parametrizing stable maps with irreducible domains. There is a unique component generically parametrizing birational stable maps to free rational curves, and the other components generically parametrize multiple covers of a $-K_S$-conic in $|-2K_S|$.*

Proof This result is a combination of several results: [Tes09, Theorem 5.1], [Tes05, Theorem 7.1.2], and [BLRT23, Theorem 1.6]. However, the statement of this form is stated only in [BLRT23, Theorem 1.5]. In [BLRT23, Theorem 1.5], the authors assume that the ground field has positive characteristic. However, this assumption is unnecessary, and their proof works in characteristic 0 too. □

Finally, we recall movable Bend and Break lemma for smooth del Pezzo surfaces. This is established by Testa in [Tes09, Lemma 1.14 and Lemma 3.2].

Proposition 8.3.13 (Movable Bend and Break for del Pezzo surfaces [Tes09, Lemma 1.14 and Lemma 3.2]) *Let S be a smooth del Pezzo surface. Let C be a general free rational curve on S such that*

$$-K_S.C \geq 4.$$

Fix two general points x_1, x_2 of C. Then we can deform C as a stable map, keeping x_1 and x_2 fixed, until it degenerates into the union of two free rational curves $C_1 \cup C_2$ such that $x_i \in C_i$ for $i = 1, 2$.

To this end, first we prove the following lemma:

Lemma 8.3.14 *Let S be a smooth del Pezzo surface. Let α be a nef class of anticanonical degree d. Let $M \subset \overline{M}_{0,0}(S, \alpha)$ be a component generically parametrizing birational stable maps to free rational curves. Let*

$$M^{(n)} \subset \overline{M}_{0,n}(S, \alpha)$$

be the component above M. Then the evaluation map $\mathrm{ev}_n \colon M^{(n)} \to S^n$ is dominant if and only if $n \leq d - 1$.

Proof Since M generically parametrizes free rational curves, M has expected dimension $d - 1$. Thus $M^{(n)}$ has dimension $d - 1 + n$. To be dominant, we need $2n \leq d - 1 + n$. Thus $n \leq d - 1$ is necessary.

Suppose $n \leq d - 1$. We prove that $\mathrm{ev}_n \colon M^{(n)} \to S^n$ is dominant based on induction on n. When $n = 1$, the claim follows from the fact that M parametrizes free rational curves. Assume the induction hypothesis for $n - 1$.

We fix general points $x_1, \ldots, x_{n-1}$ on S, then we can find a free rational curve $f: C \to S$ passing through those points. The expected dimension of the locus parametrizing rational curves passing through these points is $d - 1 - (n-1) = d - n > 0$. Since this gives a lower bound for dimension, $f: C \to S$ must deform while passing through these points. Since f is a birational stable map to the image, deformations of f must sweep out S. Thus our assertion follows. □

Proof of Proposition 8.3.13 Let $d = -K_S.C \geq 4$. We fix general points $x_1, \ldots, x_{d-2}$ on S. Then the dimension of the locus parametrizing rational curves C passes through these points is $d - 1 - (d-2) = 1$, so C deforms while fixing these points. By Corollary 8.1.6, it degenerates to the union of rational curves

$$C_1 \cup \cdots \cup C_r$$

such that $x_1 \in C_1$ and $x_2 \in C_2$. In particular, C_1 and C_2 are free rational curves. Let d_i be the anticanonical degree of C_i and a_i be the number of general points C_i passing through. Then it follows from Lemma 8.3.14 that

$$d - 2 \leq \sum_{i=1}^{r} a_i \leq \sum_{i=1}^{r} (d_i - 1) = d - r.$$

Thus r must be 2, proving the claim. □

8.3.2 Standard Models of del Pezzo Fibrations

Here we review results of Corti ([Cor96]) showing the existence of a nice model of del Pezzo fibrations, which plays an important role in our analysis. First, we introduce the following definition:

Definition 8.3.15 (Standard models) Let B be a smooth projective curve and $K = k(B)$ its function field. Let X_K be a geometrically integral smooth del Pezzo surface of degree ≥ 3 defined over K. An integral model $\pi: \mathcal{X} \to B$ of X_K is called a *standard model* if the following properties hold:

(i) $\mathcal{X}$ is a Gorenstein terminal threefold;
(ii) every fiber is reduced and irreducible; and
(iii) $-K_{\mathcal{X}}$ is π-relatively very ample.

The main result of [Cor96] is the following theorem:

Theorem 8.3.16 ([Cor96, 1.10 Theorem]) *Let B be a smooth projective curve and $K = k(B)$ its function field. Let X_K be a geometrically integral smooth del Pezzo surface of degree ≥ 3 defined over K. Then a standard model of X_K exists.*

Remark 8.3.17 [Cor96, 1.18 Theorem] shows that a standard model as defined in Definition 8.3.15 may not exist for degree-1 and degree-2 del Pezzo surfaces. However, there is a weaker notion of standard models ([Cor96, 1.13 Definition]). The existence of such models has been established in [Cor96, 1.15 Theorem] for degree 2 and in [AFK22, Theorem 1.4] for degree 1.

Remark 8.3.18 A standard model is, in general, not unique, as shown in [Cor96, 5.8 Example].

For some applications, the following models will suffice:

Definition 8.3.19 (Good nef models) Let B be a smooth projective curve and $K = k(B)$ its function field. Let X_K be a geometrically integral smooth del Pezzo surface defined over K. An integral model $\pi: \mathcal{X} \to B$ of X_K is called a *good nef model* if the following properties hold:

(i) $\mathcal{X}$ is a Gorenstein terminal threefold; and
(ii) $-K_{\mathcal{X}}$ is π-relatively nef.

Theorem 8.3.20 ([Cor96, 3.3 Theorem]) *Let B be a smooth projective curve and $K = k(B)$ its function field. Let X_K be a geometrically integral smooth del Pezzo surface defined over K. Then a good nef model of X_K exists.*

8.3.3 Non-dominant Families

The goal of this section is to establish an analogue of Corollary 7.4.4 in the setting of del Pezzo fibrations over $\mathbb{P}^1$. From now on we assume that $B = \mathbb{P}^1$.

Here is the main theorem of this section:

Theorem 8.3.21 ([LT24, Theorem 1.2]) *Let $\pi: \mathcal{X} \to \mathbb{P}^1$ be a del Pezzo fibration from a Gorenstein projective terminal threefold such that $-K_{\mathcal{X}/\mathbb{P}^1}$ is relatively nef. Then there is a proper closed subset $\mathcal{V} \subsetneq \mathcal{X}$ such that any component $M \subset \mathrm{Sec}(\mathcal{X}/\mathbb{P}^1)$ parametrizing a non-dominant family of sections will parametrize sections in $\mathcal{V}$.*

Remark 8.3.22 For any smooth del Pezzo surface over $k(t)$ it follows from Theorem 8.3.20 that there exists a good nef model satisfying the assumptions of Theorem 8.3.21.

Before discussing the proofs of Theorem 8.3.21, let us make a few remarks about this theorem. The following example shows that an analogue of Proposition 7.4.1 does not hold for sections of Fano fibrations:

Example 8.3.23 Let $\pi \colon \mathbb{F}_e \to \mathbb{P}^1$ be a Hirzebruch surface defined as

$$\mathbb{F}_e = \mathbb{P}_{\mathbb{P}^1}(O \oplus O(-e)) \to \mathbb{P}^1.$$

Then the generic fiber is isomorphic to $\mathbb{P}^1_\eta$, so in particular, there is no generic subvariety with a higher a-invariant. There is a rigid section corresponding to a natural quotient $O \oplus O(-e) \to O(-e)$, and the expected dimension of the rigid section is negative whenever $e \geq 2$. Thus there is a non-dominant family of sections which is not captured by a-invariants.

This example demonstrates that an analogue of Proposition 7.4.1 is only valid for sections of sufficiently large anticanonical degrees.

Next, the following example shows that, even when the anticanonical degree is large, a non-dominant family of sections may sweep out a surface Y whose a-invariant is same as the a-invariant of $\mathcal{X}$:

Example 8.3.24 ([LT24, Example 5.2]) Let $h \colon \mathbb{F}_2 \to \mathbb{P}^1$ be the Hirzebruch surface and let C_0 be the unique rigid section of h. We denote a fiber of h by F. Let $\mathcal{X}$ be $\mathbb{P}_{\mathbb{F}_2}(O \oplus O(-C_0 - F))$ with the structure morphism $g \colon \mathcal{X} \to \mathbb{F}_2$. We denote the composition $h \circ g \colon \mathcal{X} \to \mathbb{P}^1$ by π. Since we have $F.(-C_0 - F) = -1$, every fiber of π is isomorphic to the Hirzebruch surface $\mathbb{F}_1$. Let us denote the pullback g^*C_0 on $\mathcal{X}$ by Y. The generic fiber of π is isomorphic to $(\mathbb{F}_1)_\eta$, and the generic fiber Y_η is a fiber of $g_\eta \colon (\mathbb{F}_1)_\eta \to \mathbb{P}^1_\eta$. In particular, we have

$$a(\mathcal{X}_\eta, -K_{\mathcal{X}/\mathbb{P}^1}|_{\mathcal{X}_\eta}) = a(Y_\eta, -K_{\mathcal{X}/\mathbb{P}^1}|_{Y_\eta}) = 1.$$

On the other hand, by adjunction, we have

$$-K_{\mathcal{X}/\mathbb{P}^1}|_Y \sim -K_{Y/\mathbb{P}^1} - 2T,$$

where T is a general fiber of $Y \to \mathbb{P}^1$. Thus for any section C on Y, we have

$$-K_{\mathcal{X}/\mathbb{P}^1}.C + 2 = -K_{Y/\mathbb{P}^1}.C = (-K_{Y/\mathbb{P}^1}.C + 1) - 1.$$

This means that the expected dimension of deformations of C on $\mathcal{X}$ is smaller than the expected dimension of deformations of C on Y, showing that any dominant family of sections of Y has dimension higher than expected dimension on $\mathcal{X}$.

These examples show subtleties in the statements of Theorem 8.3.21.

Breaking Sections on Ruled Surfaces

To prove Theorem 8.3.21, we first prepare some auxiliary results on breaking sections on ruled surfaces.

Lemma 8.3.25 ([LT24, Lemma 4.3]) *Let Y be a smooth projective surface with a fibration $\pi: Y \to \mathbb{P}^1$ such that a general fiber of π is isomorphic to $\mathbb{P}^1$. Let F be a singular fiber of π with components E_is for $i = 1, \dots, r$. Suppose that E_1 is a (-1)-curve with multiplicity 1 in the fiber F. Then there is another (-1)-curve in F.*

Proof The proof is based on induction on r. When $r = 2$, F is the union of two (-1)-curves, proving the claim.

Next, we prove the inductive step. There is a sequence of contractions of (-1)-curves:

$$Y = Y_0 \to Y_1 \to \dots \to Y_{r-1},$$

which contracts all components of F but one. We denote the contraction $Y_i \to Y_{i+1}$ by ϕ_i. We may assume that ϕ_0 contracts E_1 as otherwise the statement is clear. Let E_2 be the exceptional divisor for $\phi_1 : Y_1 \to Y_2$ and, by abuse of notation, we also denote its strict transform on Y by E_2. If E_1 and E_2 do not meet on Y, then E_2 is a (-1)-curve on Y, proving the claim. Suppose that they meet. Since the multiplicity of E_1 is 1 in F, the multiplicity of E_2 in F must be 1. By the induction hypothesis, there is a component $E_j \neq E_2$ which is a (-1)-curve on Y_1. If E_1 and E_j meet, then E_1 is the exceptional divisor at an intersection of E_2 and E_j. This implies that the multiplicity of E_1 in F must be greater than or equal to 2, a contradiction. We conclude that E_1 and E_j do not meet, and E_j is a (-1)-curve on Y. Thus our assertion follows. □

Corollary 8.3.26 ([LT24, Corollary 4.4]) *Let Y be a smooth projective surface with a fibration $\pi: Y \to \mathbb{P}^1$ such that a general fiber of π is isomorphic to $\mathbb{P}^1$. Fix a section C of π. Then there is a $\mathbb{P}^1$-birational morphism to a Hirzebruch surface $\rho: Y \to \mathbb{F}_e = \mathbb{P}_{\mathbb{P}^1}(\mathcal{O} \oplus \mathcal{O}(-e))$ that is an isomorphism in an open neighborhood of C.*

Proof Suppose that π admits a reducible fiber F_0. Then F_0 must contain a (-1)-curve E_0. This curve may or may not be disjoint from C. If they meet, then the multiplicity of E_0 in F_0 must be 1, so by Lemma 8.3.25, there is another (-1)-curve disjoint from C. Thus either way we will find a (-1)-curve that is disjoint from C. We inductively contract such curves, and obtain a birational morphism to a Hirzebruch surface, proving the claim. □

Here is a breaking result for sections of ruled surfaces:

Proposition 8.3.27 ([LT24, Theorem 4.5]) *Let Y be a smooth projective surface with a fibration $\pi: Y \to \mathbb{P}^1$ such that a general fiber F of π is isomorphic to $\mathbb{P}^1$. Fix a dominant family of sections C of π with $-K_{Y/\mathbb{P}^1}.C = q$. Let*

$\rho: Y \to \mathbb{F}_e$ be a birational morphism obtained by Corollary 8.3.26. Then $e \leq q$ and $q + e$ is even. Moreover, the following hold:

(i) *there exists a section C_0 such that*

$$C \sim C_0 + \frac{q+e}{2}F + T,$$

where T is some π-vertical effective cycle; and

(ii) *there exists a dominant family of sections C_1 such that*

$$C \sim C_1 + \frac{q-e}{2}F.$$

Proof Let $\widetilde{C}$ be $\rho_* C$. This forms a moving family of sections of $\mathbb{F}_e$. Since ρ is an isomorphism in a neighborhood of C, we have $q = -K_{Y/\mathbb{P}^1}.C = -K_{\mathbb{F}_e/\mathbb{P}^1}.\widetilde{C}$. Since the minimal height of moving sections on $\mathbb{F}_e$ is e, we conclude that $e \leq q$. Moreover, since the difference between two sections on $\mathbb{F}_e$ is linearly equivalent to a multiple of F, we conclude that q and e have the same parity. On $\mathbb{F}_e$, let $\widetilde{C}_0$ be the rigid section of the minimal height $-e$ and let $\widetilde{C}_1$ be a moving section of the minimal height e. Then we have

$$\widetilde{C} \sim \widetilde{C}_0 + \frac{q+e}{2}F \sim \widetilde{C}_1 + \frac{q-e}{2}F.$$

Since we have $C = \rho^* \rho_* C$, our assertion follows from this relation. □

Proof of Theorem 8.3.21

Using Proposition 8.3.27, we obtain:

Theorem 8.3.28 ([LT24, Theorem 5.4]) *Let $\pi: \mathcal{X} \to \mathbb{P}^1$ be a del Pezzo fibration which is a good nef model of the generic fiber. Let $M \subset \mathrm{Sec}(\mathcal{X}/\mathbb{P}^1)$ be a component parametrizing sections C satisfying*

$$-K_{\mathcal{X}/\mathbb{P}^1}.C \geq \max\{-2\mathrm{neg}(\mathcal{X}, -K_{\mathcal{X}/\mathbb{P}^1}) - 1, 1\}.$$

Let Y denote the closure of the locus swept out by sections parametrized by M. Assume that either

(i) *$Y \subsetneq \mathcal{X}$; or*

(ii) *$\dim M > -K_{\mathcal{X}/\mathbb{P}^1}.C + 2$.*

Then we have $a(Y_\eta, -K_{\mathcal{X}/\mathbb{P}^1}|_{Y_\eta}) \geq a(\mathcal{X}_\eta, -K_{\mathcal{X}/\mathbb{P}^1}|_{\mathcal{X}_\eta}) = 1$.

Proof By Lemma 8.3.6, the second assumption implies the first, so we may assume the first assumption. By the height bound, the expected dimension of M is $-K_{\mathcal{X}/\mathbb{P}^1}.C + 2 \geq 3$. Hence, by Corollary 8.3.5, sections parametrized by M deform so that Y must have dimension 2. Let $\beta: \widetilde{Y} \to Y$ be a smooth resolution and $\widetilde{C}$ be the strict transform of a general deformation of C. Since dim $M \geq 3$, C deforms while fixing two points. This means that $\pi|_Y \circ \beta\colon \widetilde{Y} \to \mathbb{P}^1$ must have connected fibers. Moreover, by Bend and Break (Lemma 8.1.4), it must break off a vertical rational curve passing through a general point. This means that $\pi|_Y \circ \beta\colon \widetilde{Y} \to \mathbb{P}^1$ is a generically $\mathbb{P}^1$-bundle.

Since $\widetilde{C}$ dominates $\widetilde{Y}$, we have

$$\dim M = -K_{\widetilde{Y}/\mathbb{P}^1}.\widetilde{C} + 1.$$

So, again by Corollary 8.3.5, we must have

$$(K_{\widetilde{Y}/\mathbb{P}^1} - \beta^* K_{\mathcal{X}/\mathbb{P}^1}).\widetilde{C} \leq -1 < 0.$$

It follows from Corollary 8.3.26 that we have a birational morphism

$$\phi\colon \widetilde{Y} \to \mathbb{F}_e,$$

which is an isomorphism in a neighborhood of $\widetilde{C}$. Then by Proposition 8.3.27, there exists a section C_0 of $\pi|_Y \circ \beta$ and a vertical effective cycle T such that

$$\widetilde{C} \sim C_0 + \frac{q+e}{2} F + T,$$

where

$$q = -K_{\widetilde{Y}/\mathbb{P}^1}.\widetilde{C} > -K_{\mathcal{X}/\mathbb{P}^1}.C \geq -2\mathrm{neg}(\mathcal{X}, -K_{\mathcal{X}/\mathbb{P}^1}) - 1.$$

In particular, since $q + e$ is an integer and e is non-negative, we conclude that

$$\frac{q+e}{2} \geq -\mathrm{neg}(\mathcal{X}, -K_{\mathcal{X}/\mathbb{P}^1}).$$

Combining everything together we obtain

$$\begin{aligned}
0 &> (K_{\widetilde{Y}/\mathbb{P}^1} - \beta^* K_{\mathcal{X}/\mathbb{P}^1}).\widetilde{C} \\
&= -q - \beta^* K_{\mathcal{X}/\mathbb{P}^1}.(C_0 + \frac{q+e}{2} F + T) \\
&= e + \frac{q+e}{2}(-\beta^* K_{\mathcal{X}/\mathbb{P}^1}.F - 2) - \beta^* K_{\mathcal{X}/\mathbb{P}^1}.(C_0 + T) \\
&\geq e + \frac{q+e}{2}(-\beta^* K_{\mathcal{X}/\mathbb{P}^1}.F - 2) - K_{\mathcal{X}/\mathbb{P}^1}.\beta_* C_0.
\end{aligned} \tag{8.2}$$

Here, at the last step, we use our assumption that $-K_{\mathcal{X}/\mathbb{P}^1}$ is relatively nef. If $a(Y_\eta, -K_{\mathcal{X}/\mathbb{P}^1}|_{Y_\eta}) < a(\mathcal{X}_\eta, -K_{\mathcal{X}/\mathbb{P}^1}|_{\mathcal{X}_\eta})$, then this is equivalent to saying that

$$-\beta^* K_{\mathcal{X}/\mathbb{P}^1}.F \geq 3.$$

In turn, this implies that

$$0 > e + \frac{q+e}{2} + \mathrm{neg}(\mathcal{X}, -K_{\mathcal{X}/\mathbb{P}^1}) \geq e,$$

yielding a contradiction. □

When the equality $a(Y_\eta, -K_{\mathcal{X}/\mathbb{P}^1}|_{Y_\eta}) = a(\mathcal{X}_\eta, -K_{\mathcal{X}/\mathbb{P}^1}|_{\mathcal{X}_\eta})$ holds, we can say more:

Theorem 8.3.29 ([LT24, Theorem 5.5]) *Let $\pi\colon \mathcal{X} \to \mathbb{P}^1$ be a del Pezzo fibration that is a good nef model of the generic fiber. Let $M \subset \mathrm{Sec}(\mathcal{X}/\mathbb{P}^1)$ be a component parametrizing sections C satisfying*

$$-K_{\mathcal{X}/\mathbb{P}^1}.C \geq \max\{-2\mathrm{neg}(\mathcal{X}, -K_{\mathcal{X}/\mathbb{P}^1}) - 1, 1\}$$

and sweeping out a surface $Y \subsetneq \mathcal{X}$ satisfying

$$a(Y_\eta, -K_{\mathcal{X}/\mathbb{P}^1}|_{Y_\eta}) = a(\mathcal{X}_\eta, -K_{\mathcal{X}/\mathbb{P}^1}|_{\mathcal{X}_\eta}).$$

Then there exists a component M_1 of $\mathrm{Sec}(\mathcal{X}/\mathbb{P}^1)$ parametrizing sections C_1 satisfying

$$-K_{\mathcal{X}/\mathbb{P}^1}.C_1 < -\mathrm{neg}(\mathcal{X}, -K_{\mathcal{X}/\mathbb{P}^1}) - 1$$

and sweeping out Y.

Proof We use the setup established in the proof of Theorem 8.3.28. In our situation, we have

$$-\beta^* K_{\mathcal{X}/\mathbb{P}^1}.F = 2,$$

and this implies that

$$(K_{\widetilde{Y}/\mathbb{P}^1} - \beta^* K_{\mathcal{X}/\mathbb{P}^1}).F = 0.$$

By Proposition 8.3.27, there is a moving section C_1 on $\widetilde{Y}$ such that

$$\widetilde{C} \sim C_1 + \frac{q+e}{2} F.$$

Thus we can conclude

$$0 > (K_{\widetilde{Y}/\mathbb{P}^1} - \beta^* K_{\mathcal{X}/\mathbb{P}^1}).\widetilde{C} = (K_{\widetilde{Y}/\mathbb{P}^1} - \beta^* K_{\mathcal{X}/\mathbb{P}^1}).C_1 = -e - \beta^* K_{\mathcal{X}/\mathbb{P}^1}.C_1.$$

On the other hand, the inequality (8.2) indicates that

$$0 > e + \mathrm{neg}(\mathcal{X}, -K_{\mathcal{X}/\mathbb{P}^1}),$$

and altogether these imply that

$$-\beta^* K_{\mathcal{X}/\mathbb{P}^1}.C_1 < -\mathrm{neg}(\mathcal{X}, -K_{\mathcal{X}/\mathbb{P}^1}) - 1.$$

Finally, since we have

$$0 > (K_{\widetilde{Y}/\mathbb{P}^1} - \beta^* K_{\mathcal{X}/\mathbb{P}^1}).C_1,$$

we conclude that

$$-K_{\widetilde{Y}/\mathbb{P}^1}.C_1 + 1 \geq -\beta^* K_{\mathcal{X}/\mathbb{P}^1}.C_1 + 2.$$

If deformations of C_1 move out of Y, then they must sweep out $\mathcal{X}$. This implies that a component parametrizing deformations of C_1 in $\mathcal{X}$ must have expected dimension. The component parametrizing a general deformation of C_1 in Y also has the expected dimension. We conclude that we must have

$$-\beta^* K_{\mathcal{X}/\mathbb{P}^1}.C_1 + 2 > -K_{\widetilde{Y}/\mathbb{P}^1}.C_1 + 1,$$

which is a contradiction to the above inequality. Thus the component parametrizing deformations of C_1 in $\mathcal{X}$ must sweep out Y, proving the claim. □

We now prove the main theorem of this section:

Proof of Theorem 8.3.21 By Theorem 8.3.28, any component $M \subset \mathrm{Sec}(\mathcal{X}/\mathbb{P}^1)$ parametrizing a non-dominant family of rational curves falls into one of the following categories:

(i) M parametrizes sections C satisfying

$$-K_{\mathcal{X}/\mathbb{P}^1}.C < \max\{-2\mathrm{neg}(\mathcal{X}, -K_{\mathcal{X}/\mathbb{P}^1}) - 1, 1\};$$

(ii) M parametrizes sections C satisfying

$$-K_{\mathcal{X}/\mathbb{P}^1}.C \geq \max\{-2\mathrm{neg}(\mathcal{X}, -K_{\mathcal{X}/\mathbb{P}^1}) - 1, 1\},$$

which sweep out a surface $Y \subset \mathcal{X}$ such that

$$a(Y_\eta, -K_{\mathcal{X}/\mathbb{P}^1}|_{Y_\eta}) > a(\mathcal{X}_\eta, -K_{\mathcal{X}/\mathbb{P}^1}|_{\mathcal{X}_\eta}) = 1;$$

or

(iii) M parametrizes sections C satisfying

$$-K_{\mathcal{X}/\mathbb{P}^1}.C \geq \max\{-2\mathrm{neg}(\mathcal{X}, -K_{\mathcal{X}/\mathbb{P}^1}) - 1, 1\},$$

which sweep out a surface $Y \subset \mathcal{X}$ such that

$$a(Y_\eta, -K_{\mathcal{X}/\mathbb{P}^1}|_{Y_\eta}) = a(\mathcal{X}_\eta, -K_{\mathcal{X}/\mathbb{P}^1}|_{\mathcal{X}_\eta}) = 1.$$

For (i), it follows from Proposition 8.3.10 that there are only finitely many components M parametrizing such sections, so they sweep out a proper closed subset. For (ii), it follows from Theorem 5.1.6 that such Ys are contained in a proper closed subset. Finally, for (iii), it follows from Theorem 8.3.29 that such a Y is swept out by sections C_1, parametrized by a component M_1, satisfying

$$-K_{\mathcal{X}/\mathbb{P}^1}.C_1 < -\mathrm{neg}(\mathcal{X}, -K_{\mathcal{X}/\mathbb{P}^1}) - 1.$$

Thus by Proposition 8.3.10, again such Ys are contained in a proper closed subset, proving the claim. □

8.3.4 Movable Bend and Break

In this section, we establish movable Bend and Break for sections of del Pezzo fibrations over $\mathbb{P}^1$:

Theorem 8.3.30 ([LT24, Theorem 1.5]) *Let $\pi: \mathcal{X} \to \mathbb{P}^1$ be a del Pezzo fibration such that $\mathcal{X}$ is Gorenstein terminal and the relative anticanonical divisor $-K_{\mathcal{X}/\mathbb{P}^1}$ is relatively ample. Then there exists a constant $\mathsf{Q}(\pi)$ satisfying the following property: Suppose that $M \subset \mathrm{Sec}(\mathcal{X}/\mathbb{P}^1)$ is a component parametrizing a dominant family of sections C satisfying $-K_{\mathcal{X}/\mathbb{P}^1}.C \geq \mathsf{Q}(\pi)$. Then the closure of M in $\overline{M}_{0,0}(\mathcal{X})$ contains a point that represents a stable map whose domain is the union of two $\mathbb{P}^1$s, both mapping birationally to a free rational curve.*

Remark 8.3.31 We will make the constant $\mathsf{Q}(\pi)$ explicit. See Theorem 8.3.40.

Remark 8.3.32 Integral models satisfying the assumptions of Theorem 8.3.30 exist for any del Pezzo surface of degree ≥ 3 over $k(\mathbb{P}^1)$ by a result of Corti (Theorem 8.3.16).

Stable Maps through General Points

In this section, we analyze descriptions of stable maps going through a number of general points. First, we record the following lemma:

Lemma 8.3.33 ([LT24, Lemma 2.8]) *Let $\pi: \mathcal{X} \to \mathbb{P}^1$ be a Fano fibration such that* $\dim \mathcal{X} \leq 3$ *and $\mathcal{X}$ is Gorenstein and terminal. Let n be a positive integer. Suppose that we have a section C such that any n general points of $\mathcal{X}$ are contained in a deformation of C. Then for a general deformation C' of C, we have*

$$N_{C'/\mathcal{X}} = \oplus_i \mathcal{O}(a_i),$$

with $a_i \geq n-1$ for any i. Moreover, we can find such a C' passing through any n general points.

Conversely, if C is a section of π such that we have

$$N_{C/\mathcal{X}} = \oplus_i \mathcal{O}(a_i),$$

with $a_i \geq n-1$, then any n general points are contained in some deformation C' of C.

Proof Suppose that a deformation of C contains a set of n general points. We fix $n-1$ general points $x_1, \ldots, x_{n-1}$ on $\mathcal{X}$, and consider the locus $M(x_1, \ldots, x_{n-1})$ of $M_{0,0}(\mathcal{X})$ parametrizing deformations of C passing through these $n-1$ general points. Let $p\colon C(x_1,\dots,x_{n-1}) \to M(x_1,\dots,x_{n-1})$ be the universal family. Then our assumption implies that the evaluation map

$$\mathrm{ev}\colon C(x_1,\dots,x_{n-1}) \to \mathcal{X}$$

is dominant. Now the tangent space of $M(x_1,\dots,x_{n-1})$ at

$$[C'] \in M(x_1,\dots,x_{n-1})$$

is given by

$$H^0(C', N_{C'/\mathcal{X}}(-x_1 - \cdots - x_{n-1})),$$

so the evaluation map being dominant implies that $N_{C'/\mathcal{X}}(-x_1 - \cdots - x_{n-1})$ is generically globally generated. This implies $a_i \geq n-1$.

Conversely, suppose that C is a section with $a_i \geq n-1$. We prove our claim by induction on n. When $n = 1$, C is free, so a general deformation of C passes through a general point. Suppose that we have $a_i \geq n$ for C. We write

$$N_{C'/\mathcal{X}} \cong \oplus_i \mathcal{O}(a'_i).$$

Since we have

$$h^1(C, N_{C/\mathcal{X}}(-n-1)) = 0,$$

the upper semicontinuity of h^1 [Har77, Chapter III Theorem 12.8] implies that, for a general C', we have

$$H^1(C', N_{C'/\mathcal{X}}(-n-1)) = 0.$$

This means that $a'_i \geq n$ for any i. By the induction hypothesis, we may assume that C' passes through n general points $x_1,\dots,x_n$. Again we consider the locus $M(x_1,\dots,x_n)$ of $M_{0,0}(\mathcal{X})$ parametrizing deformations of C passing through these n general points. Since $a'_i \geq n$, the twisted normal bundle

$$N_{C'/\mathcal{X}}(-x_1 - \cdots - x_n)$$

is globally generated. It follows that the evaluation map

$$\mathrm{ev}\colon C(x_1,\dots,x_n) \to \mathcal{X}$$

is dominant. Thus C' passes through $n+1$ general points, proving the claim. □

Let $\pi\colon \mathcal{X} \to \mathbb{P}^1$ be a del Pezzo fibration where $\mathcal{X}$ is Gorenstein and terminal. Suppose that a genus 0 stable map $f\colon C \to \mathcal{X}$ passes through n general points on $\mathcal{X}$. We discuss what restrictions this condition imposes on C and f. First, we discuss the restrictions when C is irreducible:

Lemma 8.3.34 ([LT24, Lemma 6.1]) *Let $\pi\colon \mathcal{X} \to \mathbb{P}^1$ be a del Pezzo fibration such that $\mathcal{X}$ is Gorenstein and terminal. Let $M \subset \mathrm{Sec}(\mathcal{X}/\mathbb{P}^1)$ be a component such that, for any set of n general points on $\mathcal{X}$, there exists a member of M containing these general points. Then for any section C parametrized by M, we have*

$$-K_{\mathcal{X}/\mathbb{P}^1}.C \geq 2n - 2.$$

Moreover, if this is the equality, then there are only finitely many sections passing through a fixed set of n general points. If $-K_{\mathcal{X}/\mathbb{P}^1}.C = 2n - 1$, then the sublocus of M parametrizing sections passing through a fixed set of n general points is at most 1*-dimensional.*

Proof Since M parametrizes a dominant family of rational curves, it follows from Lemma 8.3.6 that we have

$$\dim M = -K_{\mathcal{X}/\mathbb{P}^1}.C + 2.$$

Let $M^{(n)} \subset M_{0,n}(\mathcal{X})$ be a component above M. Our assumption implies that the evaluation map

$$\mathrm{ev}_n\colon M^{(n)} \to \mathcal{X}^n$$

is dominant. It follows that

$$\dim M + n = \dim M^{(n)} \geq 3n.$$

Altogether we have

$$-K_{\mathcal{X}/\mathbb{P}^1}.C = \dim M - 2 \geq 2n - 2.$$

When this is equal, $\mathrm{ev}_n\colon M^{(n)} \to \mathcal{X}^n$ is generically finite. When

$$-K_{\mathcal{X}/\mathbb{P}^1}.C = 2n - 1,$$

the dimension of a general fiber of ev_n is 1-dimensional. The last two statements follow from these observations. □

Next, we discuss the case when $f\colon C \to \mathcal{X}$ is a stable map from a reducible domain.

Definition 8.3.35 ([LT24, Definition 6.2]) Let $\pi\colon \mathcal{X} \to \mathbb{P}^1$ be a del Pezzo fibration. Fix an integer d. We let $\mathrm{maxdef}(d)$ denote the maximal dimension of any component of $\mathrm{Sec}(\mathcal{X}/\mathbb{P}^1)$ parametrizing sections of anticanonical height d. When there is no section of height d, we simply set $\mathrm{maxdef}(d) = -\infty$. We also define

$$\mathrm{maxdef}(\mathcal{X}) := \max_{d<0}\{\mathrm{maxdef}(d)\},$$

and call this number the *maximum dimension of components of negative height*.

Lemma 8.3.36 ([LT24, Lemma 6.3]) *Let $\pi\colon \mathcal{X} \to \mathbb{P}^1$ be a del Pezzo fibration such that $\mathcal{X}$ is Gorenstein terminal and $-K_{\mathcal{X}/\mathbb{P}^1}$ is relatively ample. We fix a negative integer $d < 0$ and a positive integer $n \geq \operatorname{maxdef}(d)$. Suppose that there is a connected effective cycle $C = C_0 + \sum_i a_i T_i$ where C_0 is a section of the anticanonical height d and each T_i is a π-vertical rational curve. If C contains n general points of $\mathcal{X}$, then we have*

$$-K_{\mathcal{X}/\mathbb{P}^1}.C \geq d + 3n - \operatorname{maxdef}(d).$$

Proof Let $M \subset \operatorname{Sec}(\mathcal{X}/\mathbb{P}^1)$ be a component containing C_0. If M parametrizes a dominant family of sections, then its dimension must be equal to the expected dimension $d+2$. Moreover, since the universal family over M maps dominantly to $\mathcal{X}$, we conclude that $d \geq 0$, but this contradicts our assumption. Thus M cannot parametrize a dominant family, and let Y be the closure of the locus swept out by sections in M.

Assume that Y is a surface. Since Y is a proper closed subset, any general point contained in C has to be contained in some T_j. Suppose that $-K_{\mathcal{X}/\mathbb{P}^1}.T_j = 2$. Since there are only finitely many conics passing through a general point on a del Pezzo surface, the intersection of these conics with Y forms a finite set. To ensure that C is connected, C_0 has to pass through a point in this set. Since such a point is a general point on Y, the condition imposed on C_0 is independent for each T_j. Thus we can have at most $\operatorname{maxdef}(d)$ such conics T_j in the curve C. This proves that if we let m denote the number of such conics, then we have

$$-K_{\mathcal{X}/\mathbb{P}^1}.C \geq d + 2m + 3(n-m) = d + 3n - m \geq d + 3n - \operatorname{maxdef}(d).$$

When C_0 is rigid, any vertical conic passing through a general point avoids C_0. Thus we obtain

$$-K_{\mathcal{X}/\mathbb{P}^1}.C \geq d + 3n \geq d + 3n - \operatorname{maxdef}(d).$$

Thus our assertion follows. □

The following two propositions describe stable maps passing through a number of general points, and this forms a basis for our proof of movable Bend and Break for del Pezzo fibrations:

Proposition 8.3.37 ([LT24, Proposition 6.4]) *Let $\pi\colon \mathcal{X} \to \mathbb{P}^1$ be a del Pezzo fibration such that $\mathcal{X}$ is Gorenstein and terminal and $-K_{\mathcal{X}/\mathbb{P}^1}$ is relatively ample. Let n be a positive integer such that*

$$n \geq \operatorname{maxdef}(\mathcal{X}) + 2 + \max\{0, -\operatorname{neg}(\mathcal{X}, -K_{\mathcal{X}/\mathbb{P}^1})\}.$$

Suppose that $f: C \to \mathcal{X}$ is a genus 0 stable map of the anticanonical height $2n-2$ such that the intersection with a general fiber $\mathcal{X}_b$ is equal to 1. Then the following hold:

(i) *Suppose that the image of C contains n general points. Then f is a birational map to a free section.*
(ii) *Fix a general hyperplane section D in a general fiber of π, and suppose that the image of C contains $n-1$ general points and intersects D. Suppose also that the pushforward of C is not integral as a 1-cycle and at least one of general points is contained in a π-vertical component of C. Then C contains exactly two components, one maps isomorphically to a free section and the other maps birationally to a free $-K_{\mathcal{X}}$-conic or cubic in a general fiber of π.*

Proof (i) It suffices to show that the pushforward of f is integral as a 1-cycle. Suppose otherwise. We can write

$$f_*C = C_0 + \sum_{i \in I} a_i T_i,$$

where C_0 is a section of π and the anticanonical height d, T_i is a π-vertical rational curve, and a_i is a positive integer. Let $t_i = -K_{\mathcal{X}/\mathbb{P}^1}.T_i$. Then we have

$$2n - 2 = d + \sum_{i \in I} a_i t_i.$$

If C_0 contains a general point, then by Lemma 8.3.6, any component parametrizing C_0 has expected dimension $d + 2$. This shows that C_0 contains at most

$$\max\{0, \lfloor d/2 \rfloor + 1\}$$

general points. A vertical rational curve T_i can contain at most one general point, and if it does, it must be $t_i \geq 2$. Let $I' \subset I$ be a subset of i such that T_i contains a general point.
Case 1: $d \geq 0$. In this case, we have

$$\max\{0, \lfloor d/2 \rfloor + 1\} + \#I' \geq n = d/2 + 1 + \sum_i a_i t_i/2.$$

Since $d \geq 0$, the left-hand side is an upper bound for the right-hand side. Thus this must be an equality. This implies that d is even, that C_0 contains the maximum number of general points, that $a_i = 1$, and that each T_i contains one general point and $t_i = 2$. By Lemma 8.3.34, there are only finitely many possibilities for C_0, so in particular, we can ensure that this is general in the

moduli. It follows from Lemma 8.3.6 that C_0 is free. Since T_i is a vertical $-K_{\mathcal{X}}$-conic passing through a general point, there are only finitely many choices for T_i. Since $f(C)$ has to be connected, this means that $I = \emptyset$, proving the claim.
Case 2: $d < 0$. From Lemma 8.3.36, that we have

$$2n - 2 = -K_{\mathcal{X}/\mathbb{P}^1}.C \geq \operatorname{neg}(\mathcal{X}, -K_{\mathcal{X}/\mathbb{P}^1}) + 3n - \operatorname{maxdef}(\mathcal{X}).$$

However, this contradicts our assumption on n.

(ii) Again we write

$$f_*C = C_0 + \sum_{i \in I} a_i T_i,$$

where C_0 is a section of π and anticanonical height d, T_i is a π-vertical rational curve, and a_i is a positive integer. Let $t_i = -K_{\mathcal{X}/\mathbb{P}^1}.T_i$. As before, we have

$$2n - 2 = d + \sum_{i \in I} a_i t_i.$$

Case 1: $d \geq 0$ and C_0 intersects D. In this case, C_0 can have at most

$$\max\left\{0, \left\lfloor \frac{d+1}{2} \right\rfloor\right\}$$

general points. Thus we must have

$$\max\left\{0, \left\lfloor \frac{d+1}{2} \right\rfloor\right\} + \#I' \geq n - 1 = \frac{d}{2} + \sum_{i \in I} \frac{a_i t_i}{2}.$$

When d is even, we must have equalities. This means that the locus of deformations of C_0 that intersect D and contain the maximum number of general points is at most 1-dimensional, that for every i, $a_i = 1$, and that every T_i contains a general point and satisfies $t_i = 2$. Among such C_0s there are only finitely many possibilities meeting with one T_i. By Lemma 8.3.6, C_0 is free and there is only one conic T_i meeting with C_0, proving the claim.

If d is odd, then C_0 has to contain the maximum number of general points, that is, $\lfloor \frac{d+1}{2} \rfloor$. This implies that the difference between $\#I'$ and $\sum_{i \in I} \frac{a_i t_i}{2}$ is at most $1/2$. This reduces the possibilities for T_is to the following:

- every T_i has anticanonical degree 2 and contains a general point;
- every T_i but one has anticanonical degree 2 and contains a general point, and the last one has anticanonical degree 3 and contains a general point;
- every T_i but one has anticanonical degree 2 and contains a general point, and the last one has anticanonical degree 1.

Note that there are only finite many C_0 passing through $\lfloor \frac{d+1}{2} \rfloor$ general points and intersecting Z. By generality C_0 has to be free. Moreover, such a C_0 does not meet any vertical conic containing a general point nor a vertical line, proving that there is a unique vertical cubic in C.

Case 2: $d \geq 0$ and C_0 does not intersect D. The analysis for this case is similar to the one for Case 1, so we omit the proof. We refer interested readers to the proof of [LT24, Proposition 6.4].

Case 3: $d < 0$. It follows from Lemma 8.3.36 that

$$2n - 2 = -K_{\mathcal{X}/\mathbb{P}^1}.C \geq \mathrm{neg}(\mathcal{X}, -K_{\mathcal{X}/\mathbb{P}^1}) + 3(n-1) - \mathrm{maxdef}(\mathcal{X}).$$

Again this is a contradiction of our assumption on n. □

Remark 8.3.38 We can make what we mean by "general" in Proposition 8.3.37 more explicit. See [LT24, Construction 6.5] for more details.

Proposition 8.3.39 ([LT24, Proposition 6.6]) *Let $\pi: \mathcal{X} \to \mathbb{P}^1$ be a del Pezzo fibration such that $\mathcal{X}$ is Gorenstein and terminal and $-K_{\mathcal{X}/\mathbb{P}^1}$ is relatively ample. Let n be a positive integer such that*

$$n \geq \mathrm{maxdef}(\mathcal{X}) + \max\{0, -\mathrm{neg}(\mathcal{X}, -K_{\mathcal{X}/\mathbb{P}^1})\}.$$

Suppose that $f: C \to \mathcal{X}$ is a genus 0 stable map of the anticanonical height $2n - 1$ such that the intersection with a general fiber $\mathcal{X}_b$ is equal to 1. Then the following hold:

(i) *Fix a general hyperplane section D in a general fiber of π, and suppose that the image of C contains $n - 1$ general points and intersects D. Then f is a birational map to a free section.*
(ii) *Suppose that the image of C contains n general points. Suppose also that the pushforward of C is not integral as a 1-cycle and at least one of general points is contained in a π-vertical component of C. Then C contains exactly two components, one maps isomorphically to a free section and the other maps birationally to a free $-K_{\mathcal{X}}$-conic or cubic in a general fiber of π.*

Proof The proof is similar to the proof of Proposition 8.3.37. Proof is omitted. □

Proof of Movable Bend and Break

Theorem 8.3.30 is a direct conclusion of the following theorem:

Theorem 8.3.40 ([LT24, Theorem 7.2]) *Let $\pi: \mathcal{X} \to \mathbb{P}^1$ be a del Pezzo fibration such that $\mathcal{X}$ is Gorenstein terminal and the relative anticanonical* *divisor* $-K_{\mathcal{X}/\mathbb{P}^1}$ *is relatively ample. Let* $\mathsf{Q}(\pi)$ *be the constant defined by*

$$\begin{aligned}\mathsf{Q}(\pi) = \max\{3, &-2\mathrm{neg}(\mathcal{X}, -K_{\mathcal{X}/\mathbb{P}^1}) - 5,\\ &-\mathrm{neg}(\mathcal{X}, -K_{\mathcal{X}/\mathbb{P}^1}) + 3,\\ &2\mathrm{maxdef}(\mathcal{X}) - 5\mathrm{neg}(\mathcal{X}, -K_{\mathcal{X}/\mathbb{P}^1}) - 5,\\ &2\mathrm{maxdef}(\mathcal{X}) - \mathrm{neg}(\mathcal{X}, -K_{\mathcal{X}/\mathbb{P}^1}) - 3,\\ &2\mathrm{maxdef}(\mathcal{X}) + 2 + 2\max\{0, -\mathrm{neg}(\mathcal{X}, -K_{\mathcal{X}/\mathbb{P}^1})\}\}.\end{aligned}$$

Suppose that $M \subset \mathrm{Sec}(\mathcal{X}/\mathbb{P}^1)$ *is a component parametrizing a dominant family of sections* C *satisfying* $-K_{\mathcal{X}/\mathbb{P}^1}.C \geq \mathsf{Q}(\pi)$. *Then the closure of* M *in* $\overline{M}_{0,0}(\mathcal{X})$ *contains a point that represents a stable map whose domain is the union of two* $\mathbb{P}^1$*s, both mapping birationally to a free rational curve.*

This theorem prompts the following definition:

Definition 8.3.41 Let $\pi\colon \mathcal{X} \to \mathbb{P}^1$ be a del Pezzo fibration such that $\mathcal{X}$ is Gorenstein terminal and the relative anticanonical divisor $-K_{\mathcal{X}/\mathbb{P}^1}$ is relatively ample. We define $\mathrm{MBBbound}(\mathcal{X})$ to be the smallest non-negative integer r such that any component of $\overline{M}_{0,0}(\mathcal{X})$ generically parametrizing free sections of height r will contain a stable map whose domain has exactly two components each mapping birationally to a free curve.

Theorem 8.3.40 implies that we have $\mathrm{MBBbound}(\mathcal{X}) \leq \mathsf{Q}(\pi)$.

Proof of Theorem 8.3.40 Let $M \subset \mathrm{Sec}(\mathcal{X}/\mathbb{P}^1)$ be a component parametrizing a dominant family of sections C such that

$$-K_{\mathcal{X}/\mathbb{P}^1}.C \geq \mathsf{Q}(\pi).$$

In particular, a general member C of M is free, and we consider its normal bundle:

$$N_{C/\mathcal{X}} \cong \mathcal{O}(a) \oplus \mathcal{O}(b),$$

with $0 \leq a \leq b$. Our height bound implies that $b \geq 3$ when $a = 0$ and $b \geq 2$ when $a = 1$. We prove our assertion based on the splitting behavior of $N_{C/\mathcal{X}}$.
Case 1: $a \geq 1$ *and* $b \geq a+2$. We fix $a+1$ general points on $\mathcal{X}$. By Lemma 8.3.33, a general member C parametrized by M goes through these general points. As explained in [She12, Remark 2.2], the locus parametrizing sections with the generic normal bundle containing all $a+1$ general points has a unique smooth irreducible component containing C. Let Σ be the closure of the locus swept out by sections parametrized by such a component, which is an irreducible projective surface. Let $\Sigma' \to \Sigma$ be the normalization of Σ and C' be the strict

transform of C in Σ'. It follows from [She12, Corollary 2.7] that Σ' is smooth along C' and the normal bundle $N_{C'/\Sigma'}$ is isomorphic to $\mathcal{O}(b)$.

Let $\widetilde{\Sigma}$ be the minimal resolution of Σ', and $\nu\colon \widetilde{\Sigma} \to \Sigma$ be the induced birational morphism. We denote the strict transform of C' in $\widetilde{\Sigma}$ by $\widetilde{C}$. Since Σ' is smooth along C', the birational morphism $\widetilde{\Sigma} \to \Sigma'$ is locally isomorphic along C'. This implies that

$$N_{\widetilde{C}/\widetilde{\Sigma}} \cong N_{C'/\Sigma'} \cong \mathcal{O}(b).$$

Since $\widetilde{\Sigma}$ is swept out by sections, $\psi\colon \widetilde{\Sigma} \to \mathbb{P}^1$ has connected fibers. Moreover, since $b \geq 3$, $\widetilde{C}$ deforms in $\widetilde{\Sigma}$ while fixing two points. By Bend and Break II (Lemma 8.1.3) a general fiber of $\psi\colon \widetilde{\Sigma} \to \mathbb{P}^1$ is isomorphic to $\mathbb{P}^1$. Let F be a general fiber of ψ. Since F goes through a general point of $\mathcal{X}$, we conclude that

$$-K_{\mathcal{X}/\mathbb{P}^1}.F \geq 2.$$

As in Proposition 8.3.27, one can find a moving section $\widetilde{C}_1$ on $\widetilde{\Sigma}$ such that

$$\widetilde{C} \sim \widetilde{C}_1 + \frac{b-e}{2}F,$$

and another section $\widetilde{C}_0$ and a vertical effective cycle T such that

$$\widetilde{C} \sim \widetilde{C}_0 + \frac{b+e}{2}F + T.$$

We have

$$\begin{aligned} a &= -b - \nu^* K_{\mathcal{X}/\mathbb{P}^1}.\widetilde{C} \\ &= -b + \frac{b+e}{2}(-\nu^* K_{\mathcal{X}/\mathbb{P}^1}.F) - \nu^* K_{\mathcal{X}/\mathbb{P}^1}.(\widetilde{C}_0 + T) \\ &= e + \frac{b+e}{2}(-\nu^* K_{\mathcal{X}/\mathbb{P}^1}.F - 2) - \nu^* K_{\mathcal{X}/\mathbb{P}^1}.(\widetilde{C}_0 + T) \\ &\geq e + \frac{b+e}{2}(-\nu^* K_{\mathcal{X}/\mathbb{P}^1}.F - 2) - \nu^* K_{\mathcal{X}/\mathbb{P}^1}.\widetilde{C}_0, \end{aligned} \tag{8.3}$$

where the last inequality is a consequence of the fact that $-\nu^* K_{\mathcal{X}/\mathbb{P}^1}$ is relatively nef.

Proposition 8.3.27 guarantees that we have $b \geq e$. We claim that $b > e$. Assume $b = e$ for a contradiction. First, let us suppose that $-K_{\mathcal{X}/\mathbb{P}^1}.F \geq 3$. It follows from (8.3) that we have

$$a \geq 2b - \nu^* K_{\mathcal{X}/\mathbb{P}^1}.\widetilde{C}_0 \geq 2b + \operatorname{neg}(\mathcal{X}, -K_{\mathcal{X}/\mathbb{P}^1}).$$

Recall that we assume $b - 2 \geq a$. Combining two inequalities, we have

$$b \leq -2 - \operatorname{neg}(\mathcal{X}, -K_{\mathcal{X}/\mathbb{P}^1}).$$

Thus we conclude

$$-K_{\mathcal{X}/\mathbb{P}^1}.C = a + b \leq 2b - 2 \leq -6 - 2\mathrm{neg}(\mathcal{X}, -K_{\mathcal{X}/\mathbb{P}^1}),$$

which contradicts our height bound.

Next, we assume that $-K_{\mathcal{X}/\mathbb{P}^1}.F = 2$. Then (8.3) implies that

$$a - b \geq -\nu^* K_{\mathcal{X}/\mathbb{P}^1}.\widetilde{C}_0, \tag{8.4}$$

and in particular, this means we have $-2 \geq -\nu^* K_{\mathcal{X}/\mathbb{P}^1}.\widetilde{C}_0$. Recall that Σ contains $a + 1$ general points and these give a set of $a + 1$ general points on $\widetilde{\Sigma}$. Since we have $b > a + 1$, we can ensure that b fibers of ψ contain these general points. Using this observation and applying Lemma 8.3.36, we conclude that

$$-K_{\mathcal{X}/\mathbb{P}^1}.C \geq \mathrm{neg}(\mathcal{X}, -K_{\mathcal{X}/\mathbb{P}^1}) + 3(a + 1) - \mathrm{maxdef}(\mathcal{X}),$$

or $a + 1 < \mathrm{maxdef}(\mathcal{X})$. Moreover, by (8.4), we have

$$-K_{\mathcal{X}/\mathbb{P}^1}.C = a + b \leq 2a - \mathrm{neg}(\mathcal{X}, -K_{\mathcal{X}/\mathbb{P}^1}).$$

Combining these, we obtain

$$a \leq \max\{\mathrm{maxdef}(\mathcal{X}) - 2\mathrm{neg}(\mathcal{X}, -K_{\mathcal{X}/\mathbb{P}^1}) - 3, \mathrm{maxdef}(\mathcal{X}) - 2\},$$

which implies that

$$\begin{aligned} &- K_{\mathcal{X}/\mathbb{P}^1}.C \leq \\ &\max\{2\mathrm{maxdef}(\mathcal{X}) - 5\mathrm{neg}(\mathcal{X}, -K_{\mathcal{X}/\mathbb{P}^1}) - 6, 2\mathrm{maxdef}(\mathcal{X}) - 4 - \mathrm{neg}(\mathcal{X}, -K_{\mathcal{X}/\mathbb{P}^1})\}. \end{aligned}$$

However, this contradicts our height bound.

Thus we have $b > e$. This means that $\frac{b-e}{2}$ is a positive integer. When $\frac{b-e}{2} > 1$, we smooth

$$\widetilde{C}_1 + \left(\frac{b - e}{2} - 1\right) F$$

to obtain a section $\widetilde{C}_2$. When $\frac{b-e}{2} = 1$, we set $\widetilde{C}_2 = \widetilde{C}_1$. In both cases, we obtain

$$\widetilde{C} \sim \widetilde{C}_2 + F.$$

We claim that $\nu(\widetilde{C}_2)$ and $\nu(F)$ deforms to cover $\mathcal{X}$. Indeed, recall that Σ contains a general point so that it is not contained in a proper closed subset given by Theorem 8.3.21. Since $\widetilde{C}_2$ and F deform to cover $\widetilde{\Sigma}$, our claim follows. Finally note that F avoids the singular locus of $\mathcal{X}$ by ensuring that F is sitting in a smooth fiber of π. Now the preimage of the singular locus of $\mathcal{X}$ on $\widetilde{\Sigma}$ is the union of points and ψ-vertical curves. Since $\widetilde{C}$ and $\widetilde{C}_2$ have the same intersections with ψ-vertical divisors, the fact that C avoids the singular locus implies that $\nu_*\widetilde{C}_2$ avoids the singular locus of $\mathcal{X}$. Thus one may ensure that $\nu_*\widetilde{C}_2$ is free.

Case 2: $a = 0$ and $b \geq 3$. The proof for this case is similar to Case 1. So a proof is omitted and we refer interested readers to the proof in [LT24, Theorem 7.2].
Case 3: $a \geq 1$ and $b - a \leq 1$:. First, let us assume that $a = b$. In this case, our height bound implies that

$$a \geq \mathrm{maxdef}(\mathcal{X}) + 1 + \max\{0, -\mathrm{neg}(\mathcal{X}, -K_{\mathcal{X}/\mathbb{P}^1})\}.$$

We fix a general points and a general hyperplane section D in a general fiber of π. Then there is a one-parameter family of sections C in M containing these general points and meeting D. By Corollary 8.1.6, sections C degenerate to $C_0 + T$ where

- C_0 is a section; and
- T is a vertical effective cycle such that at least one of general points is contained in the support of T.

By Proposition 8.3.37, $C_0 + T$ has the desired form.

When $b = a + 1$, one can give a similar proof. We consider a one-parameter family of sections containing $a+1$ general points, and we apply Corollary 8.1.6 and Proposition 8.3.39. Thus our assertion follows. □

Consequences of Movable Bend and Break

In this section, we explore several consequences of movable Bend and Break (Theorem 8.3.40).

Various Versions of Movable Bend and Break. First, we introduce the following definition:

Definition 8.3.42 A tree C of rational curves is called a *comb* if it consists of a unique central rational curve C_0 and $r \geq 0$ rational *teeth* $T_1, \ldots, T_r$ such that each T_i intersects C_0 at a single point, and there are no other intersection.

A stable map $f : C \to X$ from a comb to a smooth projective variety X is called a *free comb* if f realizes every component of C as a free rational curve on X.

The first consequence of Theorem 8.3.40 is the following:

Corollary 8.3.43 ([LT24, Corollary 7.6]) *Let $\pi : \mathcal{X} \to \mathbb{P}^1$ be a del Pezzo fibration such that $\mathcal{X}$ is Gorenstein terminal and $-K_{\mathcal{X}/\mathbb{P}^1}$ is relatively ample. Let $C \subset \mathcal{X}$ be a section of π such that*

- *$-K_{\mathcal{X}/\mathbb{P}^1}.C \geq \mathrm{MBBbound}(\mathcal{X})$; and*
- *C is free in $\mathcal{X}$.*

Then C deforms as a stable map to $f\colon C' \to \mathcal{X}$ whose domain C' is a comb $C_0 \cup T_1 \cup \cdots \cup T_r$ satisfying the following:

(i) *f realizes each component of C' as a free rational curve;*
(ii) *C_0 is a section of π satisfying $-K_{\mathcal{X}/\mathbb{P}^1}.C_0 < \mathrm{MBBbound}(\mathcal{X})$; and*
(iii) *each T_i is a free π-vertical rational curve.*

Proof Let $\overline{M} \subset \overline{M}_{0,0}(\mathcal{X})$ be a component that parametrizes a stable map $f\colon C \to \mathcal{X}$ such that C is a comb, the central curve C_0 maps isomorphically to a free section, and the teeth $T_1, \ldots, T_r$ are mapped birationally to π-vertical free rational curves. (Our assumption starts this with $r = 0$.) Assume that

$$-K_{\mathcal{X}/\mathbb{P}^1}.C_0 \geq \mathrm{MBBbound}(\mathcal{X}).$$

We will show that $\overline{M}$ contains a comb C' of the same form such that the central curve C_0' has strictly smaller height.

Let $\overline{M}_i$ be components of $\overline{M}_{0,0}(\mathcal{X})$ containing C_0 for $i = 0$ or T_i for $i \geq 1$. As before, we denote the unique component of $\overline{M}_{0,n}(\mathcal{X})$ above $\overline{M}_i$ by $\overline{M}_i^{(n)}$. Since our stable map $f\colon C \to \mathcal{X}$ is a smooth point of $\overline{M}_{0,0}(\mathcal{X})$, there exists a component $\widetilde{M}$ of

$$\overline{M}_0^{(r)} \times_{\mathcal{X}^r} \left(\prod_{i \geq 1} \overline{M}_i^{(1)} \right)$$

that contains f and maps into $\overline{M}$. Note that on the locus parametrizing free curves, the evaluation maps

$$\overline{M}_0^{(r)} \to \mathcal{X}, \quad \prod_{i \geq 1} \overline{M}_i^{(1)} \to \mathcal{X}^r$$

are flat, and this implies that $\widetilde{M}$ maps surjectively to $\overline{M}_0^{(r)}$ and $\overline{M}_i^{(1)}$ for every $i \geq 1$.

By our movable Bend and Break (Theorem 8.3.40), $\overline{M}_0^{(r)}$ contains in its smooth locus a general point of a component $\overline{N}$ of

$$\overline{N}_0^{(r+1)} \times_{\mathcal{X}} \overline{N}_{r+1}^{(1)},$$

where $\overline{N}_0$ is a component parametrizing free sections of smaller height and $\overline{N}_{r+1}$ is a component parametrizing free π-vertical rational curves. Now $\overline{N}$ has codimension 1 in $\overline{M}_0^{(r)}$ and $\widetilde{M} \to \overline{M}_0^{(r)}$ is flat over the locus parametrizing free combs. We conclude that a general point in the preimage of $\overline{N}$ in $\widetilde{M}$ yields the desired free comb, proving the claim. □

Another version of movable Bend and Break is the following:

Corollary 8.3.44 ([LT24, Theorem 7.7]) *Let $\pi\colon \mathcal{X} \to \mathbb{P}^1$ be a del Pezzo fibration such that $\mathcal{X}$ is Gorenstein terminal and $-K_{\mathcal{X}/\mathbb{P}^1}$ is relatively ample. Let $C \subset \mathcal{X}$ be a section of π such that*

(i) $-K_{\mathcal{X}/\mathbb{P}^1}.C \geq \mathrm{MBBbound}(\mathcal{X})$*; and*
(ii) *C is free in $\mathcal{X}$.*

Then C deforms as a stable map to the union of a free section C_0 with $-K_{\mathcal{X}/\mathbb{P}^1}.C_0 < \mathrm{MBBbound}(\mathcal{X})$ and a π-vertical free rational curve.

We prove this corollary by applying Corollary 8.3.43 then gluing and smoothing teeth:

Proof Let $\overline{M} \subset \overline{M}_{0,0}(\mathcal{X})$ be the unique component containing C. By Corollary 8.3.43, $\overline{M}$ contains a comb consisting of $r+1$ free rational curves such that the central curve C_0 satisfies

$$-K_{\mathcal{X}/\mathbb{P}^1}.C_0 < \mathrm{MBBbound}(\mathcal{X}).$$

We prove our claim based on induction on r. When $r = 1$, there is nothing to prove.

For the induction step, suppose that we have teeth $T_1, \ldots, T_r$. Since all curves in the comb are free, this comb represents a smooth point on $\overline{M}_{0,0}(\mathcal{X})$. In particular, we may deform our comb so that $C_0 \cap T_1$ is general such that any π-vertical irreducible curve of bounded degree passing through this point is free. Next, we deform T_2 along C_0 so that the resulting degeneration curve T_2' intersects $C_0 \cap T_1$. Moreover, since this point is a general point in $\mathcal{X}$ we may ensure that T_2' is an irreducible free rational curve. Then the resulting stable map $f'\colon C' \to \mathcal{X}$ admits a component E contracted to the intersection of C_0 and T_1 so that C_0, T_1, T_2' are attached to E in C'. This is still a smooth point of $\overline{M}_{0,0}(\mathcal{X})$. Then, deforming T_2' along T_1, we obtain an irreducible free deformation T_2'' of T_2 such that T_2'' is attached to T_1 at a point different from $C_0 \cap T_1$. Note that again this stable map is a smooth point of $\overline{M}_{0,0}(\mathcal{X})$. Then using [Kol96, II.7.6.1 Theorem], we smooth $T_1 \cup T_2''$ while fixing the intersection $C_0 \cap T_1$, and we obtain a free comb $C_0 \cup T_* \cup \cdots \cup T_r$. Then we may apply the induction hypothesis, proving our assertion. □

We will use the following version for our later applications:

Corollary 8.3.45 ([LT24, Proposition 7.8]) *Let $\pi\colon \mathcal{X} \to \mathbb{P}^1$ be a del Pezzo fibration such that $\mathcal{X}$ is Gorenstein terminal and $-K_{\mathcal{X}/\mathbb{P}^1}$ is relatively ample. Let $C \subset \mathcal{X}$ be a section of π such that*

(i) $-K_{\mathcal{X}/\mathbb{P}^1}.C \geq \mathrm{MBBbound}(\mathcal{X})$*; and*

(ii) *C is free in $\mathcal{X}$.*

Then C deforms as a stable map to the union of a free section C_0 and a π-vertical free rational curve T of anticanonical degree ≤ 3.

Proof Let $C \subset \mathcal{X}$ be a free section. By movable Bend and Break (Theorem 8.3.40), it degenerates to the union of a free section C_0 and a π-vertical free rational curve T. We claim that this broken curve can deform as a stable map to a curve in our statement. We prove this claim based on induction on the anticanonical degree of T. If the anticanonical degree of T is less than or equal to 3, then there is nothing to prove.

Suppose that $-K_{\mathcal{X}/\mathbb{P}^1}.T \geq 4$. Since the degeneration $C_0 \cup T$ represents a smooth point of $\overline{M}_{0,0}(\mathcal{X})$, after deforming this stable map, we may assume that the intersection $C_0 \cap T$ maps to a general point x_1 of $\mathcal{X}$. Moreover, since we assume that $-K_{\mathcal{X}/\mathbb{P}^1}.T \geq 4$, we may assume that T goes through a general point x_2 in the smooth fiber containing T. Using Proposition 8.3.13, T degenerates to the union of two free vertical rational curves $T_1 \cup T_2$ while fixing two points x_1 and x_2 so that $x_i \in T_i$ for $i = 1, 2$. Then, again, $C_0 \cup T_1 \cup T_2$ is a smooth point of $\overline{M}_{0,0}(\mathcal{X})$. Then using [Kol96, II.7.6.1 Theorem], we smooth $C_0 \cup T_1$ while fixing $T_1 \cap T_2$ and we obtain a free section C_1. Now we apply the induction hypothesis to $C_1 \cup T_2$, and we conclude our claim. □

Fibers of the Evaluation Maps: Next, we prove an analogue of Proposition 7.4.12 in the setting of del Pezzo fibrations using movable Bend and Break. First, we record the following lemma:

Lemma 8.3.46 ([LT24, Lemma 7.9]) *Let $\pi\colon \mathcal{X} \to \mathbb{P}^1$ be a del Pezzo fibration such that $\mathcal{X}$ is Gorenstein terminal and $-K_{\mathcal{X}/\mathbb{P}^1}$ is relatively ample. Let $M \subset$ $\mathrm{Sec}(\mathcal{X}/\mathbb{P}^1)$ be a dominant component. Let $p\colon \mathcal{C} \to M$ be the universal family and* $\mathrm{ev}\colon \mathcal{C} \to \mathcal{X}$ *be the evaluation map.*

Suppose that we have a component N of $\overline{M}_{0,0}(\mathcal{X})$ generically parametrizing a dominant family of π-vertical rational curves T with $-K_{\mathcal{X}/\mathbb{P}^1}.T \geq 3$. Assume that a general curve parametrized by N appears as a component of some stable map in the closure $\overline{M} \subset \overline{M}_{0,0}(\mathcal{X})$. Then a general fiber of ev *is irreducible.*

Proof Let $\overline{p}\colon \overline{\mathcal{C}} \to \overline{M}$ be the universal family. After taking a smooth resolution, we assume that we have a smooth projective variety $\overline{\mathcal{C}}'$ with morphisms $\overline{p}'\colon \overline{\mathcal{C}}' \to \overline{M}$ and $\overline{\mathrm{ev}}\colon \overline{\mathcal{C}}' \to \mathcal{X}$. Let

$$\overline{\mathcal{C}}' \to \mathcal{Y} \to \mathcal{X}$$

be the Stein factorization of $\overline{\mathrm{ev}}\colon \overline{\mathcal{C}}' \to \mathcal{X}$ where the first morphism is denoted by $\overline{\mathrm{ev}}_{\mathcal{Y}}$ and the second morphism is denoted by h. Our goal is to show that a finite morphism $h: \mathcal{Y} \to \mathcal{X}$ is birational.

After taking resolutions, we may assume that $\mathcal{Y}$ is smooth and we still have the corresponding family $\overline{p}': \overline{C}' \to \overline{M}$ and the evaluation map $\overline{\mathrm{ev}}_{\mathcal{Y}}: \overline{C}' \to \mathcal{Y}$. By the construction, $\mathcal{Y}$ is dominated by sections of $\psi := \pi \circ h: \mathcal{Y} \to \mathbb{P}^1$, so we conclude that $\psi: \mathcal{Y} \to \mathbb{P}^1$ is an algebraic fiber space.

Let $\overline{M}_{\mathcal{Y}} \subset \overline{M}_{0,0}(\mathcal{Y})$ be a component generically parametrizing $f: C \to \mathcal{Y}$ where C is a general fiber of $\overline{p}': \overline{C}' \to \overline{M}$ and f is induced by $\overline{\mathrm{ev}}_{\mathcal{Y}}: \overline{C}' \to \mathcal{Y}$. Then we have a surjective generically finite morphism $\overline{M}_{\mathcal{Y}} \to \overline{M}$. Our assumption implies that there is a dominant component $N_{\mathcal{Y}}$ of $\overline{M}_{0,0}(\mathcal{Y})$ generically parametrizing ψ-vertical rational curves T mapping to general members parametrized by N. Then we have a dominant generically finite map $N_{\mathcal{Y}} \to N$, so the expected dimensions of $N_{\mathcal{Y}}$ and N coincide. This implies that

$$(K_{\mathcal{Y}/\mathbb{P}^1} - f^*K_{\mathcal{X}/\mathbb{P}^1}).T = 0.$$

By the ramification formula, $K_{\mathcal{Y}/\mathbb{P}^1} - f^*K_{\mathcal{X}/\mathbb{P}^1}$ is relatively pseudo-effecitve. Since T is a ψ-vertical moving curve, we conclude that

$$a(\mathcal{Y}_\eta, -f^*K_{\mathcal{X}/\mathbb{P}^1}|_{\mathcal{Y}_\eta}) = a(\mathcal{X}_\eta, -K_{\mathcal{X}/\mathbb{P}^1}|_{\mathcal{Y}_\eta}) = 1.$$

It follows from Corollary 6.4.9 that either h is birational or the Iitaka dimension of $K_{\mathcal{Y}/\mathbb{P}^1}|_{\mathcal{Y}_\eta} - f^*K_{\mathcal{X}/\mathbb{P}^1}|_{\mathcal{Y}_\eta}$ is 1.

Suppose that h is not birational. This means that after applying a relative MMP with respect to $K_{\mathcal{Y}/\mathbb{P}^1} - f^*K_{\mathcal{X}/\mathbb{P}^1}$ over $\mathbb{P}^1$, we obtain a birational contraction $\phi: \mathcal{Y} \dashrightarrow \mathcal{Y}'$ such that $K_{\mathcal{Y}'/\mathbb{P}^1} - \phi_* f^*K_{\mathcal{X}/\mathbb{P}^1}$ is relatively semi-ample, that is, we have a morphism $\mathcal{Y}' \to \mathcal{S} \to \mathbb{P}^1$ with $\mathcal{S}$ a normal projective surface such that $K_{\mathcal{Y}'/\mathbb{P}^1} - \phi_* f^*K_{\mathcal{X}/\mathbb{P}^1}$ is the pullback of a relatively ample $\mathbb{Q}$-divisor on $\mathcal{S}$. Then the strict transform T' of a general T in $N_{\mathcal{Y}}$ is identified with a fiber of $\mathcal{Y}' \to \mathcal{S}$. This means that $-f^*K_{\mathcal{X}/\mathbb{P}^1}.T = 2$, but this contradicts our assumption. Hence h must be birational. □

Corollary 8.3.47 ([LT24, Corollary 7.10]) *Let $\pi: \mathcal{X} \to \mathbb{P}^1$ be a del Pezzo fibration such that $\mathcal{X}$ is Gorenstein terminal and $-K_{\mathcal{X}/\mathbb{P}^1}$ is relatively ample. Let $M \subset \mathrm{Sec}(\mathcal{X}/\mathbb{P}^1)$ be a dominant component parametrizing sections C such that $-K_{\mathcal{X}/\mathbb{P}^1}.C \geq \mathrm{MBBbound}(\mathcal{X}) + 2$. Let $p: \mathcal{C} \to M$ be the universal family and $\mathrm{ev}: \mathcal{C} \to \mathcal{X}$ be the evaluation map. Then a general fiber of* ev *is irreducible.*

Proof Let $\overline{M} \subset \overline{M}_{0,0}(\mathcal{X})$ be the closure of M. By Corollary 8.3.44, $\overline{M}$ contains a stable map consisting of a free section and a π-vertical free rational curve T of anticanonical degree ≥ 3. Since such a stable map is a smooth point of $\overline{M}_{0,0}(\mathcal{X})$, a general deformation of T is also a component of some stable map parametrized by $\overline{M}$. Thus our assertion follows from Lemma 8.3.46. □

8.3.5 Examples

In this section, we analyze the space of sections for certain del Pezzo fibrations.

Improving Bounds

Under some stronger assumptions, one can improve the bound in Theorem 8.3.40:

Lemma 8.3.48 ([LT24, Lemma 8.1]) *Let $\pi\colon \mathcal{X} \to \mathbb{P}^1$ be a del Pezzo fibration such that $\mathcal{X}$ is Gorenstein terminal and $-K_{\mathcal{X}/\mathbb{P}^1}$ is relatively ample. Assume the following properties:*

(i) *for every $d < 0$, we have*

$$\mathrm{maxdef}(d) - d \leq 2;$$

(ii) *the generic fiber $\mathcal{X}_\eta$ does not admit a geometrically integral rational curve of the anticanonical degree 2.*

Then we have

$$\mathrm{MBBbound}(\mathcal{X}) \leq 3.$$

Proof This essentially follows from the proof of Theorem 8.3.40 with minor improvements. Under the first assumption, one may prove Lemma 8.3.36 with the following conclusion:

$$-K_{\mathcal{X}/\mathbb{P}^1}.C \geq 3n - 2.$$

With this improved lemma, Proposition 8.3.37 can be proved under the assumption that

$$n \geq 4.$$

Proposition 8.3.39 is also true assuming

$$n \geq 2.$$

Then, with these improved propositions, one might follow the proof of Theorem 8.3.40. Note that, under our second assumption, the case of $-K_{\mathcal{X}/\mathbb{P}^1}.F = 2$ is excluded. These improvements establish movable Bend and Break assuming the anticanonical height ≥ 3 except when the anticanonical height is 4.

Finally, when the anticanonical height is 4, the only place where we need an improvement is Case 3 of the proof of Theorem 8.3.40 with $a = b = 2$. In this case, we consider sections passing through two general points and meeting a general hyperplane section D in a general fiber of π. Such sections form a one-parameter family and they sweep out a surface with a general fiber that is a rational curve of height ≥ 3. Then we can improve Lemma 8.3.36 using the fact that T_i containing a general point has anticanonical degree ≥ 3. This improves the conclusion as

$$-K_{\mathcal{X}/\mathbb{P}^1}.C \geq d + 3n.$$

In turn, Proposition 8.3.37 now holds with $n = 3$. Hence our assertion follows. □

Next, we improve Corollary 8.3.45 by showing that one can always break off a vertical free conic:

Theorem 8.3.49 ([LT24, Theorem 8.2]) *Let $\pi\colon \mathcal{X} \to \mathbb{P}^1$ be a del Pezzo fibration such that $\mathcal{X}$ is Gorenstein and terminal and $-K_{\mathcal{X}/\mathbb{P}^1}$ is relatively ample. Suppose that a general fiber S of π is a del Pezzo surface with $2 \leq (-K_S)^2 \leq 8$ and every surface swept out by π-vertical lines in smooth fibers is a big $\mathbb{Q}$-Cartier divisor on $\mathcal{X}$.*

Let $C \subset \mathcal{X}$ be a free section of height $d \geq$ MBBbound$(\mathcal{X})$. Then C deforms as a stable map to the union of a free section C_0 and a π-vertical free conic.

Proof It follows from Corollary 8.3.45 that C deforms as a stable map to the union of a free section C_0 and a vertical free conic or cubic T. In the former case, there is nothing to prove, so we assume the latter. Since this union is a smooth point of $\overline{M}_{0,0}(\mathcal{X})$, we can ensure that T is sitting in a smooth fiber of π. Then there are two possibilities for T:

- T is the pullback of a hyperplane class on $\mathbb{P}^2$; or
- T is the pullback of a rational member in the anticanonical system on a smooth cubic surface.

In the first case, let E be a (-1)-curve that does not intersect T. In the second case, let E be a (-1)-curve such that $E.T = 1$. Let $S \subset \mathcal{X}$ be the surface swept out by deformations of E. Since this is a big $\mathbb{Q}$-Cartier divisor, a movable curve C_0 must intersect S. We may assume that an intersection x of C_0 and S is general.

Now we deform T along C_0 so that the degeneration T' passes through x. Then, by the construction, T' has to be the union of a line E' and a conic F such that $F^2 = 0$ in $\mathcal{X}_{\pi(x)}$. While deforming T, we may ensure that T meets a section that passes through a general point in the smooth fiber containing x. This will guarantee that F is a smooth conic. Then we apply [Kol96, Chapter II 7.9 Theorem] to smooth $C_0 \cup E'$ to C_1. While smoothing, we may deform F to F' so that $C_1 \cup F'$ is connected. Thus our assertion follows. □

Pencils of Hyperplane Sections on del Pezzo Threefolds

Let X be a del Pezzo threefold with $\mathrm{Pic}(X) = \mathbb{Z}H$ and $-K_X \sim 2H$. We consider a general pencil $\mathfrak{p} \subset |H|$ of hyperplane sections and we denote its base locus by Z, which is a smooth projective curve of genus 1. Let $\beta: \mathcal{X} \to X$ be the blowup of X along Z which comes with the exceptional divisor E and the projection $\pi: \mathcal{X} \to \mathbb{P}^1$ induced by the pencil $\mathfrak{p}$. Then $\mathcal{X}$ is a smooth Fano threefold, and $\pi: \mathcal{X} \to \mathbb{P}^1$ is a standard model of a del Pezzo fibration.

We denote the space of sections of anticanonical height d by

$$\mathrm{Sec}(\mathcal{X}/\mathbb{P}^1, d).$$

Regarding this space, we have the following theorem:

Theorem 8.3.50 ([LT24, Section 8.2]) *Assume that $H^3 \geq 3$ and $\mathfrak{p}$ is general. Then the following statements are true:*

(i) $\mathrm{neg}(\mathcal{X}, -K_{\mathcal{X}/\mathbb{P}^1}) = -1$*;*
(ii) *for each $d \geq -1$, $\mathrm{Sec}(\mathcal{X}/\mathbb{P}^1, d)$ is non-empty, irreducible, and has expected dimension; and*
(iii) *when $d \geq 0$, $\mathrm{Sec}(\mathcal{X}/\mathbb{P}^1, d)$ parametrizes a dominant family of sections.*

Proof We give a proof when $H^3 = 3$. First, one can deduce the theorem over an algebraically closed field k of characteristic 0 from the corresponding theorem over $\mathbb{C}$. So we may assume that our ground field is $\mathbb{C}$.

Analysis of π-vertical rational curves of low height. One can prove that the monodromy action on the Néron–Severi space of a general fiber of π is maximal, that is, it is the full Weyl group $W(E_6)$. (For details, readers should consult [LT24, Example 8.3].) Since $W(E_6)$ acts on the set of classes of (-1)-curves (resp. smooth conics) transitively we may conclude that for π-vertical rational curves of height 1 (resp. 2), there is a unique family of rational curves which dominates the base $\mathbb{P}^1$.

Computation of a proper closed subset. Since $\mathcal{X}$ is Fano, for any section C, we have

$$-K_{\mathcal{X}/\mathbb{P}^1}.C \geq -1.$$

In particular, any component of $\mathrm{Sec}(\mathcal{X}/\mathbb{P}^1)$ has positive expected dimension, so there is no rigid section of π. Thus if we have a non-dominant family of sections, then it must sweep out a surface Y. Since there is no geometrically integral line and conic on $\mathcal{X}_\eta$, the anticanonical degree of Y_η is greater than or equal to 3. This implies that we have

$$a(Y_\eta, -K_{\mathcal{X}/\mathbb{P}^1}|_{Y_\eta}) < a(\mathcal{X}_\eta, -K_{\mathcal{X}_\eta}) = 1.$$

By Theorem 8.3.28, any non-dominant family of sections parametrizes sections of height ≤ 0. Below we will show that there is only one non-dominant family of sections of height 0, and it parametrizes sections in E. In particular, a proper closed subset $\mathcal{V}$ in Theorem 8.3.21 is given by E.

Sections of low height. We now study sections of low height.
Sections of height -1. We have

$$-K_{\mathcal{X}/\mathbb{P}^1} \sim E,$$

so any sections of height -1 are contained in $E \cong Z \times \mathbb{P}^1$. Such sections are parametrized by Z.
Sections of height 0. Since we have $\mathcal{X}_b \sim H - E$, such sections are the strict transforms of H-lines on X that do not meet Z. It follows from Theorem 8.2.7 that $\mathrm{Sec}(\mathcal{X}/\mathbb{P}^1, 0)$ is irreducible and 2-dimensional. It also parametrizes a dominant family of sections.
Sections of height 1. These sections are strict transforms of H-conics on X meeting Z once. Let $p\colon \mathcal{C} \to M$ be the universal family of H-conics on X. Then, by Proposition 8.2.10, a general fiber of the evaluation map $\mathrm{ev}\colon \mathcal{C} \to X$ is irreducible and 2-dimensional. Thus, assuming Z is general, we conclude that $\mathrm{Sec}(\mathcal{X}/\mathbb{P}^1, 1)$ is irreducible and 3-dimensional. This space parametrizes a dominant family of sections.
Sections of height 2. These are strict transforms of H-cubics on X that meet Z twice. As we observed in Lemma 8.2.14, general H-cubics are very free, so they are smooth by Lemma 7.2.14. In particular, such H-cubics are twisted cubics in $\mathbb{P}^4$. Thus such an H-cubic C is contained in a unique hyperplane $H' \subset \mathbb{P}^4$. Then $S = X \cap H'$ is a smooth cubic surface and $C \subset S$ is the pullback of a hyperplane class on $\mathbb{P}^2$. Let $U \subset (\mathbb{P}^4)^\vee$ be the Zariski open subset parametrizing hyperplane sections H' such that $S = X \cap H'$ is smooth. By Theorem 8.2.1, the space of twisted cubics on X is irreducible. This implies that the monodromy action of $\pi_1(U)$ on the set of classes of smooth rational cubics on S is transitive.

Now observe that we have a natural morphism

$$\Phi\colon \mathrm{Sec}(\mathcal{X}/\mathbb{P}^1, 2) \to \mathrm{Sym}^{(2)}(Z), \quad [C] \mapsto [C \cap Z].$$

If we fix two points x_1, x_2 on Z, then we have the line $\ell \subset \mathbb{P}^4$ passing through x_1, x_2. We consider the 2-dimensional family P of H' containing ℓ. Then, by the Lefschetz hyperplane theorem, the monodromy of $\pi_1(U \cap P)$ on classes of smooth rational cubics is still transitive, so a general fiber of Φ is irreducible and 2-dimensional. Hence we conclude that $\mathrm{Sec}(\mathcal{X}/\mathbb{P}^1, 2)$ is irreducible and 4-dimensional, proving the claim.

Inductive step: By Lemma 8.3.48, we have $\mathrm{MBBbound}(\mathcal{X}) \leq 3$. We prove that for $d \geq 3$ the space

$$\mathrm{Sec}(\mathcal{X}/\mathbb{P}^1, d)$$

is non-empty, irreducible, and has the expected dimension. First, by gluing a π-vertical conic and smoothing it, we can prove that $\mathrm{Sec}(\mathcal{X}/\mathbb{P}^1, d)$ is non-empty by induction on d. We have already shown that this space parametrizes a dominant family of sections. Thus it must have the expected dimension.

It suffices to show that $\mathrm{Sec}(\mathcal{X}/\mathbb{P}^1, d)$ is irreducible by assuming that

$$\mathrm{Sec}(\mathcal{X}/\mathbb{P}^1, d-2)$$

is irreducible. By Theorem 8.3.49, any free section C of height d deforms to the union of a free section C_0 of height $d - 2$ and a π-vertical free conic T. Then since the monodromy on the set of classes of conics is transitive, the space of vertical conics T that meet C_0 is also irreducible. This implies that $\mathrm{Sec}(\mathcal{X}/\mathbb{P}^1, d)$ is irreducible, which proves the claim.

□

8.4 Further Topics: Higher Genus Base

So far we have discussed the moduli space of rational curves on a Fano variety. One might also develop a similar story for arbitrary smooth projective curves of genus ≥ 1. Let B be a smooth projective curve of genus $g = g(B)$ defined over an algebraically closed field k of characteristic 0. Let X be a Fano variety defined over k. The protagonist here is the morphism scheme

$$\mathrm{Mor}(B, X).$$

One may equally study the dimension and the number of components of this moduli space. As observed in the previous section, it is quite useful to have a projective compactification of such a moduli space, and this is given by the *moduli stack of stable maps of genus g with n marked points*:

$$\overline{\mathcal{M}}_{g,n}(X),$$

and its coarse moduli space

$$\overline{M}_{g,n}(X).$$

There is one remark regarding this compactification. While the morphism scheme $\mathrm{Mor}(B, X)$ parametrizes morphisms $f : B \to X$, the moduli stack $\overline{\mathcal{M}}_{g,n}(X)$ parametrizes stable maps from any genus g curve. Thus in general $\overline{\mathcal{M}}_{g,n}(X)$ does not give a compactification of $\mathrm{Mor}(B, X)$ even up to automorphism. To this end, we have the following forgetting map:

$$\overline{\mathcal{M}}_{g,n}(X) \to \overline{\mathcal{M}}_{g,n},$$

where $\overline{\mathcal{M}}_{g,n}$ is the *moduli stack of stable curves of genus g with n marked points*. Then one might think that the fiber of this map at $[B]$ is a compactification of $\mathrm{Mor}(B, X)$ up to automorphism. See, for example, [FP97] for more information on these moduli spaces.

When we have a Fano fibration $\pi : \mathcal{X} \to B$, we are interested in the space of sections:

$$\mathrm{Sec}(\mathcal{X}/B).$$

In the following subsections, we will briefly discuss some recent results on this space for del Pezzo fibrations [LT22] and for Fano fibrations in general [LRT25c]. Note that one may view $\overline{\mathcal{M}}_{g(B),n}(\mathcal{X})$ as a compactification for the space of sections $\mathrm{Sec}(\mathcal{X}/B)$.

8.4.1 Del Pezzo Fibrations

Let B be a smooth projective curve of genus g. Let $\pi : \mathcal{X} \to B$ be a del Pezzo fibration. Here we discuss the main results of [LT22]. First, we introduce the following definition:

Definition 8.4.1 Let $\pi : \mathcal{X} \to B$ be a Fano fibration. A section $C \subset \mathcal{X}$ is *generically free* if the following are true:

(i) C is contained in the smooth locus of $\mathcal{X}$; and
(ii) the normal bundle $N_{C/\mathcal{X}}$ satisfies (a) $H^1(C, N_{C/\mathcal{X}}) = 0$ and (b) $N_{C/\mathcal{X}}$ is generically globally generated, that is,

$$H^0(C, N_{C/\mathcal{X}}) \otimes \mathcal{O}_C \to N_{C/\mathcal{X}},$$

is surjective at the generic point.

We say C is *relatively free* if additionally $N_{C/\mathcal{X}}$ is globally generated.

With this definition, components of $\mathrm{Sec}(\mathcal{X}/B)$ can be classified into the following three categories:

(1) non-dominant families of sections;

(2) dominant families of sections for which a general section is not generically free; or
(3) dominant families of sections for which a general section is generically free.

Note that components in (3) will have expected dimension equal to

$$-K_{\mathcal{X}/B}.C + 2(1 - g(B)).$$

[LT22] obtained results for each category of components. The first theorem we discuss is for non-dominant families of sections. This is an analogue of Theorem 8.3.21:

Theorem 8.4.2 ([LT22, Theorem 1.3]) *Let $\pi: \mathcal{X} \to B$ be a del Pezzo fibration from a Gorenstein projective terminal threefold such that $-K_{\mathcal{X}/B}$ is relatively nef. Then there is a proper closed subset $\mathcal{V} \subsetneq \mathcal{X}$ such that any component $M \subset$ $\mathrm{Sec}(\mathcal{X}/B)$ parametrizing a non-dominant family of sections will parametrize sections contained in $\mathcal{V}$.*

The next theorem is unique in the higher genus setting, and is useful in understanding components in category (2):

Theorem 8.4.3 ([LT22, Theorem 1.6]) *Let $\pi: \mathcal{X} \to B$ be a del Pezzo fibration such that $\mathcal{X}$ is Gorenstein terminal and $-K_{\mathcal{X}/B}$ is relatively nef. Then there is a bounded family of surfaces $Y \subset \mathcal{X}$ whose intersections with a general fiber are rational curves of anticanonical degree 2, and a positive constant $\mathsf{D}(\pi)$ with the following property: let $M \subset \mathrm{Sec}(\mathcal{X}/B)$ be a dominant component parametrizing sections of height $\geq \mathsf{D}(\pi)$ which are not generically free. Then a general section parametrized by M is contained in some surface Y in our bounded family.*

Remark 8.4.4 When $\mathcal{X}_\eta$ has Picard rank 1 and $(-K_{\mathcal{X}_\eta})^2 \geq 3$, there is no geometrically integral rational conic on $\mathcal{X}_\eta$. In particular, Theorem 8.4.3 implies that any dominant family of sections of height $\geq \mathsf{D}(\pi)$ generically parametrizes a generically free section.

Finally, we have an analogue of Theorem 8.3.30 that establishes movable Bend and Break in this setting:

Theorem 8.4.5 ([LT22, Theorem 1.7]) *Let $\pi: \mathcal{X} \to B$ be a del Pezzo fibration such that $\mathcal{X}$ is Gorenstein terminal and the relative anticanonical divisor $-K_{\mathcal{X}/B}$ is relatively ample. Then there exists a constant $\mathsf{Q}(\pi)$ satisfying the following property: Suppose that $M \subset \mathrm{Sec}(\mathcal{X}/B)$ is a component parametrizing a dominant family of generically free sections C satisfying $-K_{\mathcal{X}/B}.C \geq \mathsf{Q}(\pi)$.*

Then the closure of M in $\overline{M}_{g,0}(\mathcal{X})$ contains a point that represents a stable map whose domain is the union of a generically free section and a π-vertical free rational curve.

Combining the above theorems, [LT22] establishes geometric Manin's conjecture II when the generic fiber $\mathcal{X}_\eta$ is split:

Theorem 8.4.6 ([LT22, Theorem 1.15]) *Let $\pi\colon \mathcal{X} \to B$ be a del Pezzo fibration such that the following are true:*

(i) *$\mathcal{X}$ is Gorenstein, $\mathbb{Q}$-factorial, and terminal;*
(ii) *every fiber of π is a del Pezzo surface with at worst canonical singularities; and*
(iii) *$\mathcal{X}_\eta$ is realized as the blowup of $\mathbb{P}^2_\eta$ at finitely many $k(B)$-rational points.*

Then there is a section C such that for every integral numerical class $\alpha \in [C] + \mathrm{Nef}_1(\mathcal{X}_\eta)$, there is exactly one component of $\mathrm{Sec}(\mathcal{X}/B, \alpha)$ *parametrizing generically free sections.*

Remark 8.4.7 [LT22, Theorem 1.15] assumes that every fiber of π is a del Pezzo surface S with $(-K_S)^2 \geq 2$. The proof relies on the main theorems of [LT22] as well as results from [Tes09] and [Tes05] which established geometric Manin's conjecture II for del Pezzo surfaces. However, in [Tes05], results for degree 1 del Pezzo surfaces assume that the surface itself is general in its moduli. This was why in [LT22, Theorem 1.15] we imposed the above assumption. Since geometric Manin's conjecture II for degree 1 del Pezzo surfaces was fully established in [BLRT23, Theorem 1.6], we state Theorem 8.4.6 for degree 1 del Pezzo surfaces as well.

Example 8.4.8 Natural examples satisfying the assumptions of Theorem 8.4.6 are the trivial families

$$\mathcal{X} = S \times B,$$

where S is a smooth del Pezzo surface defined over k.

8.4.2 Fano Fibrations

In this section, we briefly discuss results for higher dimensional Fano fibrations that were obtained in [LRT25c]. These are higher dimensional generalizations of Theorems 8.4.2 and 8.4.3. For technical reasons, we assume that our ground field is $\mathbb{C}$ in this section.

The first theorem shows that components parametrizing sections that are not relatively free come from maps characterized by a-invariants:

Theorem 8.4.9 ([LRT25c, Theorem 1.3]) *Let $\pi: \mathcal{X} \to B$ be a smooth model of a Fano fibration. There is a constant $\xi = \xi(\pi)$ such that the following holds: let $M \subset \mathrm{Sec}(\mathcal{X}/B)$ be a component parametrizing a family of non-relatively free sections C that satisfy $-K_{\mathcal{X}/B}.C \geq \xi$. Let C' be the normalization of the universal family $\mathcal{C}$ over M with the evaluation map $\mathrm{ev}: \mathcal{C}' \to \mathcal{X}$. Then either*

(i) ev *is not dominant, in which case the subvariety $\mathcal{Y}$ swept out by sections in M satisfies*

$$a(\mathcal{Y}_\eta, -K_{\mathcal{X}/B}|_{\mathcal{Y}_\eta}) \geq a(\mathcal{X}_\eta, -K_{\mathcal{X}/B}|_{\mathcal{X}_\eta}) = 1; \textit{or}$$

(ii) ev *is dominant, in which case if $h: \mathcal{Y} \to \mathcal{X}$ is the finite part of the Stein factorization of* ev, *then we have*

$$a(\mathcal{Y}_\eta, -h^*K_{\mathcal{X}/B}|_{\mathcal{Y}_\eta}) = a(\mathcal{X}_\eta, -K_{\mathcal{X}/B}|_{\mathcal{X}_\eta}) = 1.$$

Furthermore, there is a dominant rational map $\phi: \mathcal{Y} \dashrightarrow \mathcal{Z}$ over B with rationally connected general fibers such that $\dim \mathcal{Z} \geq 2$ *and the following properties hold: let C' be a general section of $\mathcal{Y} \to B$ parametrized by M and let $\mathcal{W}' \subset \mathcal{Y}$ be the unique main component of $\phi^{-1}(\phi(C'))$ mapping dominantly to $\phi(C')$. There is a smooth resolution $\psi: \mathcal{W} \to \mathcal{W}'$ such that the locus, where ψ^{-1} is well defined, intersects C' and ψ has the following properties:*

(a) *we have*

$$a(\mathcal{W}_\eta, -\psi^* h^* K_{\mathcal{X}/B}|_{\mathcal{W}_\eta}) = a(\mathcal{X}_\eta, -K_{\mathcal{X}/B}|_{\mathcal{X}_\eta}) = 1;$$

(b) *the pair $(\mathcal{W}_\eta, -\psi^* h^* K_{\mathcal{X}/B}|_{\mathcal{W}_\eta})$ is adjoint rigid;*

(c) *a general deformation of the strict transform of C' in $\mathcal{W}$ is relatively free in $\mathcal{W}$; and*

(d) *there is a constant $T = T(\pi)$ depending only on π, but not on M, such that the sublocus of M parametrizing deformations of the strict transform of C' in $\mathcal{W}$ has codimension at most T in M.*

This theorem motivates us to call a generically finite B-morphism $h: \mathcal{Y} \to \mathcal{X}$ with

$$a(\mathcal{Y}_\eta, -h^*K_{\mathcal{X}/B}|_{\mathcal{Y}_\eta}) \geq a(\mathcal{X}_\eta, -K_{\mathcal{X}/B}|_{\mathcal{X}_\eta}) = 1$$

an *accumulating map*. Theorem 8.4.9 claims that accumulating maps are sources of components parametrizing non-relatively free sections. In particular, combining Theorem 8.4.9 with Theorem 5.1.4 from the BAB conjecture and Lemma 5.3.5 from [LST22, Lemma 8.3], [LRT25c] established that non-relatively free sections of sufficiently large height come from a bounded family of accumulating maps:

Theorem 8.4.10 ([LRT25c, Theorem 1.7]) *Let $\pi: \mathcal{X} \to B$ be a smooth model of a Fano fibration.*

(i) *There is a proper closed subset $\mathcal{R} \subset \mathcal{X}$ such that if $M \subset \mathrm{Sec}(\mathcal{X}/B)$ is a component parametrizing a non-dominant family of sections, then sections parametrized by M are contained in $\mathcal{R}$.*
(ii) *There is a constant $\xi = \xi(\pi)$, a proper closed subset $\mathcal{V} \subset \mathcal{X}$, and a bounded family of smooth projective B-varieties $\mathcal{Y}_t$ parametrized by $t \in \mathfrak{T}$ with B-morphisms $h_t: \mathcal{Y}_t \to \mathcal{X}$ with the following properties:*
  (a) $\dim \mathcal{Y}_t < \dim \mathcal{X}$ *and $h_t: \mathcal{Y}_t \to \mathcal{X}$ is generically finite onto the image;*
  (b) *we have*

$$a(\mathcal{Y}_{t,\eta}, -h_t^* K_{\mathcal{X}/B}|_{\mathcal{Y}_{t,\eta}}) = a(\mathcal{X}_\eta, -K_{\mathcal{X}/B}|_{\mathcal{X}_\eta}) = 1,$$

  and a pair $(\mathcal{Y}_{t,\eta}, -h_t^* K_{\mathcal{X}/B}|_{\mathcal{Y}_{t,\eta}})$ is adjoint rigid for any $t \in \mathfrak{T}$; and
  (c) *if $M \subset \mathrm{Sec}(\mathcal{X}/B)$ is a component parametrizing non-relatively free sections C of height $\geq \xi$, then for a general section C parametrized by M, we have either*
    - *$C \subset \mathcal{V}$; or*
    - *for some $h_t: \mathcal{Y}_t \to \mathcal{X}$, there is a relatively free section C' of $\mathcal{Y}_t/B$ such that $h_t(C') = C$.*

One may consider the above theorem a geometric analogue of the thinness of exceptional sets in Manin's conjecture. In particular, using Theorem 8.4.10 (i), one can give an arithmetic application that is analogous to Theorem 7.5.2 in the setting of Fano fibrations ([LRT25c, Theorem 1.11]). We also note that [LRT25c, Section 4.3] proposes a formulation of geometric Manin's conjecture in the setting of Fano fibrations over a higher genus curve; this conjecture awaits a further extensive study. Other applications of results in [LRT25c] are explored in [LRT25b] and [LRT25a].

Finally, we note that a generalization of movable Bend and Break for Fano threefolds (Theorem 8.2.12) is still open, but [Fen25] has established the first instance of such a theorem for elliptic curves on cubic threefolds. Moreover, using movable Bend and Break, [Fen25] has proved an analogue of geometric Manin's conjecture II in such a setting.

References

[Abr09] D. Abramovich, *Birational geometry for number theorists*, Arithmetic geometry, Clay Math. Proc., vol. 8, Amer. Math. Soc., Providence, RI, 2009, pp. 335–373.

[ADHL15] I. Arzhantsev, U. Derenthal, J. Hausen, and A. Laface, *Cox rings*, Cambridge Studies in Advanced Mathematics, vol. 144, Cambridge University Press, Cambridge, 2015.

[AFK22] H. Abban, M. Fedorchuk, and I. Krylov, *Stability of fibrations over one-dimensional bases*, Duke Math. J. **171** (2022), no. 12, 2461–2518.

[AGP02] M. Aguilar, S. Gitler, and C. Prieto, *Algebraic topology from a homotopical viewpoint*, Universitext, Springer-Verlag, New York, 2002, Trans. Stephen Bruce Sontz.

[AK77] A. B. Altman and S. L. Kleiman, *Foundations of the theory of Fano schemes*, Compositio Math. **34** (1977), no. 1, 3–47.

[AK19] F. Ambro and J. Kollár, *Minimal models of semi-log-canonical pairs*, Moduli of K-stable varieties, Springer INdAM Ser., vol. 31, Springer, Cham, 2019, pp. 1–13.

[Ale94] V. Alexeev, *Boundedness and K^2 for log surfaces*, Internat. J. Math. **5** (1994), no. 6, 779–810.

[Ara10] C. Araujo, *The cone of pseudo-effective divisors of log varieties after Batyrev*, Math. Z. **264** (2010), no. 1, 179–193.

[AV02] D. Abramovich and A. Vistoli, *Compactifying the space of stable maps*, J. Amer. Math. Soc. **15** (2002), no. 1, 27–75.

[AW97] D. Abramovich and J. Wang, *Equivariant resolution of singularities in characteristic* 0, Math. Res. Lett. **4** (1997), no. 2–3, 427–433.

[Bat88] V. V. Batyrev, *Distribution of rational points of bounded height*, Lecture at Math. Inst. Berlin, Thursday July 21, 1988.

[Bat92] ______, *The cone of effective divisors of threefolds*, Proceedings of the International Conference on Algebra, Part 3 (Novosibirsk, 1989), Contemp. Math., vol. 131, Amer. Math. Soc., Providence, RI, 1992, pp. 337–352.

[BB92] A. A. Borisov and L. A. Borisov, *Singular toric Fano three-folds*, Mat. Sb. **183** (1992), no. 2, 134–141.

[BC13] A. Bertram and I. Coskun, *The birational geometry of the Hilbert scheme of points on surfaces*, Birational geometry, rational curves, and arithmetic, Simons Symp., Springer, Cham, 2013, pp. 15–55.

[BCHM10] C. Birkar, P. Cascini, C. D. Hacon, and J. McKernan, *Existence of minimal models for varieties of log general type*, J. Amer. Math. Soc. **23** (2010), no. 2, 405–468.

[BDPP13] S. Boucksom, J. P. Demailly, M. Paun, and T. Peternell, *The pseudo-effective cone of a compact Kähler manifold and varieties of negative Kodaira dimension*, J. Algebraic Geom. **22** (2013), no. 2, 201–248.

[Beh97] K. Behrend, *Gromov-Witten invariants in algebraic geometry*, Invent. Math. **127** (1997), no. 3, 601–617.

[BEL91] A. Bertram, L. Ein, and R. Lazarsfeld, *Vanishing theorems, a theorem of Severi, and the equations defining projective varieties*, J. Amer. Math. Soc. **4** (1991), no. 3, 587–602.

[Bel25] P. Belmans, *Fanography*, https://fanography.info, 2025.

[BF97] K. Behrend and B. Fantechi, *The intrinsic normal cone*, Invent. Math. **128** (1997), no. 1, 45–88.

[BHB17] T. D. Browning and R. Heath-Brown, *Forms in many variables and differing degrees*, J. Eur. Math. Soc. (JEMS) **19** (2017), no. 2, 357–394.

[BHB20] ———, *Density of rational points on a quadric bundle in* $\mathbb{P}^3 \times \mathbb{P}^3$, Duke Math. J. **169** (2020), no. 16, 3099–3165.

[Bil23] M. Bilu, *Motivic Euler products and motivic height zeta functions*, Mem. Amer. Math. Soc. **282** (2023), no. 1396, v+185.

[Bir62] B. J. Birch, *Forms in many variables*, Proc. Roy. Soc. Ser. A **265** (1961/1962), 245–263.

[Bir07] C. Birkar, *Ascending chain condition for log canonical thresholds and termination of log flips*, Duke Math. J. **136** (2007), no. 1, 173–180.

[Bir19] ———, *Anti-pluricanonical systems on Fano varieties*, Ann. of Math. (2) **190** (2019), no. 2, 345–463.

[Bir21] ———, *Singularities of linear systems and boundedness of Fano varieties*, Ann. of Math. (2) **193** (2021), no. 2, 347–405.

[BJ22] A. Burke and E. Jovinelly, *Geometric Manin's Conjecture for Fano* 3*-folds*, arXiv:2209.05517, 2022.

[BK94] M. Brion and F. Knop, *Contractions and flips for varieties with group action of small complexity*, J. Math. Sci. Univ. Tokyo **1** (1994), no. 3, 641–655.

[BK13] R. Beheshti and N. M. Kumar, *Spaces of rational curves on complete intersections*, Compos. Math. **149** (2013), no. 6, 1041–1060.

[BL17] T. D. Browning and D. Loughran, *Varieties with too many rational points*, Math. Z. **285** (2017), no. 3, 1249–1267.

[BLRT22] R. Beheshti, B. Lehmann, E. Riedl, and S. Tanimoto, *Moduli spaces of rational curves on Fano threefolds*, Adv. Math. **408** (2022), Paper No. 108557, 60.

[BLRT23] ———, *Rational curves on del Pezzo surfaces in positive characteristic*, Trans. Amer. Math. Soc. Ser. B **10** (2023), 407–451.

[BM90] V. V. Batyrev and Yu. I. Manin, *Sur le nombre des points rationnels de hauteur borné des variétés algébriques*, Math. Ann. **286** (1990), no. 1–3, 27–43.

[BM96] K. Behrend and Yu. I. Manin, *Stacks of stable maps and Gromov-Witten invariants*, Duke Math. J. **85** (1996), no. 1, 1–60.

[Bor96] M. Borovoi, *The Brauer-Manin obstructions for homogeneous spaces with connected or abelian stabilizer*, J. Reine Angew. Math. **473** (1996), 181–194.

[Bou11] D. Bourqui, *Asymptotic behavior of rational curves*, arXiv:1107.3824, 2011.

[Bou12] ______, *Moduli spaces of curves and Cox rings*, Michigan Math. J. **61** (2012), no. 3, 593–613.

[Bou16] ______, *Algebraic points, non-anticanonical heights and the Severi problem on toric varieties*, Proc. Lond. Math. Soc. (3) **113** (2016), no. 4, 474–514.

[BR15] M. Beck and S. Robins, *Computing the continuous discretely*, 2nd ed., Undergraduate Texts in Mathematics, Springer, New York, 2015.

[Bro09] T. D. Browning, *Quantitative arithmetic of projective varieties*, Progress in Mathematics, vol. 277, Birkhäuser Verlag, Basel, 2009.

[Bro10] ______, *Recent progress on the quantitative arithmetic of del Pezzo surfaces*, Number theory, Ser. Number Theory Appl., vol. 6, World Sci. Publ., Hackensack, NJ, 2010, pp. 1–19.

[Bro21] ______, *Cubic forms and the circle method*, Progress in Mathematics, vol. 343, Birkhäuser/Springer, Cham, 2021.

[BS83] S. Bloch and V. Srinivas, *Remarks on correspondences and algebraic cycles*, Amer. J. Math. **105** (1983), no. 5, 1235–1253.

[BS15] B. Bhatt and P. Scholze, *The pro-étale topology for schemes*, Astérisque (2015), no. 369, 99–201.

[BS20] T. D. Browning and W. Sawin, *A geometric version of the circle method*, Ann. of Math. (2) **191** (2020), no. 3, 893–948.

[BT96a] V. V. Batyrev and Y. Tschinkel, *Height zeta functions of toric varieties*, J. Math. Sci. **82** (1996), no. 1, 3220–3239.

[BT96b] ______, *Rational points on some Fano cubic bundles*, C. R. Acad. Sci. Paris Sér. I Math. **323** (1996), no. 1, 41–46.

[BT98a] ______, *Manin's conjecture for toric varieties*, J. Algebraic Geom. **7** (1998), no. 1, 15–53.

[BT98b] ______, *Tamagawa numbers of polarized algebraic varieties*, Astérisque (1998), no. 251, 299–340.

[BT99] F. A. Bogomolov and Y. Tschinkel, *On the density of rational points on elliptic fibrations*, J. Reine Angew. Math. **511** (1999), 87–93.

[BV17] T. D. Browning and P. Vishe, *Rational curves on smooth hypersurfaces of low degree*, Algebra Number Theory **11** (2017), no. 7, 1657–1675.

[Cam04] F. Campana, *Orbifolds, special varieties and classification theory*, Ann. Inst. Fourier (Grenoble) **54** (2004), no. 3, 499–630.

[Cas04] A.-M. Castravet, *Rational families of vector bundles on curves*, Internat. J. Math. **15** (2004), no. 1, 13–45.

[CF10] J. W. S. Cassels and A. Fröhlich (eds.), *Algebraic number theory*, London Mathematical Society, London, 2010, Papers from the conference held at the University of Sussex, Brighton, September 1–17, 1965, Including a list of errata.

[CG72] C. H. Clemens and P. A. Griffiths, *The intermediate Jacobian of the cubic threefold*, Ann. of Math. (2) **95** (1972), 281–356.

[CG94] P. C. Craighero and R. Gattazzo, *Quintic surfaces of* $\mathbf{P}^3$ *having a non-singular model with* $q = p_g = 0$, $P_2 \neq 0$, Rend. Sem. Mat. Univ. Padova **91** (1994), 187–198.

[Che04] I. A. Cheltsov, *Regularization of birational automorphisms*, Mat. Zametki **76** (2004), no. 2, 286–299.

[CJS00] R. L. Cohen, J. D. S. Jones, and G. B. Segal, *Stability for holomorphic spheres and Morse theory*, Geometry and topology: Aarhus (1998), Contemp. Math., vol. 258, Amer. Math. Soc., Providence, RI, 2000, pp. 87–106.

[CLT02] A. Chambert-Loir and Y. Tschinkel, *On the distribution of points of bounded height on equivariant compactifications of vector groups*, Invent. Math. **148** (2002), no. 2, 421–452.

[CLT10] ———, *Igusa integrals and volume asymptotics in analytic and adelic geometry*, Confluentes Math. **2** (2010), no. 3, 351–429.

[Cor96] A. Corti, *Del Pezzo surfaces over Dedekind schemes*, Ann. of Math. (2) **144** (1996), no. 3, 641–683.

[Cox95] D. A. Cox, *The homogeneous coordinate ring of a toric variety*, J. Algebraic Geom. **4** (1995), no. 1, 17–50.

[CPPZ21] I. Cheltsov, J. Park, Yu. G. Prokhorov, and M. Zaidenberg, *Cylinders in Fano varieties*, EMS Surv. Math. Sci. **8** (2021), no. 1–2, 39–105.

[CS09] I. Coskun and J. Starr, *Rational curves on smooth cubic hypersurfaces*, Int. Math. Res. Not. IMRN (2009), no. 24, 4626–4641.

[CT] J.-L. Colliot-Thélène, *Letter to T. Ekedahl*, 9/21/1988, unpublished.

[CTS21] J.-L. Colliot-Thélène and A. N. Skorobogatov, *The Brauer-Grothendieck group*, Ergebnisse der Mathematik und ihrer Grenzgebiete. 3. Folge. A Series of Modern Surveys in Mathematics, vol. 71, Springer, Cham, 2021.

[Cut86] S. D. Cutkosky, *Zariski decomposition of divisors on algebraic varieties*, Duke Math. J. **53** (1986), no. 1, 149–156.

[DC17] G. Di Cerbo, *On Fujita's spectrum conjecture*, Adv. Math. **311** (2017), 238–248.

[Deb01] O. Debarre, *Higher-dimensional algebraic geometry*, Universitext, Springer-Verlag, New York, 2001.

[Del68] P. Deligne, *Théorème de Lefschetz et critères de dégénérescence de suites spectrales*, Inst. Hautes Études Sci. Publ. Math. (1968), no. 35, 259–278.

[Del80] ———, *La conjecture de Weil. II*, Inst. Hautes Études Sci. Publ. Math. (1980), no. 52, 137–252.

[dFH11] T. de Fernex and C. D. Hacon, *Deformations of canonical pairs and Fano varieties*, J. Reine Angew. Math. **651** (2011), 97–126.

[DlBB11] R. De la Bretèche and T. D. Browning, *Manin's conjecture for quartic del Pezzo surfaces with a conic fibration*, Duke Math. J. **160** (2011), no. 1, 1–69.

[DLTT25] R. Das, B. Lehmann, S. Tanimoto, and P. Tosteson, *Homological stability and Manin's conjecture for rational curves on quartic del Pezzo surfaces*, submitted, arXiv:2506.17071, 2025.

[DT24] R. Das and P. Tosteson, *Homology of spaces of curves on blowups*, arXiv:2405.12968, 2024.

[Dwo60] B. Dwork, *On the rationality of the zeta function of an algebraic variety*, Amer. J. Math. **82** (1960), 631–648.

[Eis95] D. Eisenbud, *Commutative algebra*, Graduate Texts in Mathematics, vol. 150, Springer-Verlag, New York, 1995, : With a view toward algebraic geometry.

[EJ06] A.-S. Elsenhans and J. Jahnel, *The asymptotics of points of bounded height on diagonal cubic and quartic threefolds*, Algorithmic number theory, Lecture Notes in Comput. Sci., vol. 4076, Springer, Berlin, 2006, pp. 317–332.

[Eke90] T. Ekedahl, *An effective version of Hilbert's irreducibility theorem*, Séminaire de Théorie des Nombres, Paris 1988–1989, Progr. Math., vol. 91, Birkhäuser Boston, MA, 1990, pp. 241–249.

[ELM$^+$06] L. Ein, R. Lazarsfeld, M. Mustaţă, M. Nakamaye, and M. Popa, *Asymptotic invariants of base loci*, Ann. Inst. Fourier (Grenoble) **56** (2006), no. 6, 1701–1734.

[Els11] A.-S. Elsenhans, *Rational points on some Fano quadratic bundles*, Exp. Math. **20** (2011), no. 4, 373–379.

[EP12] L. Ein and M. Popa, *Extension of sections via adjoint ideals*, Math. Ann. **352** (2012), no. 2, 373–408.

[ETW17] J. S. Ellenberg, T. Tran, and C. Westerland, *Fox-Neuwirth-Fuks cells, quantum shuffle algebras, and Malle's conjecture for function fields*, arXiv:1701.04541, 2017.

[EVW16] J. S. Ellenberg, A. Venkatesh, and C. Westerland, *Homological stability for Hurwitz spaces and the Cohen-Lenstra conjecture over function fields*, Ann. of Math. (2) **183** (2016), no. 3, 729–786.

[Fen25] E. Feng, *Moduli space of genus one curves on cubic threefold*, arXiv:2506.16686, 2025.

[FK88] E. Freitag and R. Kiehl, *étale cohomology and the Weil conjecture*, Ergebnisse der Mathematik und ihrer Grenzgebiete (3), vol. 13, Springer-Verlag, Berlin, 1988, Trans. Betty S. Waterhouse and William C. Waterhouse, with an historical introduction by J. A. Dieudonné.

[FLS18] C. Frei, D. Loughran, and E. Sofos, *Rational points of bounded height on general conic bundle surfaces*, Proc. Lond. Math. Soc. (3) **117** (2018), no. 2, 407–440.

[FM17] C. Frei and M. Madritsch, *Forms of differing degrees over number fields*, Mathematika **63** (2017), no. 1, 92–123.

[FMT89] J. Franke, Yu. I. Manin, and Y. Tschinkel, *Rational points of bounded height on Fano varieties*, Invent. Math. **95** (1989), no. 2, 421–435.

[Fou98] É. Fouvry, *Sur la hauteur des points d'une certaine surface cubique singulière*, Astérisque (1998), no. 251, 31–49.

[FP97] W. Fulton and R. Pandharipande, *Notes on stable maps and quantum cohomology*, Algebraic geometry—Santa Cruz 1995, Proc. Sympos. Pure Math., vol. 62, Part 2, Amer. Math. Soc., Providence, RI, 1997, pp. 45–96.

[FS04] O. Fujino and H. Sato, *Introduction to the toric Mori theory*, Michigan Math. J. **52** (2004), no. 3, 649–665.

[Fu11] L. Fu, *Etale cohomology theory*, Nankai Tracts in Mathematics, vol. 13, World Scientific, Hackensack, NJ, 2011.

[Fuj87] T. Fujita, *On polarized manifolds whose adjoint bundles are not semipositive*, Algebraic geometry, Sendai, 1985, Adv. Stud. Pure Math., vol. 10, North-Holland, Amsterdam, 1987, pp. 167–178.

[Fuj92] ———, *On Kodaira energy and adjoint reduction of polarized manifolds*, Manuscripta Math. **76** (1992), no. 1, 59–84.

[Fuj96] ———, *On Kodaira energy of polarized log varieties*, J. Math. Soc. Japan **48** (1996), no. 1, 1–12.

[Fuj97] ———, *On Kodaira energy and adjoint reduction of polarized threefolds*, Manuscripta Math. **94** (1997), no. 2, 211–229.

[Ful84] W. Fulton, *Intersection theory*, Ergebnisse der Mathematik und ihrer Grenzgebiete (3), vol. 2, Springer-Verlag, Berlin, 1984.

[Gao23] R. Gao, *A Zariski dense exceptional set in Manin's conjecture: dimension 2*, Res. Number Theory **9** (2023), no. 2, Paper No. 42, 15.

[Gao24a] ———, *The geometric exceptional set in Manin's conjecture for Batyrev and Tschinkel's example*, Eur. J. Math. **10** (2024), no. 4, no. 60.

[Gao24b] ———, *Quasi-étale covers of Du Val del Pezzo surfaces and Zariski dense exceptional sets in Manin's conjecture*, arXiv:2406.16240, 2024.

[Gea72] A. Grothendieck et al., *Groupes de monodromie en géométrie algébrique. I*, Lecture Notes in Mathematics, Vol. 288, Springer-Verlag, Berlin-New York, 1972, Séminaire de Géométrie Algébrique du Bois-Marie 1967–1969 (SGA 7 I), Dirigé par A. Grothendieck. Avec la collaboration de M. Raynaud et D. S. Rim.

[GHS03] T. Graber, J. Harris, and J. Starr, *Families of rationally connected varieties*, J. Amer. Math. Soc. **16** (2003), no. 1, 57–67.

[GKP16] D. Greb, S. Kebekus, and T. Peternell, *Étale fundamental groups of Kawamata log terminal spaces, flat sheaves, and quotients of abelian varieties*, Duke Math. J. **165** (2016), no. 10, 1965–2004.

[GMO08] A. Gorodnik, F. Maucourant, and H. Oh, *Manin's and Peyre's conjectures on rational points and adelic mixing*, Ann. Sci. Éc. Norm. Supér. (4) **41** (2008), no. 3, 383–435.

[Gro95] A. Grothendieck, *Techniques de construction et théorèmes d'existence en géométrie algébrique. IV. Les schémas de Hilbert*, Séminaire Bourbaki, Vol. 6, Soc. Math. France, Paris, 1995, pp. 249–276.

[Gue95] M. A. Guest, *The topology of the space of rational curves on a toric variety*, Acta Math. **174** (1995), no. 1, 119–145.

[GW20] U. Görtz and T. Wedhorn, *Algebraic geometry I. Schemes—with examples and exercises*, 2nd ed., Springer Studium Mathematik—Master, Springer Spektrum, Wiesbaden, 2020.

[Har77] R. Hartshorne, *Algebraic geometry*, Graduate Texts in Mathematics, vol. 52, Springer-Verlag, New York-Heidelberg, 1977.

[Hir64] H. Hironaka, *Resolution of singularities of an algebraic variety over a field of characteristic zero. I, II*, Ann. of Math. (2) **79** (1964), 109–203; ibid. (2) **79** (1964), 205–326.

[HJ17] C. D. Hacon and C. Jiang, *On Fujita invariants of subvarieties of a uniruled variety*, Algebr. Geom. **4** (2017), no. 3, 304–310.

[HK00] Y. Hu and S. Keel, *Mori dream spaces and GIT*, Michigan Math. J. **48** (2000), 331–348, Dedicated to William Fulton on the occasion of his 60th birthday.

[HL20] J. Han and Z. Li, *On Fujita's conjecture for pseudo-effective thresholds*, Math. Res. Lett. **27** (2020), no. 2, 377–396.

[HL25] M. Hase-Liu, *A converse to geometric Manin's conjecture for general low degree hypersurfaces*, arXiv:2501.12506, 2025.

[HM07] C. D. Hacon and J. McKernan, *On Shokurov's rational connectedness conjecture*, Duke Math. J. **138** (2007), no. 1, 119–136.

[HM10] ______, *Existence of minimal models for varieties of log general type. II*, J. Amer. Math. Soc. **23** (2010), no. 2, 469–490.

[HMX13] C. D. Hacon, J. McKernan, and C. Xu, *On the birational automorphisms of varieties of general type*, Ann. of Math. (2) **177** (2013), no. 3, 1077–1111.

[HMX14] ______, *ACC for log canonical thresholds*, Ann. of Math. (2) **180** (2014), no. 2, 523–571.

[Hör10] A Höring, *The sectional genus of quasi-polarised varieties*, Arch. Math. (Basel) **95** (2010), no. 2, 125–133.

[HRS04] J. Harris, M. Roth, and J. Starr, *Rational curves on hypersurfaces of low degree*, J. Reine Angew. Math. **571** (2004), 73–106.

[HS00] M. Hindry and J. H. Silverman, *Diophantine geometry: An introduction*, Graduate Texts in Mathematics, vol. 201, Springer-Verlag, New York, 2000.

[HT00] J. Harris and Y. Tschinkel, *Rational points on quartics*, Duke Math. J. **104** (2000), no. 3, 477–500.

[HT06] B. Hassett and Y. Tschinkel, *Weak approximation over function fields*, Invent. Math. **163** (2006), no. 1, 171–190.

[HTT15] B. Hassett, S. Tanimoto, and Y. Tschinkel, *Balanced line bundles and equivariant compactifications of homogeneous spaces*, Int. Math. Res. Not. IMRN (2015), no. 15, 6375–6410.

[Huy06] D. Huybrechts, *Fourier-Mukai transforms in algebraic geometry*, Oxford Mathematical Monographs, The Clarendon Press, Oxford University Press, Oxford, 2006.

[IM07] A. Iliev and L. Manivel, *Prime Fano threefolds and integrable systems*, Math. Ann. **339** (2007), no. 4, 937–955.

[IP99] V. A. Iskovskikh and Yu. G. Prokhorov, *Fano varieties*, Algebraic geometry, V, Encyclopaedia Math. Sci., vol. 47, Springer, Berlin, 1999, pp. 1–247.

[Isk77] V. A. Iskovskih, *Fano threefolds. I*, Izv. Akad. Nauk SSSR Ser. Mat. **41** (1977), no. 3, 516–562, 717.

[Isk78] ______, *Fano threefolds. II*, Izv. Akad. Nauk SSSR Ser. Mat. **42** (1978), no. 3, 506–549.

[Isk79] ______, *Anticanonical models of three-dimensional algebraic varieties*, Current problems in mathematics, Vol. 12 (Russian), Itogi Nauki i

Tekhniki, Akad. Nauk SSSR, Vsesoyuz. Inst. Nauchn. i Tekhn. Inform., Moscow, 1979, pp. 59–157, 239 (loose errata).

[Jah14] J. Jahnel, *Brauer groups, Tamagawa measures, and rational points on algebraic varieties*, Mathematical Surveys and Monographs, vol. 198, American Mathematical Society, Providence, RI, 2014.

[JLT25] E. Jovinelly, B. Lehmann, and S. Tanimoto, *Corrigenda to rational curves on prime Fano threefolds of index* 1, J. Algebraic Geom. **34** (2025), no. 3, 535–542.

[JO24] E. Jovinelly and F. Okamura, *Rational curves on coindex* 3 *Fano varieties*, arXiv:2409.00834, 2024.

[Kaw84] Y. Kawamata, *The cone of curves of algebraic varieties*, Ann. of Math. (2) **119** (1984), no. 3, 603–633.

[Kaw87] ———, *The Zariski decomposition of log-canonical divisors*, Algebraic geometry, Bowdoin, 1985 (Brunswick, Maine, 1985), Proc. Sympos. Pure Math., vol. 46, Amer. Math. Soc., Providence, RI, 1987, pp. 425–433.

[Kaw24] M. Kawakita, *Complex algebraic threefolds*, Cambridge Studies in Advanced Mathematics, vol. 209, Cambridge University Press, Cambridge, 2024.

[Kle66] S. L. Kleiman, *Toward a numerical theory of ampleness*, Ann. of Math. (2) **84** (1966), 293–344.

[KM92] J. Kollár and Sh. Mori, *Classification of three-dimensional flips*, J. Amer. Math. Soc. **5** (1992), no. 3, 533–703.

[KM98] J. Kollár and Sh. Mori, *Birational geometry of algebraic varieties*, Cambridge Tracts in Mathematics, vol. 134, Cambridge University Press, Cambridge, 1998, With the collaboration of C. H. Clemens and A. Corti, Translated from the 1998 Japanese original.

[KMM87] Y. Kawamata, K. Matsuda, and K. Matsuki, *Introduction to the minimal model problem*, Algebraic geometry, Sendai, 1985, Adv. Stud. Pure Math., vol. 10, North-Holland, Amsterdam, 1987, pp. 283–360.

[KMM92a] J. Kollár, Y. Miyaoka, and Sh. Mori, *Rational connectedness and boundedness of Fano manifolds*, J. Differential Geom. **36** (1992), no. 3, 765–779.

[KMM92b] J. Kollár, Y. Miyaoka, and Sh. Mori, *Rationally connected varieties*, J. Algebraic Geom. **1** (1992), no. 3, 429–448.

[Kol96] J. Kollár, *Rational curves on algebraic varieties*, Ergebnisse der Mathematik und ihrer Grenzgebiete. 3. Folge. A Series of Modern Surveys in Mathematics, vol. 32, Springer-Verlag, Berlin, 1996.

[Kol97] ———, *Singularities of pairs*, Algebraic geometry—Santa Cruz 1995, Proc. Sympos. Pure Math., vol. 62, Amer. Math. Soc., Providence, RI, 1997, pp. 221–287.

[Kol13] ———, *Singularities of the minimal model program*, Cambridge Tracts in Mathematics, vol. 200, Cambridge University Press, Cambridge, 2013, With a collaboration of Sándor Kovács.

[Kon95] M. Kontsevich, *Enumeration of rational curves via torus actions*, The moduli space of curves (Texel Island, 1994), Progr. Math., vol. 129, Birkhäuser Boston, MA, 1995, pp. 335–368.

[KP01] B. Kim and R. Pandharipande, *The connectedness of the moduli space of maps to homogeneous spaces*, Symplectic geometry and mirror symmetry (Seoul, 2000), World Sci. Publ., River Edge, NJ, 2001, pp. 187–201.

[KPS18] A. G. Kuznetsov, Y. G. Prokhorov, and C. A. Shramov, *Hilbert schemes of lines and conics and automorphism groups of Fano threefolds*, Jpn. J. Math. **13** (2018), no. 1, 109–185.

[Kur14] Y. Kurumadani, *Pencils of cubics curves and rational elliptic surfaces*, RIMS preprint 1800, 2014.

[Laz04] R. Lazarsfeld, *Positivity in algebraic geometry. I*, Ergebnisse der Mathematik und ihrer Grenzgebiete. 3. Folge. A Series of Modern Surveys in Mathematics, vol. 48, Springer-Verlag, Berlin, 2004, Classical setting: line bundles and linear series.

[Leh12] B. Lehmann, *A cone theorem for nef curves*, J. Algebraic Geom. **21** (2012), no. 3, 473–493.

[LR19] C. Le Rudulier, *Points algébriques de hauteur bornée sur une surface*, Bull. Soc. Math. France **147** (2019), no. 4, 705–748.

[LRT25a] B. Lehmann, E. Riedl, and S. Tanimoto, *Codimension of jumping loci*, J. Algebraic Geom. **34** (2025), no. 4, 755–814.

[LRT25b] ———, *Non-free curves on Fano varieties*, Osaka J. Math. **62** (2025), no. 1, 19–50.

[LRT25c] ———, *Non-free sections of Fano fibrations*, Mem. Amer. Math. Soc. (2025), in press, arXiv:2301.01695.

[LS24] D. Loughran and T. Santens, *Malle's conjecture and Brauer groups of stacks*, arXiv:2412.04196, 2024.

[LST22] B. Lehmann, A. K. Sengupta, and S. Tanimoto, *Geometric consistency of Manin's Conjecture*, Compos. Math. **158** (2022), no. 6, 1375–1427.

[LT17] B. Lehmann and S. Tanimoto, *On the geometry of thin exceptional sets in Manin's conjecture*, Duke Math. J. **166** (2017), no. 15, 2815–2869.

[LT19a] ———, *Geometric Manin's conjecture and rational curves*, Compos. Math. **155** (2019), no. 5, 833–862.

[LT19b] ———, *On exceptional sets in Manin's conjecture*, Res. Math. Sci. **6** (2019), no. 1, Paper No. 12, 41.

[LT21] ———, *Rational curves on prime Fano threefolds of index* 1, J. Algebraic Geom. **30** (2021), no. 1, 151–188.

[LT22] ———, *Classifying sections of del Pezzo fibrations, II*, Geom. Topol. **26** (2022), no. 6, 2565–2647.

[LT24] ———, *Classifying sections of del Pezzo fibrations, I*, J. Eur. Math. Soc. (JEMS) **26** (2024), no. 1, 289–354.

[LTT18] B. Lehmann, S. Tanimoto, and Y. Tschinkel, *Balanced line bundles on Fano varieties*, *J. Reine Angew. Math.* **743** (2018), 91–131.

[LWZB24] Y. Liu, M. M. Wood, and D. Zureick-Brown, *A predicted distribution for Galois groups of maximal unramified extensions*, Invent. Math. **237** (2024), no. 1, 49–116.

[Mat02] K. Matsuki, *Introduction to the Mori program*, Universitext, Springer-Verlag, New York, 2002.

[MM83] Sh. Mori and Sh. Mukai, *On Fano 3-folds with $B_2 \geq 2$*, Algebraic varieties and analytic varieties (Tokyo, 1981), Adv. Stud. Pure Math., vol. 1, North-Holland, Amsterdam, 1983, pp. 101–129.

[MM03] ______, *Erratum: "Classification of Fano 3-folds with $B_2 \geq 2$"*, Manuscripta Math. **110** (2003), no. 3, 407.

[MM82] ______, *Classification of Fano 3-folds with $B_2 \geq 2$*, Manuscripta Math. **36** (1981/82), no. 2, 147–162.

[Mor82] Sh. Mori, *Threefolds whose canonical bundles are not numerically effective*, Ann. of Math. (2) **116** (1982), no. 1, 133–176.

[Mor88] ______, *Flip theorem and the existence of minimal models for 3-folds*, J. Amer. Math. Soc. **1** (1988), no. 1, 117–253.

[MTiB25] Y. Mustopa and M. Teixidor i Bigas, *Rational curves on moduli spaces of vector bundles*, Internat. J. Math. **36** (2025), no. 3, Paper No. 2450080.

[Nak00] M. Nakamaye, *Stable base loci of linear series*, Math. Ann. **318** (2000), no. 4, 837–847.

[Nam97] Y. Namikawa, *Smoothing Fano 3-folds*, J. Algebraic Geom. **6** (1997), no. 2, 307–324.

[New68] P. E. Newstead, *Stable bundles of rank 2 and odd degree over a curve of genus 2*, Topology **7** (1968), 205–215.

[Oh10] H. Oh, *Orbital counting via mixing and unipotent flows*, Homogeneous flows, moduli spaces and arithmetic, Clay Math. Proc., vol. 10, Amer. Math. Soc., Providence, RI, 2010, pp. 339–375.

[Oka24] F. Okamura, *The irreducibility of the spaces of rational curves on del Pezzo manifolds*, Int. Math. Res. Not. IMRN **2024** (2024), no. 12, 9893–9909.

[Oka25a] ______, *Moduli spaces of rational curves on Artin-Mumford double solids*, arXiv:2501.09269, 2025.

[Oka25b] ______, *Rational curves on Fano threefolds with Gorenstein terminal singularities*, Eur. J. Math. **11** (2025), no. 2, Article: 38.

[Ols16] M. Olsson, *Algebraic spaces and stacks*, American Mathematical Society Colloquium Publications, vol. 62, American Mathematical Society, Providence, RI, 2016.

[Pet12] T. Peternell, *Varieties with generically nef tangent bundles*, J. Eur. Math. Soc. (JEMS) **14** (2012), no. 2, 571–603.

[Pey95] E. Peyre, *Hauteurs et mesures de Tamagawa sur les variétés de Fano*, Duke Math. J. **79** (1995), no. 1, 101–218.

[Pey03] ______, *Points de hauteur bornée, topologie adélique et mesures de Tamagawa*, J. Théor. Nombres Bordeaux **15** (2003), no. 1, 319–349.

[Pey17] ______, *Liberté et accumulation*, Doc. Math. **22** (2017), 1615–1659.

[Pey21] ______, *Beyond heights: slopes and distribution of rational points*, Arakelov geometry and Diophantine applications, Lecture Notes in Math., vol. 2276, Springer, Cham, 2021.

[Poo17] B. Poonen, *Rational points on varieties*, Graduate Studies in Mathematics, vol. 186, American Mathematical Society, Providence, RI, 2017.

[Pro07] Yu. G. Prokhorov, *The degree of $\mathbb{Q}$-Fano threefolds*, Mat. Sb. **198** (2007), no. 11, 153–174.

[Rei72] M. Reid, *The Complete Intersection of Two or More Quadrics*, Ph.D. thesis, University of Cambridge, 1972.

[Rei88] I. Reider, *Vector bundles of rank 2 and linear systems on algebraic surfaces*, Ann. of Math. (2) **127** (1988), no. 2, 309–316.

[RTU17] J. Rana, J. Tevelev, and G. Urzúa, *The Craighero-Gattazzo surface is simply connected*, Compos. Math. **153** (2017), no. 3, 557–585.

[RY19] E. Riedl and D. Yang, *Kontsevich spaces of rational curves on Fano hypersurfaces*, J. Reine Angew. Math. **748** (2019), 207–225.

[Sal98] P. Salberger, *Tamagawa measures on universal torsors and points of bounded height on Fano varieties*, Astérisque (1998), no. 251, 91–258.

[San95] T. Sano, *On classifications of non-Gorenstein* **Q**-*Fano 3-folds of Fano index* 1, J. Math. Soc. Japan **47** (1995), no. 2, 369–380.

[Sch79] S. H. Schanuel, *Heights in number fields*, Bull. Soc. Math. France **107** (1979), no. 4, 433–449.

[Seg79] G. Segal, *The topology of spaces of rational functions*, Acta Math. **143** (1979), no. 1-2, 39–72.

[Sen19] A. K. Sengupta, *Manin's b-constant in families*, Algebra Number Theory **13** (2019), no. 8, 1893–1905.

[Sen21] ———, *Manin's conjecture and the Fujita invariant of finite covers*, Algebra Number Theory **15** (2021), no. 8, 2071–2087.

[Ser92] J. P. Serre, *Topics in Galois theory*, Research Notes in Mathematics, vol. 1, Jones and Bartlett Publishers, Boston, MA, 1992, Lecture notes prepared by Henri Damon, with a foreword by Darmon and the author.

[SGA72a] *Théorie des topos et cohomologie étale des schémas. Tome 1: Théorie des topos*, Lecture Notes in Mathematics, vol. Vol. 269, Springer-Verlag, Berlin-New York, 1972, Séminaire de Géométrie Algébrique du Bois-Marie 1963–1964 (SGA 4), Dirigé par M. Artin, A. Grothendieck, et J. L. Verdier. Avec la collaboration de N. Bourbaki, P. Deligne et B. Saint-Donat.

[SGA72b] *Théorie des topos et cohomologie étale des schémas. Tome 2*, Lecture Notes in Mathematics, vol. Vol. 270, Springer-Verlag, Berlin-New York, 1972, Séminaire de Géométrie Algébrique du Bois-Marie 1963–1964 (SGA 4), Dirigé par M. Artin, A. Grothendieck et J. L. Verdier. Avec la collaboration de N. Bourbaki, P. Deligne et B. Saint-Donat.

[SGA73] *Théorie des topos et cohomologie étale des schémas. Tome 3*, Lecture Notes in Mathematics, vol. Vol. 305, Springer-Verlag, Berlin-New York, 1973, Séminaire de Géométrie Algébrique du Bois-Marie 1963–1964 (SGA 4), Dirigé par M. Artin, A. Grothendieck et J. L. Verdier. Avec la collaboration de P. Deligne et B. Saint-Donat.

[She12] M. Shen, *On the normal bundles of rational curves on Fano 3-folds*, Asian J. Math. **16** (2012), no. 2, 237–270.

[Shi92] T. Shioda, *Existence of a rational elliptic surface with a given Mordell-Weil lattice*, Proc. Japan Acad. Ser. A Math. Sci. **68** (1992), no. 9, 251–255.

[Sho85] V. V. Shokurov, *A nonvanishing theorem*, Izv. Akad. Nauk SSSR Ser. Mat. **49** (1985), no. 3, 635–651.

[Som86] A. J. Sommese, *On the adjunction theoretic structure of projective varieties*, Complex analysis and algebraic geometry (Göttingen, 1985), Lecture Notes in Math., vol. 1194, Springer, Berlin, 1986, pp. 175–213.

[ST16] J. Shalika and Y. Tschinkel, *Height zeta functions of equivariant compactifications of unipotent groups*, Comm. Pure Appl. Math. **69** (2016), no. 4, 693–733.

[ST22] N. Shimizu and S. Tanimoto, *The spaces of rational curves on del Pezzo threefolds of degree one*, Eur. J. Math. **8** (2022), no. 1, 291–308.

[Sta] The Stacks Project Authors, *Stacks Project*, https://stacks.math.columbia.edu.

[Sta00] J. Starr, *Rational curves on hypersurfaces in projective n-space*, ProQuest LLC, Ann Arbor, MI, 2000, Ph. D. Thesis–Harvard University.

[STBT07] J. Shalika, R. Takloo-Bighash, and Y. Tschinkel, *Rational points on compactifications of semi-simple groups*, J. Amer. Math. Soc. **20** (2007), no. 4, 1135–1186.

[Tan21] S. Tanimoto, *An introduction to Geometric Manin's conjecture*, Proceedings of Miyako-no-Seihoku Algebraic Geometry Symposium 2021, "Positivity of tangent bundles and related topics", 2021, pp. 102–118.

[Tes05] D. Testa, *The Severi problem for rational curves on del Pezzo surfaces*, ProQuest LLC, Ann Arbor, MI, 2005, Ph. D. Thesis–Massachusetts Institute of Technology.

[Tes09] ———, *The irreducibility of the spaces of rational curves on del Pezzo surfaces*, J. Algebraic Geom. **18** (2009), no. 1, 37–61.

[Tho98] J. F. Thomsen, *Irreducibility of* $\overline{M}_{0,n}(G/P,\beta)$, Internat. J. Math. **9** (1998), no. 3, 367–376.

[Tik80] A. S. Tikhomirov, *Geometry of the Fano surface of a double* $\mathbf{P}^3$ *branched in a quartic*, Izv. Akad. Nauk SSSR Ser. Mat. **44** (1980), no. 2, 415–442, 479.

[Tik81] ———, *The Fano surface of the Veronese double cone*, Izv. Akad. Nauk SSSR Ser. Mat. **45** (1981), no. 5, 1121–1197, 1199.

[Tju71] A. N. Tjurin, *The geometry of the Fano surface of a nonsingular cubic* $F \subset P^4$*, and Torelli's theorems for Fano surfaces and cubics*, Izv. Akad. Nauk SSSR Ser. Mat. **35** (1971), 498–529.

[Tsc03] Y. Tschinkel, *Fujita's program and rational points*, Higher dimensional varieties and rational points (Budapest, 2001), Bolyai Soc. Math. Stud., vol. 12, Springer, Berlin, 2003, pp. 283–310.

[Tsc09] ———, *Algebraic varieties with many rational points*, Arithmetic geometry, Clay Math. Proc., vol. 8, Amer. Math. Soc., Providence, RI, 2009, pp. 243–334.

[VE10] A. Venkatesh and J. S. Ellenberg, *Statistics of number fields and function fields*, Proceedings of the International Congress of Mathematicians. Volume II, Hindustan Book Agency, New Delhi, 2010, pp. 383–402.

[Voi02] C. Voisin, *Hodge theory and complex algebraic geometry. I*, Cambridge Studies in Advanced Mathematics, vol. 76, Cambridge University Press, Cambridge, 2002, Trans. Leila Schneps.

[Voi06] ———, *On integral Hodge classes on uniruled or Calabi-Yau threefolds*, Moduli spaces and arithmetic geometry, Adv. Stud. Pure Math., vol. 45, Math. Soc. Japan, Tokyo, 2006, pp. 43–73.

[Voi07] ______, *Hodge theory and complex algebraic geometry. II*, Cambridge Studies in Advanced Mathematics, vol. 77, Cambridge University Press, Cambridge, 2007, Trans. Leila Schneps.

[Wiś09] J. Wiśniewski, *Rigidity of the Mori cone for Fano manifolds*, Bull. Lond. Math. Soc. **41** (2009), no. 5, 779–781.

Index

For EU product safety concerns, contact us at Calle de José Abascal, 56–1°,
28003 Madrid, Spain or eugpsr@cambridge.org.

www.ingramcontent.com/pod-product-compliance
Ingram Content Group UK Ltd.
Pitfield, Milton Keynes, MK11 3LW, UK
UKHW020216290926
15470UKWH00001B/219

* 9 7 8 1 0 0 9 6 1 1 6 3 3 *